JIGS AND FIXTURES

SECOND EDITION

William E. Boyes
Editor

Published by:

Society of Manufacturing Engineers
Marketing Services Department
One SME Drive
P.O. Box 930
Dearborn, Michigan 48128

JIGS AND FIXTURES

SME wishes to express its acknowledgement and appreciation to the following publications for supplying the various articles reprinted within the contents of this book.

Control Engineering
Technical Publishing
1301 South Grove Avenue
Barrington, IL 60010

Manufacturing Engineering
Society of Manufacturing Engineers
One SME Drive
P.O. Box 930
Dearborn, MI 48128

Modern Machine Shop
Gardner Publications Inc.
600 Main Street
Cincinnati, OH 45202

Power Transmission Design
A Penton/IPC Publication
614 Superior Avenue West
Cleveland, OH 44113

Production Engineering
A Penton/IPC Publication
Penton Plaza
Cleveland, OH 44114

Stanki I Instrument
(Machines and Tooling)
The Production Engineering Research
　Association of Great Britain
Melton Mowbray
Leicestershire LE13 OPB
England

The Carbide Journal
Society of Carbide and Tool Engineers
Box 347
Bridgeville, PA 15017

Tooling & Production
Huebner Publications, Inc.
5821 Harper Road
Solon, OH 44139

Grateful acknowledgement is also extended to:

Bendix Corporation
P.O. Box 1159
Kansas City, MO 64141

Moore Products
Spring House, PA 19477

OTC Hytec
Division of Owatonna Tool Company
Owatonna, MN 55060

Positrol Inc.
3890 Virginia Avenue
Cincinnati, OH 45227

Powerhold, Inc.
Old Indian Trail
P.O. Box 447
Middlefield, CT 06455

Cover illustration courtesy of MP Tool & Engineering Company, Roseville, Michigan.

PREFACE

This edition of the Manufacturing Update Series—a collection of state-of-the-art information on the design and utilization of jigs and fixtures—is a response to SME members who have requested more information on the topic.

Jigs and fixtures allow interchangeability of parts by providing a relationship between work and the tool operating on the work. The fixture is a device for holding and locating work while operations are being performed, where as the jig is not only a device for holding and locating work, but it also guides the tool performing the operation.

The earliest jigs and fixtures were simple devices to hold parts and to guide tools for handworking or simple machining. Later, mass production dictated that parts be interchangeable in assemblies. More elaborate jigs and fixtures were required to precisely position parts and guide tools during machining. Many of these were designed and built at considerable expense.

Lately, however, universal fixturing and building block components have become available commercially. This provides the designer with economical and time-saving methods of assembling these components, sometimes with slight modifications, into practical jigs and fixtures. Additionally, much of the design and construction of jigs and fixtures is being sent to specialists familiar with varied and complicated machining operations.

Today, many jigs and fixtures are being designed and built by computer aided equipment. This trend will increase exponentially in the future.

This second edition includes an even wider range of articles discussing good workholding. A special emphasis was placed on raising productivity by decreasing downtime. There is also more material on numerical control and on industrial robots. A section in toolholding has also been included.

Novice engineers will find useful suggestions regarding sound design principles including: selection of locating points, basics of clamping and chucking, estimating costs and designing with standard components. Experienced engineers now have a detailed reference book regarding applications of hydraulics, electronics, air and vacuum for the handling, loading, workholding and indexing of parts. The volume also presents methods of fixturing for automation and NC application, including the automatic feeding and high-speed transfer of parts during machining operation.

Chapter One is primarily concerned with basic design principles, methods and formulae utilized in the design of jigs and fixtures.

Chapter Two covers methods and equipment for holding and clamping parts and assemblies.

Chapter Three contains new ideas for indexing and positioning parts for machining.

Chapter Four is devoted to the efficient feeding and handling of parts.

Chapter Five presents useful methods of fixturing for common machining and assembly operations while Chapter Six presents methods of fixturing for automation and NC.

Chapter Seven contains updated information on toolholding devices.

Each of the articles was written by an expert in the field and many of their names are included. Grateful acknowledgement is expressed to these authors.

I also wish to express my gratitude to the following publications who supplied much of the material in this volume: *Control Engineering, Manufacturing Engineering, Modern Machine Shop, Power Transmission Design, Production Engineering, Stanki I Instrument, The Carbide Journal* and *Tooling & Production.* Grateful acknowledegment is also extended to Bendix Corporation, Moore Products, OTC Hytec, Positrol Inc. and Powerhold, Inc.

Finally, my thanks is extended to the staff of the SME Marketing Services Department for their assistance in developing this book.

William E. Boyes
Section Manager
Metrology and Quality Control Engineering
Mack Trucks, Inc.
Editor

W.E. Boyes
William E. Boyes is currently the Section Manager of the
Metrology and Quality Control Engineering Departments of Mack
Trucks, Inc., Powertrain Division, Hagerstown, Maryland. He has
worked in various quality control positions for Mack Trucks over
the past 20 years.

Mr. Boyes is both a Certified Manufacturing Engineer and a
Registered Professional Engineer. He is an active member of
many engineering societies and associations including the
American Society of Quality Control, the Precision Measurements
Association, the American Society for Non-destructive Testing
and the American Defense Preparedness Association. Mr. Boyes
is also a member delegate to the National Conference of
Standards Laboratories.

An SME member since 1958, Mr. Boyes has served on SME's
Tool Engineering Council in many capacities. He was chairman of
the council from 1974 to 1978 and prior to that served as the
chairman of its Gage Division. Mr. Boyes is currently a member of
the Tool Engineering Council's Quality Assurance Division.

SME

The informative volumes of the Manufacturing Update Series are part of the Society of Manufacturing Engineers' effort to keep its Members better informed on the latest trends and developments in engineering.

With 60,000 members, SME provides a common ground for engineers and managers to share ideas, information and accomplishments.

An overwhelming mass of available information requires engineers to be concerned about keeping up-to-date, in other words, continuing education. SME Members can take advantage of numerous opportunities, in addition to the books of the Manufacturing Update Series, to fulfill their continuing educational goals. These opportunities include:

- Chapter programs through the over 200 chapters which provide SME Members with a foundation for involvement and participation.
- Educational programs including seminars, clinics, programmed learning courses and videotapes.
- Conferences and expositions which enable engineers to see, compare, and consider the newest manufacturing equipment and technology.
- Publications including Manufacturing Engineering, the SME Newsletter, Technical Digest and a wide variety of books including the Tool and Manufacturing Engineers Handbook.
- SME's Manufacturing Engineering Certification Institute formally recognizes manufacturing engineers and technologists for their technical expertise and knowledge acquired through years of experience.

In addition, the Society works continuously with the American National Standards Institute, the International Standards Organization and other organizations to establish the highest possible standards in the field.

SME Members have discovered that their membership broadens their knowledge throughout their career.

In a very real sense, it makes SME the leader in disseminating and publishing technical information for the manufacturing engineer.

MANUFACTURING UPDATE SERIES

Published by the Society of Manufacturing Engineers, the Manufacturing Update Series provides significant, up-to-date information on a variety of topics relating to the manufacturing industry. This series is intended for engineers working in the field, technical and research libraries, and also as reference material for educational institutions.

The information contained in this volume doesn't stop at merely providing the basic data to solve practical shop problems. It also can provide the fundamental concepts for engineers who are reviewing a subject for the first time to discover the state-of-the-art before undertaking new research or application. Each volume of this series is a gathering of journal articles, techical papers and reports that have been reprinted with expressed permission from the various authors, publishers or companies identified within the book.

SME technical committees, which are made up of educators, engineers and managers working within industry, are responsible for the selection of material in this series.

We sincerely hope that the information collected in this publication will be of value to you and your company. If you feel there is a shortage of technical information on a specific manufacturing area, please let us know. Send your thoughts to the Manager of Educational Resources, Marketing Services Department at SME. Your request will be considered for possible publication by SME—the leader in disseminating and publishing technical information for the engineer.

TABLE OF CONTENTS

———— CHAPTERS ————

1 DESIGN AIDS

2 WORKHOLDING AND CLAMPING

3 INDEXING AND POSITIONING DEVICES

4 PARTS FEEDING AND HANDLING

5 MACHINING AND ASSEMBLY FIXTURES

6 FIXTURING FOR AUTOMATION AND NC

7 TOOL HOLDING DEVICES

CHAPTER ONE

DESIGN AIDS

Reprinted from Modern Machine Shop, April 1977

Tool Designer's Notebook

The make or break of any manufacturing operation often depends upon the way it is tooled—especially the design of jigs and fixtures.

By J. ROBERT HAMNER, Manufacturing Engineer
C. A. Norgren Co.
Littleton, Colorado

It is very easy to forget some of the basics that have grown up over the years—basics that should become second nature to anyone designing jigs and fixtures. At one extreme, jigs and fixtures can be simple, functional, and a decisive aid to productivity. At the other end they can be complicated, expensive, and do little to aid a smooth flowing production schedule.

The design and cost principles in Mr. Hamner's notebook are neither radical nor necessarily new. Many have appeared in such outstanding text and handbooks as the Tool Engineers Handbook and Tool Design by Donaldson and LeCain. But they all have served Mr. Hamner well in creating practical, efficient, and economical tooling for his company. These principles are also the ones he recommends to people moving from the engineering classroom to the real world of manufacturing.

Tool Cost

What is the maximum justifiable tool cost? A formula that is particularly applicable when tooling for small, fixed production rates follows:

$$C = \frac{N[a(1 + t)]-S}{G}$$

Where:

C = Dollar cost of tool—including design, drafting, manufacture and overhead.

N = Minimum number of pieces that will be produced per year.

a = Dollar savings in direct labor cost per unit.

t = Percentage of overhead charged to direct labor.

S = Yearly cost of setting up tools and machine.

G = Annual total allowance (percent) for interest, taxes, depreciation, repairs and storage.

Assume that 500 parts are needed per year. It is estimated that the tool will save $.60 in labor per part. The overhead charged to direct labor is 50 percent of the labor cost. It cost $30 to set the tool up for a run. Overhead costs are 9 percent for interest, 4 percent for fixed charges, taxes, insurance, and rent, 50 percent for depreciation, and 7 percent for repairs—a total of 70 percent. How much can we allow for the new tool?

$$C = \frac{N[a(1 + t)]-S}{G}$$

$$C = \frac{500\ [.6(1 + .5)] -30}{.7}$$

$$C = \$600$$

This is the maximum total amount that should be spent on the new tool of which about $200 or 33 percent will probably be spent on design. In general, the cost of designing fixtures is about 33 percent of the total cost of the tool, about 25 percent of the cost of die sets, and about 15 percent of the cost of gages. If more than one tool of identical design is needed, the percentage of manufacturing cost charged to design will be lowered.

ABOUT THE AUTHOR

Mr. Hamner studied engineering at Wichita State University and the University of Colorado. His 19 years of industrial experience includes a variety of assignments in tool design. He has designed tooling for high-production machinery, nuclear weapons, ICBM's, oil field equipment, and military aircraft. Bob also taught courses in manufacturing orientation and has authored a manual on Tool Design Standards. His present responsibilities as Group Leader in the Process Engineering Department of C. A. Norgren include planning, estimating, and directing the activities of process engineers and tool designers.

Lathe Fixture

Machining a long flanged workpiece may require a support, such as a pilot. Additional support may be obtained by adding clamps to each chucking jaw.

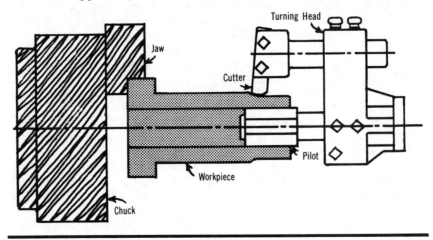

4

Drill Bushings

It is best to locate the bushing directly on the part. Any chips produced will follow the helix of the drill up and out of the drill jig. Also, be aware of drill rotation; locating points and clamps must counteract this force.

Any chip entanglement that may occur is outside the jig where the operator can remove it. It also provides maximum drill guiding effect.

However, zero chip clearance has two disadvantages: (1) The abrasive action of chips in the bushing causes excessive wear, and (2) a deep-hole effect is produced, making it difficult for the chips to come up through the drill flutes.

If there is a space between the workpiece and the drill bushing, the clearance should be 1 to 1½ times the drill diameter. This method should be avoided on continuous chip operations because the chips may become entangled and prevent removal of the drill.

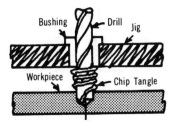

Chip tangle caused by the use of chip clearance when producing continuous chips.

Zero chip clearance causes chip tangle to occur outside the drill jig when producing continuous chips.

Locating

When locating the workpiece, use 50 percent of the engineering part tolerance; don't arbitrarily assign ±.001 inch to everything.

Locate from the same surfaces as those from which the hole is dimensioned on the engineering print. That's how inspection will check it.

Don't locate on parting lines, gates, and overflows. If the hole or feature is dimensioned from such a surface, resolve this problem before actual production begins.

Use adjustable stops for locating sand castings. Workpieces made from sand castings can vary.

Locate on cored surfaces rather than cavity surfaces.

Maintain the same locating points for sequential operations.

Clamping

Here is the right way and the wrong way to clamp a workpiece in a milling fixture. The wrong way is when the direction of the feed and the rotation of the cutter both force the work against the clamp screw.

The right way is when the workpiece is firmly clamped against a solid shoulder of the fixture, which absorbs the thrust of the cutter and there is no possibility of chatter. Locate the workpiece as low as possible to reduce vibration.

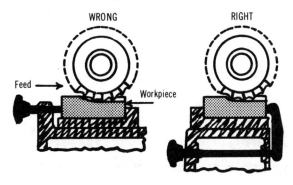

Chip Removal

Chips must be removed from the jig before each new workpiece is located. Always leave plenty of room around the workpiece so that chips can be easily washed, brushed, or blown away.

Corner relief and area relief are required to eliminate the accumulation of chips and dirt in locating corners. Also, corner relief will provide burr clearance if the workpiece has been milled in a previous operation.

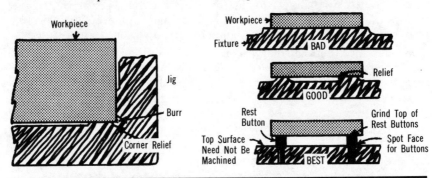

Chip Control

There are two basic types of chips: discontinuous and continuous. Discontinuous chips are small segments such as those produced in machining cast iron and certain forms of cast brass. Discontinuous chips are also produced when milling, grinding and shaping most materials.

Continuous chips are produced by turning and drilling operations. The long, stringy chips take the form of coils and spirals.

Obviously, it is advantageous to create discontinuous chips because they are easier to remove. This can be accomplished by intermittent feed of the drill or by grinding chip breakers into the cutting tools.

Hook-Bolt Clamp

Hook-bolt clamps may be used whenever other direct methods of clamping cannot be used. Remember to use a backup to prevent bending of the clamp and to maintain alignment.

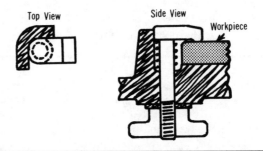

Where to Clamp

Clamping force should be directly over, or in line with, locating points—especially on light work. This will prevent distortion of the workpiece and will maintain accuracy.

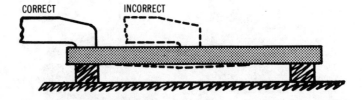

Wear on Locating Surfaces

Wear on locating surfaces can affect accuracy of high-production fixtures. If the locator is made slightly smaller in diameter there is no danger of wearing a pocket and adversely affecting accuracy.

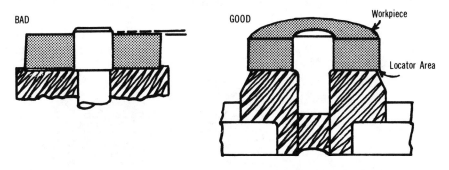

Standard Fixture Width

Try to maintain standard fixture widths, which will reduce setup time on the drill press by the use of guides. Also, the guides can save production time in progressive drilling, reaming and tapping operations.

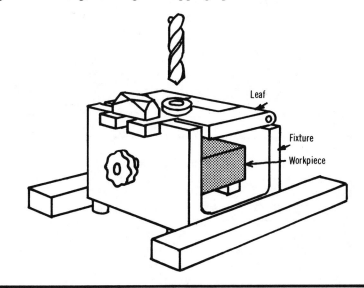

Burr Relief

The burr raised on the workpiece at the start of a cut is called a minor burr, while the burr caused at the end of a cut is termed a major burr. Always provide relief for burrs so that the workpiece can be removed easily. For example, a slot can be provided in the fixture to prevent the minor burr from holding the part in the fixture.

A major burr is formed when the drill breaks through the workpiece. Use a fixture that allows the workpiece to be lifted straight up to prevent binding by the major burr. Be aware of the location of these burrs.

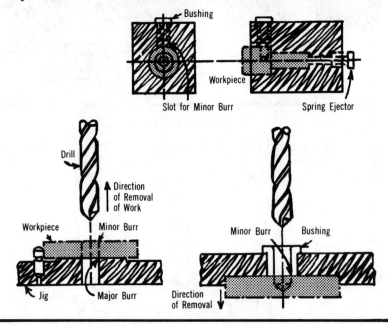

Jig and Fixture Fundamentals

1. Loading and unloading of small parts should not exceed two seconds for each operation.

2. If a fixture has removable parts, such as a drill bushing and a ream bushing for one hole, provide a hole and setscrew to retain the bushing not being used. This prevents loss in the tool crib.

3. If locating points are covered, provide sight or peep holes for the operator's benefit.

4. Tools that weigh over 30 pounds must have lifting eyes.

5. Take a look at previous tools designed for a similar operation.

6. Use standard, off-the-shelf components wherever possible.

7. Use cold rolled steel to eliminate the machining of hot rolled steel.

8. If springs are subject to chip accumulation, shield them or position them on raised bosses. Don't ask the operator to continually pick or blow chips out of a spring.

9. Don't demand unnecessary tolerances on noncritical dimensions. Base plates can often be ± ⅛ inch.

10. Stress simplicity. It is very easy—almost natural—to allow a fixture to become complicated. The real challenge is to design something simple.

11. Design around stock sizes.

12. Foolproofing is a must for any fixture. It is defined as the incorporation of certain design features that do not interfere with the loading and locating of the workpiece, yet make it impossible to place the part in an improper positon.

13. Safety is a prime responsibility of the tool designer. He must design the jig or fixture so that the operator's hands will not be too close to cutting tools and so that clamps and levers can be operated without danger of injury. Although it is fairly easy to incorporate safety ideas in an original design, considerable time and money are often spent after the tool is built and someone is injured. Keep the operator's movements within normal range—not maximum. This reduces fatigue, and increases production as well as safety.

Leaf Jigs

A leaf jig opens and closes over the top of the workpiece. Quarter-turn screws are frequently used in leaf jigs. Want to save production time? Cut the underside of the quarter-turn screw away and place a pin in the post. The operator can rotate the screw to the correct position very easily for opening the leaf.

Another method utilizes a bent pin pressed into the leaf.

A cam lock is a quick, easy way to open and close a leaf jig; however, check the cam frequently for wear.

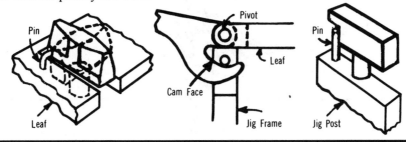

Ejectors

It is just as important to eject some types of workpieces from fixtures as it is to locate and clamp them, especially when located on a stud, plug or ring. Don't pry the part off with a screwdriver; use spring ejectors or pin ejectors.

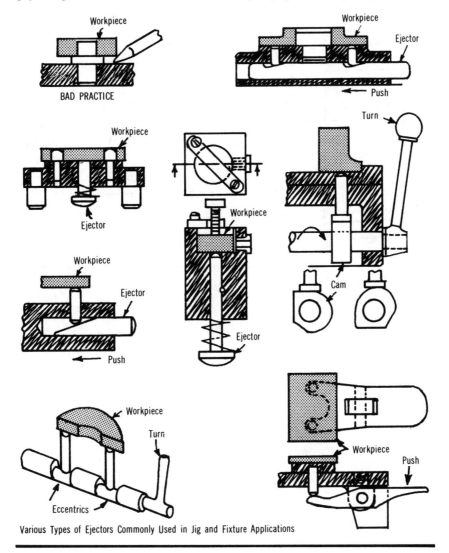

Various Types of Ejectors Commonly Used in Jig and Fixture Applications

Jig Feet

Jig feet are used to provide clearance for an accumulation of chips under the fixture. If they are pressed into place be sure to provide a knockout hole. This is true for dowels, pins and bushings. Anything that is a press fit should be provided with a knockout hole.

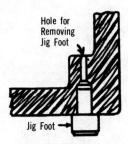

Pump Jigs

Commercially available fixtures, such as pump jigs, can reduce design and fabrication costs. They have the advantages of rigidity, ample chip clearance, and ease of operation. With the addition of jaws, locators, bushings, and so on, they are suitable for drilling, reaming and tapping small parts. Also, by making locating details interchangeable, the pump jig can be used for more than one workpiece.

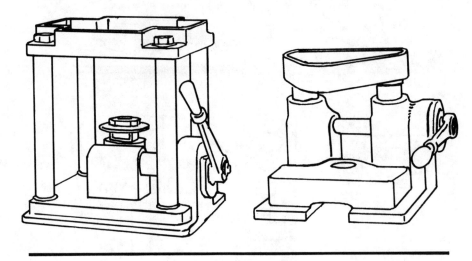

Getting a grip on
NEW VISE DEVELOPMENTS

Reprinted from Tooling & Production, December 1980

by John Lenz
Vice President
Industrial Products Div
Kurt Mfg Co
Minneapolis, MN

In many machine shops the precision machine vise is the basic workholding fixture that is universally applied to repetitive positioning of parts on the tables of milling and drilling machines of wide variety. The single most critical factor in selecting a vise is immobility—the ability of the vise to hold a part immobile not only while it is being clamped but also during the machining operation.

The well-known Kurt patented Anglock design is such that for each pound of force applied horizontally in the direction of clamping there is one-half pound of force vertically pushing the movable jaw downward and seating it on the vise bed.

To this basic principle we have added some new ideas that make this universal tool even more useful. ■

Designated Model APD60-6", this development features hydraulic operation powered by a fast and reliable Kurt-built air-over-oil accumulator. Shop air at 100 lb pressure provides 9000 lb clamping force. The hydraulics have a $\frac{3}{16}$" power stroke. Setup is facilitated by the hand knob which gives easy and rapid jaw-opening adjustment. The base casting has a coolant trough to keep things tidy, and can be mounted on a standard swivel base for angular positioning.

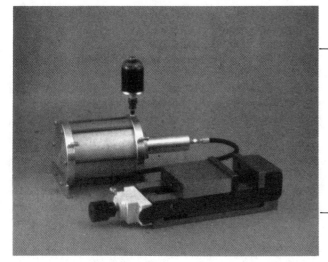

This development is similar to the one above but is built on a narrow base so that a series of these vises can be anchored snugly side by side on an NC machine bed. That way a number of parts in a production run can be clamped and machined at the same time with minimum center to center travel distance. Yoke plates on sides provide zero stationary jaw deflection. This is Model XXD60-6".

The movable jaw of this vise has a patent pending force amplification and intelligence system. The linkage on the hand lever applies high clamping pressure with minimal operator effort. The hand knob enables easy and rapid adjustment of the jaw opening. The optional intelligence attachment stops the machine tool if preset clamping force is not applied. This prevents machine and tool damage caused by loosely clamped parts. Model HPA60-6".

As the precision machine vise has become more of a production fixture in NC shops we have recognized a need to tie the machine tool and the fixture together with some control intelligence so that the machine does not cycle if the parts are not firmly clamped in the vise. So we developed this machine-control relay box that accepts the hydraulic pressure signal from any of the three vise assemblies on the previous page and relays that to the NC control panel.

In many NC machine applications it becomes appropriate to turn a vise up on its side in what we call the vertical position. For most vises this would require a T-slotted angle block to mount the vise with any degree of stability. To accommodate this situation we developed the Model VNC60-6" NC and vertical vise. Both the base and sides have T slots for easy clamping.

Not every setup that requires angular positioning of the vise can use the swivel-base unit, so we developed a simple protractor kit for the 6" standard Anglock vise. This allows the vise to be mounted directly to the machine table. It is used to accurately set the vise angle in relation to the machine table T slots. While the vise is clamped to the table, the protractor can be removed and used for setting other vises. Model D60.

No swivel bases are available for setting NC vises at an angle, so we developed the new NCD60 vise protractor clamp kit which allows the vise to be accurately set at any angle in relation to the machine table T slots and also serves as the clamp for one side of the vise. This unit is not removed from the setup once the vise is positioned, as is the case with the unit shown above. We have patents pending on both of the protractor kits.

Presented at SME's Jigs and Fixtures Seminar, December 1975

Modular Standard Components For Jigs, Fixtures, And Positioners

By Nihad Hamed
Lapeer Manufacturing Company

MODULAR PREFABRICATED COMPONENTS FOR JIGS & FIXTURES

A Jig or Fixture is a piece of equipment containing several parts, assembled together in a particular arrangement to perform such processes as welding, machining, fitting, etc., in order to hold, provide, or force a fixed relationship between the different parts.

Mechanical forces are necessary in a certain magnitude, direction, and point of action to act in the space of the Jig or Fixture.

A wide variety of Toggle-Action Clamps; manually, air or hydraulically operated are on the market and can provide the required forces.

To bring the action of these clamps or forces to the needed points of action, a clamp has to be seated and secured on different levels, angles, and elevations.

A modular system of parts and components could significantly save engineering time, technical know-how, machining, fitting, and valuable time in design and building Jigs & Fixtures.

Such modular components could be custom made for a particular application or purchased from their manufacturer.

The main components of such modular lines are described as follows:

(1-1)   To seat a clamp at low levels, we start by using shims of a selected standard thickness (i.e., 1/8 & 1/4 inch). These shims may have the same standard hole pattern as the regular support mentioned in (1-2). If there is enough thickness, we screw the clamp onto the shimplate; otherwise, we use shims identical to the clamp base.

(Fig.1)

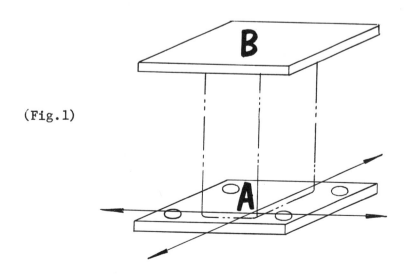

(1-2)   A simple support component of two plates-lower plate "A" with a
         fixed hole pattern and size, and a similar upper plate drilled
         and tapped to accommodate the selected clamp.

         The required height may now be made up by inserting the proper
         cross-section modulus which will withstand the stress exerted
         by the clamping action. This section could be made solid or
         hollow using square, rectangular or round tubing. For clamp-
         ing purposes a hollow rectangular cross-section is preferable.
         The tubing could be made up by welding two channels or angles
         together to form a rectangle, (Fig. 2) or suitable rectangular
         tubing can be ordered from any steel firm.

(Fig.2)

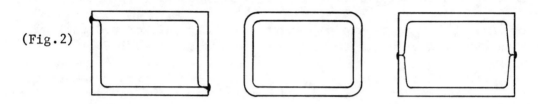

As one tubular cross-section could not cover the wide range of stresses
exerted by the different categories of clamps, three sizes of tubing may
be selected for light, medium, or heavy use covering most clamps of a
manufacturer's line, in order to get the right support for the right
clamp to avoid inadequate selections as (Fig. 3) and to get the proper
selection as in (Fig. 4).

(Fig. 3)

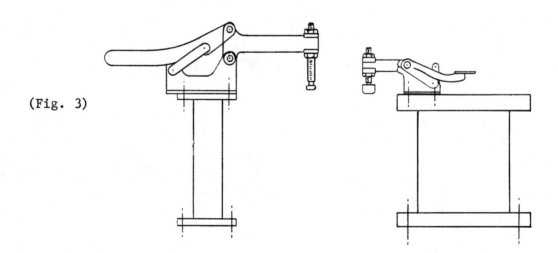

To standardize the height of the support, the rectangular tubing could be made up in inch or centimeter increments, etc., then the spindle on the clamp can be adjusted, in most cases, for final adjustment. (Fig. 5)

(Fig. 4)

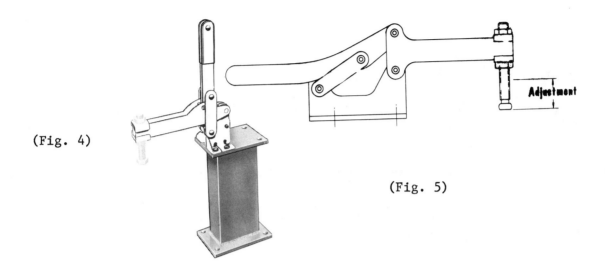

(Fig. 5)

To apply a vertical force "F" at point "A" (See Fig. 6), we select the proper size Toggle Clamp that will give us the necessary force. Then, we select the proper support that gives the proper height "H".

(Fig 6)

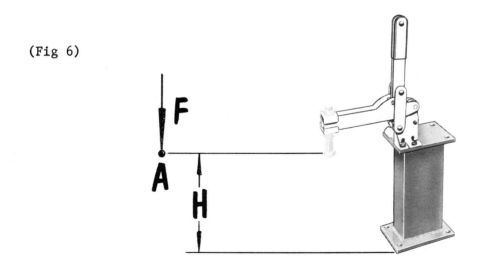

(Fig. 7)

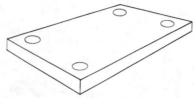

To bring the action of the clamp to point "A" without any measuring, drilling, or tapping holes at that point, we use an adapter plate. The adapter plate is identical to the lower plate mentioned in (1-1) with the only difference being that the holes are tapped transfer holes to fasten the clamp support onto. (Fig 7 & 8).

The advantages of using the adapter plate are:

(a) To eliminate the need for accurate measuring, including compensation for possible deviation caused by distortion and contraction due to welding.

(b) To eliminate drilling and tapping on the job or under the drill. Therefore, just by putting our clamp in its proper relationship to point "A" and by tack welding, we have our clamp in its proper place. Obviously, there is no special skill needed to get this procedure completed.

This plate can be removed later by cutting the welds.

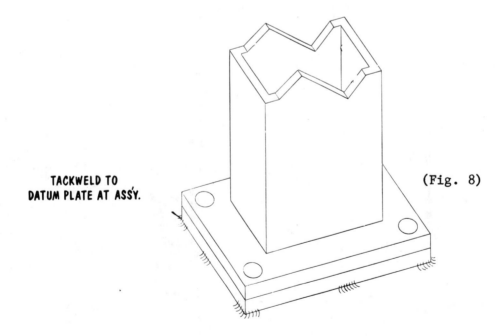

TACKWELD TO
DATUM PLATE AT ASS'Y.

(Fig. 8)

Secondary Plate (drilled and tapped) to be welded on the fixture in order not to weld the support permanently. This plate can be removed later by cutting the welds.

ADJUSTABLE SUPPORT

If the clamp spindle is in a horizontal position and therefore not vertically adjustable, an adjustable support may be used on the upper surface
of the standard support. This adjustable device consists of 2 spindles
and plates which could be raised or lowered as little as one-thousandth
of an inch - or up to 2 inches. (Fig. 9 & 10).

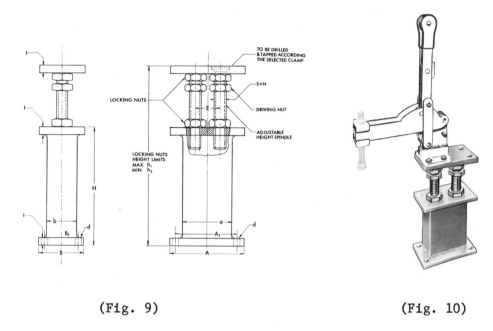

(Fig. 9)   (Fig. 10)

THE ADJUSTABLE INCLINED SUPPORT

Inclined clamping surfaces need special bases with a
special calculated inclination. An inclined support
with a tilting base and properly selected standard
height will bring the Toggle Action Clamp to the
point of action required. (Fig. 11 & 12).

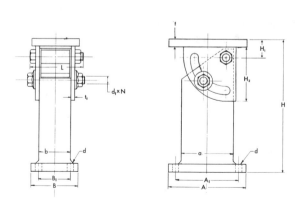

(Fig. 11)

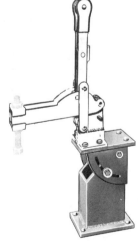

(Fig. 12)

(4) A clamp on a horizontal angle usually needs special preparation and
     requires added set-up time.

     However, once the proper height and size Rotary Support is selected
     and the clamp is screwed to the upper plate, the whole unit (with
     adapter weld plate) can easily be placed in its required position.
     Then, to secure the required horizontal angle, we lock the clamp
     position with the locking screw provided. (Fig. 13 & 14).

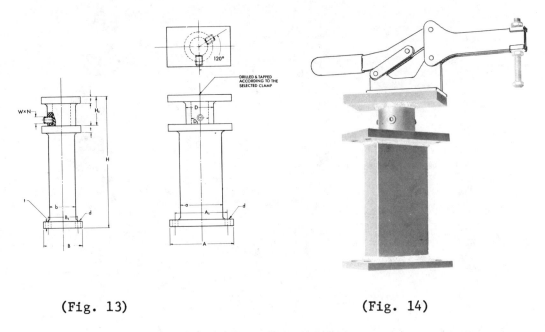

(Fig. 13)   (Fig. 14)

THE INCLINED SUPPORT

(5) The inclined supports are used when there is an obstruction to the
     support. Because their capacities are reduced due to the increase
     in their total reach, we use a higher strength support if needed.

     The inclined supports may have the upper plate
     horizontal or vertical to cover more needs.
     (Fig. 14, 15, 16, & 17)

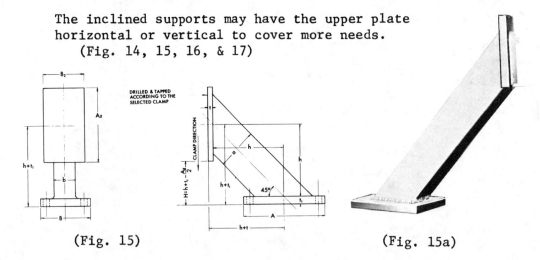

(Fig. 15)   (Fig. 15a)

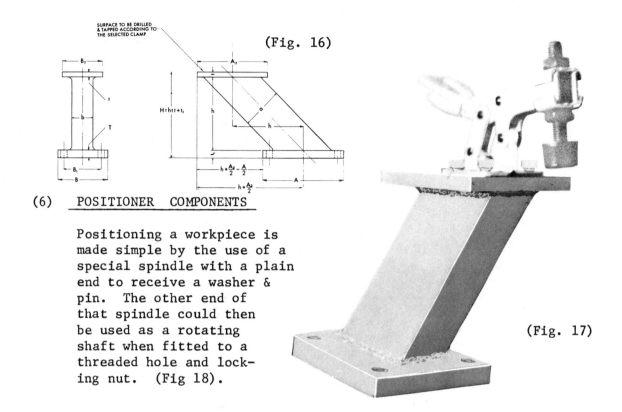

SURFACE TO BE DRILLED
& TAPPED ACCORDING TO
THE SELECTED CLAMP

(Fig. 16)

$H = h + t + t_1$

(Fig. 17)

(6)   POSITIONER   COMPONENTS

Positioning a workpiece is made simple by the use of a special spindle with a plain end to receive a washer & pin. The other end of that spindle could then be used as a rotating shaft when fitted to a threaded hole and locking nut. (Fig 18).

It is then attached to the threaded spindle and acts as the end of the shaft. Now the center of gravity is located by adjusting the screw in the radial slot. (Fig. 19).

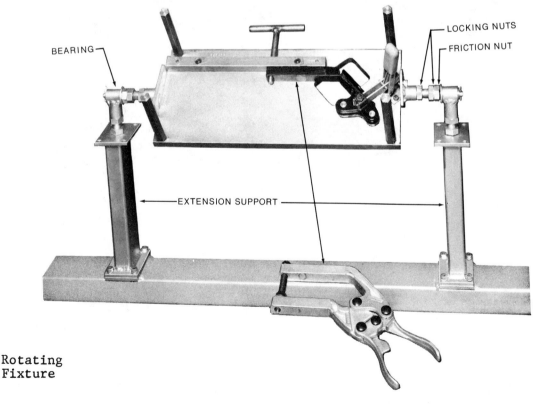

LOCKING NUTS

FRICTION NUT

BEARING

EXTENSION SUPPORT

Rotating
Fixture

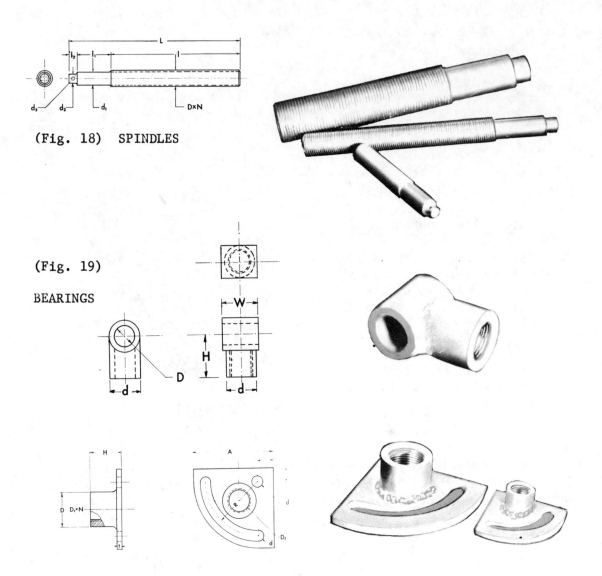

(Fig. 18) SPINDLES

(Fig. 19)

BEARINGS

(Fig. 20) OFF-SET PLATES

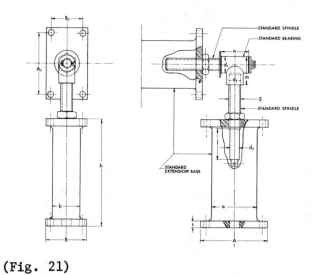

If the work piece is not
symmetrical, a special
off-set plate can be used
as a counter-balance.
(Fig. 20)

STANDARD SPINDLE
STANDARD BEARING
STANDARD SPINDLE
STANDARD EXTENSION BASE

(Fig. 21)

PIPE & CYLINDER POSITIONER

(7)   Positioning pipe or any cylindrical object can easily be achieved
       by using special casters mounted on plain supports for required
       height as shown. (Fig. 22)

(Fig. 22)

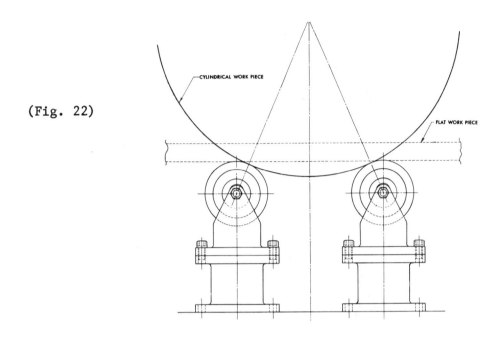

TURNING TABLES

(8)   Turn Tables consist of a Base Plate and a tubular axis to fit the
       plain end of the spindle which is mounted to a standard extension
       base by the use of a tapped hole and locking nuts as shown in ...
       (Fig. 23).

(Fig. 23)

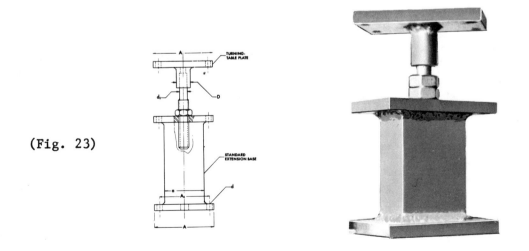

CANTILEVERS

(8)   Cantilevers can be fitted to any of the plain supports and act as a second support as shown in (Fig. 24). Cantilevers can be fixed at any level desired and provide the following functions:

    (a)   A lower jaw for clamping

    (b)   A lower clamp support

    (c)   An adjustable surface to carry the workpiece or job

    (d)   A positioner mount (by using the central tapped hole) when upper clamping is needed.

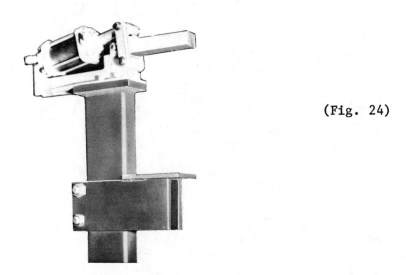

(Fig. 24)

(Fig. 24a)

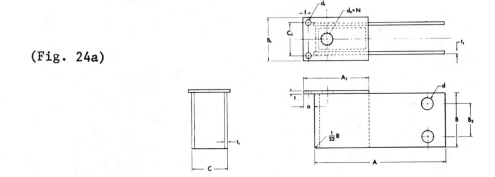

As a clamp is exerting a force, that force acts as a vertical or horizontal inclined line of action at a certain distance from the X-X axis at the lower plate. Needless to say, if we have a rectangular support cross-section, the most appropriate position for force "F" (at a distance "r") is in the plane Y-Y, because the modulus of the section around axis Y-Y is greater than around axis X-X. (Fig. 25)

(Fig. 25)

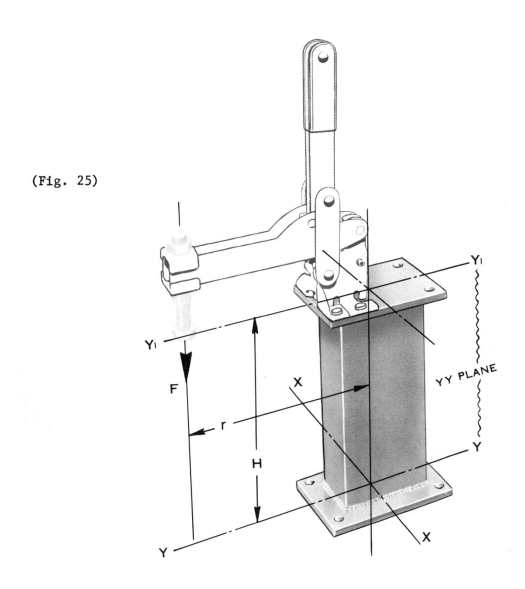

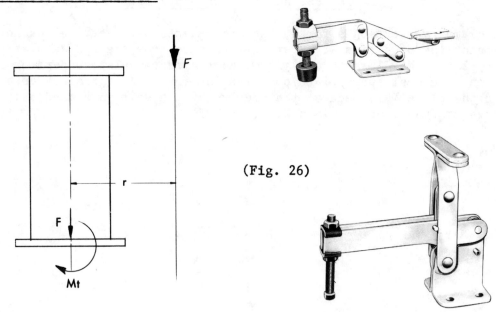

(Fig. 26)

A force "F" acting at a distance "r" gives, at the base of the support, an equivalent force "F" and a bending moment (Mt). The height, in this example, will not effect the value of "F" and "Mt". (Fig. 26)

HORIZONTAL
ACTION
CLAMPS

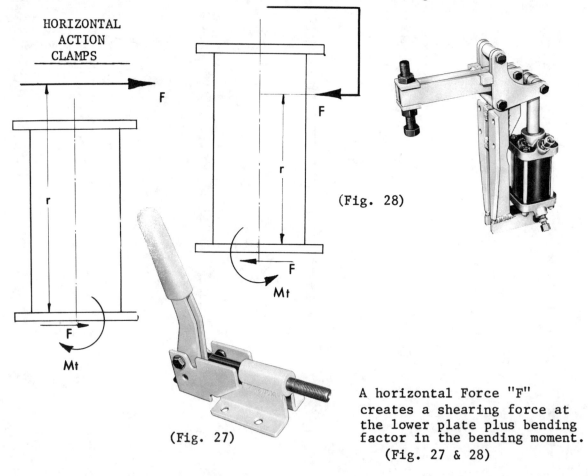

(Fig. 28)

(Fig. 27)

A horizontal Force "F" creates a shearing force at the lower plate plus bending factor in the bending moment. (Fig. 27 & 28)

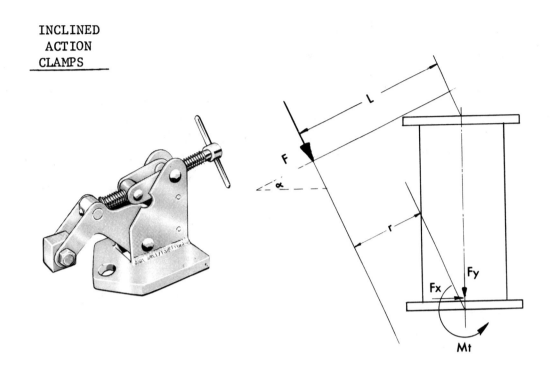

An inclined force acting at a distance "r" gives 2 components, Fx and Fy, plus the illustrated bending moment. (Fig 29 & 30)

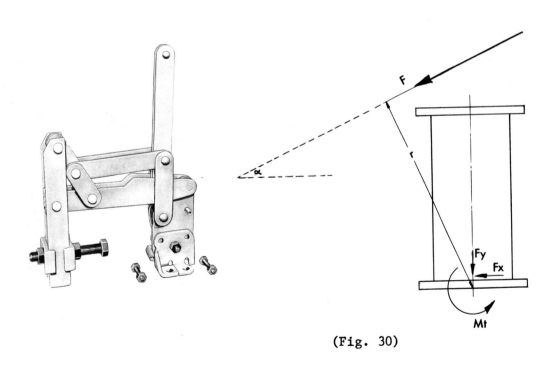

(Fig. 30)

25

SUGGESTED APPLICATIONS

Illustrated here are applications
of special components used in a
heavy turning fixture and a moveable fixture. There are endless
combinations in the use of these
standard components necessary for
specific design and manufacturing
needs.

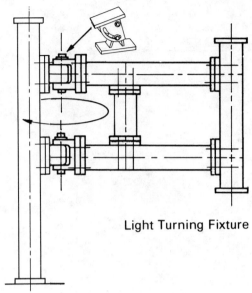

Light Turning Fixture

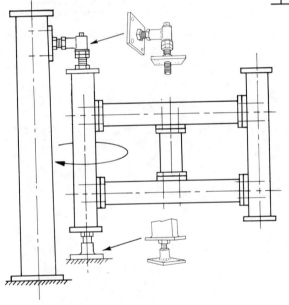

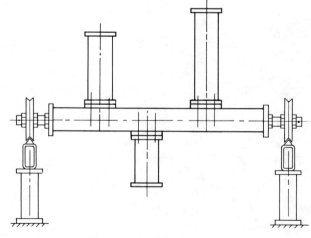

Reprinted from:
Manufacturing Engineering, March 1977

A Subplate System for Speedy Setups

– it provides a high-speed approach to fixture building

Precisely matched hole patterns and associated pull dowels are key features contributing to fast, accurate setups with a new line of subplate tooling components designed for use on NC and conventional milling machines. Hardened and ground angle plates and parallels as well as vises and indexers are quickly attached and accurately aligned with the *X* and *Y*-machine axes using screws and pull dowels. According to the manufacturer, Stevens Engineering Inc., Phoenix, AZ, these components eliminate the need for rigging and indicating and cut a typical half-hour setup job down to around five minutes. Making setups without the need for specialized tooling also provides significant cost savings.

The use of tooling plates, *Figure* 1, in conjunction with the subplate and other accessories provides a fast, inexpensive approach to building a wide variety of holding fixtures. The pull dowels locate in accurately jig-bored holes in the tooling plate and in corresponding bushed holes in the angle plate, parallels, or the subplate. Tapped holes in these components accept the 1/2″ (13 mm)-13 x 1 1/2″ (38-mm) long screws. Setups made with the system are unusually strong and rigid, an important factor with respect to safety and the use of high feedrates in machining.

Subplate Design. Subplates are made from stabilized carbon steel

1. DOWEL HOLE and screw hole patterns in the tooling plate match precisely with those of other subplate system components.

ground flat and parallel to within 0.0005″ (0.013 mm). Fully hardened units are optional. Precision 1/2″ (13-mm) ID bushings are located in jig-bored holes at 5″ (127 mm) intervals. These bushings are used for locating and aligning subplate accessories which are then attached using holes located at 1 1/4″ (31.8 mm) intervals between the bushings.

The subplate fastens to the machine table T-slots with T-nuts and capscrews. A coolant trough around the

on production parts. The pull dowel setup technique helps assure precise repetition from one run to the next. A special threaded-end tool simplifies insertion and extraction of the 1/2″ (13-mm) diameter pull dowels.

Versatile Vise. One of the most versatile accessories in the system is a force control vise, *Figure* 2, that has subplate perimeter channels cutting fluid to screened outlets at each end.

One of the advantages of the system is the uniformly accurate results

a friction clutch in the leadscrew assembly which can be set to any torque level desired. The clamping force established is constant and repeatable. Because the fixed jaw is the primary reference locator for parts machined in a vise, deflection of the jaw under load must be constant from part to part for precision machining. The Model 400 vise accomplishes this automatically without the need for time-consuming settings with a torque wrench. Fragile parts are protected

2. BOTH FIXED and movable jaws of the force control vise are quickly and accurately aligned.

from damage caused by overtightening by setting the friction clutch for the desired clamping force.

One problem with many vises is that the base tends to bow up in proportion to the clamping force applied. In this vise the subplate itself is the base, assuring solid setups with minimum jaw deflection.

Alignment of the fixed and movable jaw assemblies with the machine axes is immediate and automatic when the pull dowels are inserted through the jaw members into the subplate. Two screws threaded into the subplate through each jaw complete the setup. Jaws can be located on the subplate in separation increments of 2 1/2″ (64 mm) up to a maximum depending on subplate length. Range of movement of the movable jaw is 3″ (76 mm).

Savings Potential. In a typical application, machining operations on a pump body require three setups on a mill. The first setup, *Figure* 3, requires a bridge fixture for drilling, tapping, milling, and boring. The second and third setups require an angle plate and a flat plate holding fixture, respectively.

The conventional approach to machining this part involves building two fixtures — a combination bridge and angle plate holding fixture for the first and second setups, and a plate holding fixture for the third setup. Approximate total investment in

3. TYPICAL BRIDGE FIXTURE arrangement using a tooling plate and two parallels.

tooling with this approach is $417.

Using the new subplate components, a register diameter is bored in the tooling plate, the plate is then drilled and tapped for clamps, and a removable diamond pin is installed. Total tooling cost is $155 for a fixture suitable for all three setups. That's a total savings of $262 in tooling expense compared to the conven-

tional method.

Savings in setup costs are also attractive. An estimated 115 minutes would be required to make the three setups conventionally. This compares to just 18 minutes when using the Stevens subplate tooling system. At a shop rate of $25 per hour, this time differential translates into a savings of over $40 every time the job is run. ∎

Reprinted from Stanki I Instrument, Vol. 45, Issue 2, 1974

STRENGTH OF TEE-SLOTS IN BASEPLATES

STANKI I INSTRUMENT, VOL. 45, ISSUE 2, 1974, pp. 22

A.S. SHATS et al

U.D.C. 621.9.06-229.3-115-182.7

A method of determining maximum permissible stresses in Tee slots of fixture baseplates and machine-tool tables, allowing for slot length and design features.

Tee slots in baseplates of universal composite fixtures take considerable loads when threaded joints are tightened up on assembly of the fixture, when workpieces are clamped to the fixture and when cutting forces are exerted during machining. With minimum slot depth, the design and dimensions of the slot must ensure good strength and life of the fixture. If deep slots are required, the baseplate dimensions must be increased so as to maintain the required stiffness, and this is often undesirable and sometimes impossible.

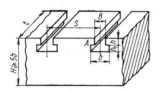

Fig. 1.
Baseplate with Tee slots.

Strength calculations based on mechanics and theory of elasticity methods are very complex and give only a very approximate picture of the stresses in the slot elements. It is simpler and more accurate to assess slot strength experimentally by photoelasticity methods[1].

The stressed state of the slot is determined by tearing forces P_{work} acting normally to the plane of the baseplate. On the basis of previous research[2] it is known that maximum stresses in the slot section occur at point A (Fig. 1). As a rule, in most baseplates the slots are in two mutually perpendicular directions. With a dimension of $t = (1,5-2)\ell_b$, therefore, where ℓ_b is the length of the bolt head, the three-dimensional stressed state can be reduced to a two-dimensional stressed state.

To assess slot strength, photoelasticity tests were performed on plane models of ED−5−P optically active transparent material. The models were made with different slot element dimension ratios $S/B, h_1/B, b/B$ and h/B (see Fig. 1) in 1:1 scale. The basic slot was taken to be one with $h = 15$mm, $b = 24$mm, $S = 60$mm and $h_1 = 8,5$mm.

The effect of the parameters on slot strength was studied on the basis of Tee slot designs used in existing fixtures and machine-tool tables, and allowing for GOST Standard 1574−62 for Tee slot designs. The load was applied by loading device 3 (Fig. 2). The stresses at point A (see Fig. 1) were measured by the compensation method using a KSP−5 co-ordinate-

Table 1

Test conditions	C_v	yv	m	zv	K_c for drills					K_1 with ℓ/d		
					Milled	Ground from solid with ℓ/d				4—7	7—12	12—16
						4—7	7—12	12—16				
Normal......................	16,3	0,57	0,25	0,4	1	1	1,07	1,19		1,14	1	0,77
Arduous	12,1	0,57	0,25	0,4	1	1	1,21	1,41		1,47	1	0,72

Note: In the first approximation the test conditions are considered normal when empirical coefficient a assumed and checked for drills with d = 5—8mm equals $a = vs/d \leqslant 0,85$mm, and arduous when $a > 0,85$.

hatched columns P values; the horizontal lines indicate theoretical values, while the broken and continuous lines and numbers 1—3 have the same meaning as in Fig. 1.

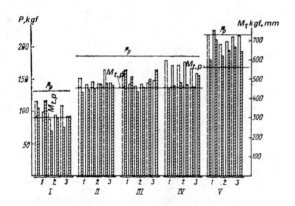

Fig. 2.
Effect of drill production technology and length on force and torque in drilling: I—V— drilling rates.

Analysis of the results showed good coincidence of the experimental results and the results calculated in accordance with the standards[1,2]. $M_{t.e}$ and P_e are practically unaffected by the value of ratio ℓ/d on the drill production technology, i.e. the anticipated difference between the drilling conditions with milled and ground drills was not observed[3]. With increasing stiffness of the machine, force P_e tends to increase, and torque $M_{t.e}$ to decrease (see Table 2). This is also confirmed by the oscillograms in Fig. 3, which show that the greater the stiffness of the machine, the more clearly defined the peaks on the P_e and $M_{t.e}$ curves. The test rig had a static stiffness 3—4 times greater than that of the basic drilling machine.

Figure 4 shows the variation in drill wear h_{C1} (with ground flutes) plotted against drilling rates (I—V): rate I corresponds to a feed of 131mm/min, II to 159mm/ min, III to 223mm/min, IV to 279mm/min and V to 288mm/min. It will be noted that wear h_{C1} depends on the feed per minute: the greater this feed, the more rapid the clearance-face wear.

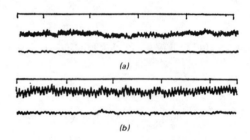

Fig. 3.
Oscillograms of variation in axial force (upper curve) and torque (lower curve) when drilling in Model 2A125 machine (a) and in rig (b) (time marks at 0,01s intervals).

To investigate the effect of drill length and production technology on drilled hole quality, holes of 8mm diameter were drilled in sandwich blanks in a Model 2A450 jig-boring machine. After drilling, the triple-layer blanks were dismantled and the hole displacement and diameter were gauged through the entire hole depth. Maximum co-ordinate read-off error was 0,002mm, and drill land eccentricity at setting-up did not exceed 0,05mm. Fig. 5(a) shows the results of tests with s = 0,16mm/rev using 80mm long drills (curves 1), 115mm long (curves 2) and 165mm long (curves 3), while Fig. 5(b) shows results with 165mm long drills with a hole depth of h = 1mm (curves 4), h = 25mm (curves 5) and h = 41mm (curves 6).

Table 2

Machine model	Drill type			s (mm/rev)	v (m/min)	P_e(kgf)	P_p (kgf)	$M_{t.e}$ (kgf.mm)	$M_{t.p}$ (kgf.mm)
	d	GOST	Production method						
2A125	8			0,17	32,9	152	185	450	456
Rig	8			0,17	32,9	170	185	426	456
2A150	8	10902—64	Ground from solid	0,19	35,0	172	199	308	499
2A125	9			0,22	33,9	243	248	802	700
Rig	9			0,22	33,9	252	248	704	700
2A125	13			0,28	29,9	480	426	1650	1770
Rig	13			0,28	31,1	570	426	1060	1770
2A125	18	10903—64	Milled	0,28	29,3	530	590	3340	3890
Rig	18			0,28	26,6	501	590	2610	3890
2A150	18			0,28	29,3	587	590	3890	3890

Notes: 1. P_e and $M_{t.e}$ are experimental values of axial force and torque; P_p and $M_{t.p}$ are theoretical values.
2. Test rig based on Model 2A125 machine with the column, table and base joined by a single reinforced concrete bed.

Presented at SME's Jigs and Fixtures Seminar, December 1975

Toggle Action Clamp and Its Application
By Nihad Hamed
Lapeer Manufacturing Company

WHAT IS A TOGGLE ACTION CLAMP

It is a mechanical leverage device that increases a force exerted by human effort or through the effort of oil or air pressure.

The leverage ratio could go up theoretically to infinity and practically to a hundred times.

HOW TOGGLE ACTION MECHANISM WORKS:

A toggle action mechanism is a two link mechanism that hinges as shown.

When a force (P) is acting at the hinge (A), this force could be analyzed on the two axis (AB & AC) and we get the force (P1 & P2) acting along the two center lines of the two links.

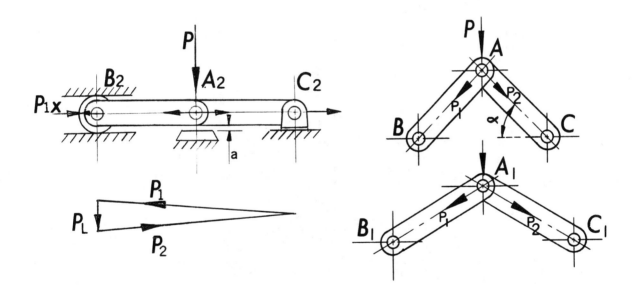

The triangle of force for that link is as shown.

If (C1) becomes a fixed hinge and (B1) a horinzontally guided roller, and if we assume that the roller is moving against a horizontal resistance (-Plx).

It is clear from the 2nd force triangle that (Plx) is increasing as long as the force (P) is constant and moving down to the flat positions (B2 A2 C2).

31

At that position, if we remove (P), we get an unstable
equilibrium and point (A) tends to regain its upper
position or goes to a lower position.

To provide a permanent lock at that position, (A) has to
travel down an additional distance (a) and to stop against
the locking stop (usually a few thousandths of an inch).

This will create a locking force (PL) and the whole joint
is now in a stable equilibrium.

If the clamp should now be de-clamped, another force (PL)
acting in opposite directions is able to unlock the clamp.

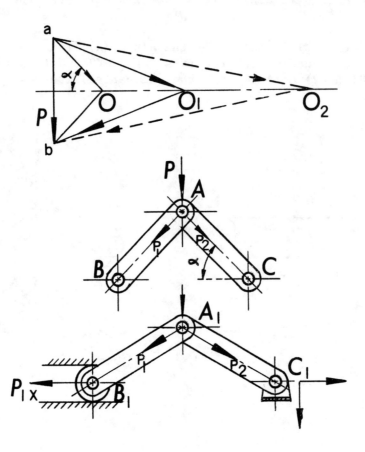

Plotting the triangle of force of Oab, O_1ab and O_2ab, it
is clear that with the same force P, the forces P_1 and P_2
increases. And when O is infinity, P and P equal infinity.
In practice, the buckling, bending, and deviation of the
link provide a determined value of P_1 and P_2.

HOW THE CLAMP HANDLE ACTS:

In the illustrated drawing (P1) acting at the edge of the
hinged lever at (a) the reaction (P) at (b) is equal.

$$P = P1 \times \frac{L1}{L}$$

That means, besides the toggle action multiplication of
force, we are still able to increase the mechanical effort
applied by means of the clamp handle.

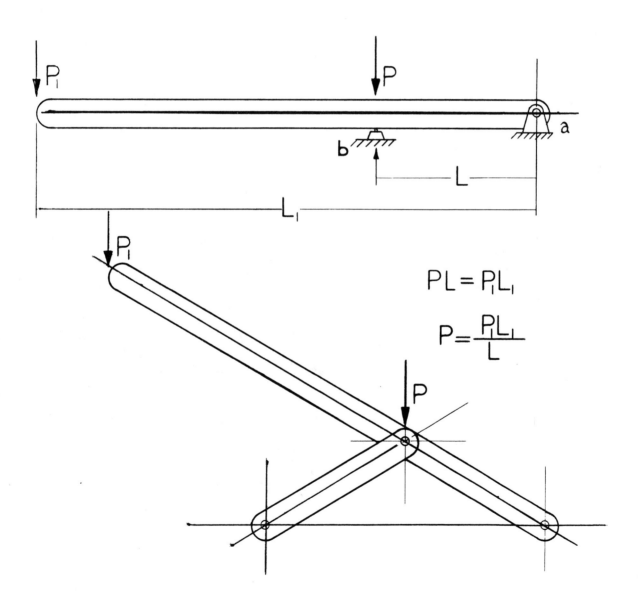

$$PL = P_1 L_1$$

$$P = \frac{P_1 L_1}{L}$$

REVERSED TOGGLE ACTION:

In this case the two links are not equal and the distance
between (B) and (C) is smaller as shown.

The triangle of force becomes (abo).

If the angle (α) becomes smaller keeping the action and the
magnitude of force (P), the triangle becomes (abo) with the
result of increasing (P1x) as shown.

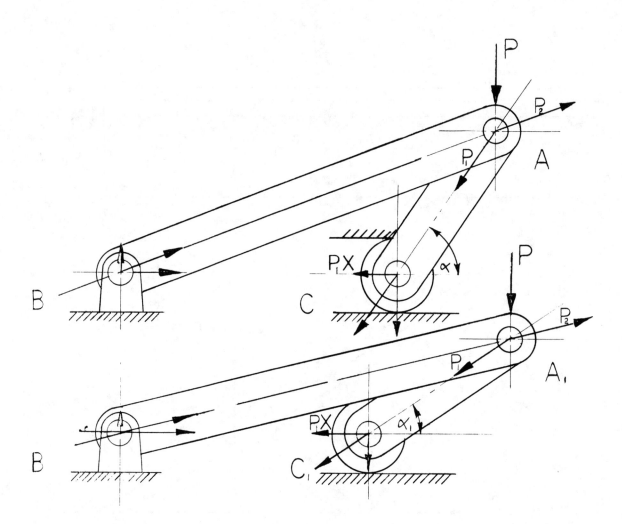

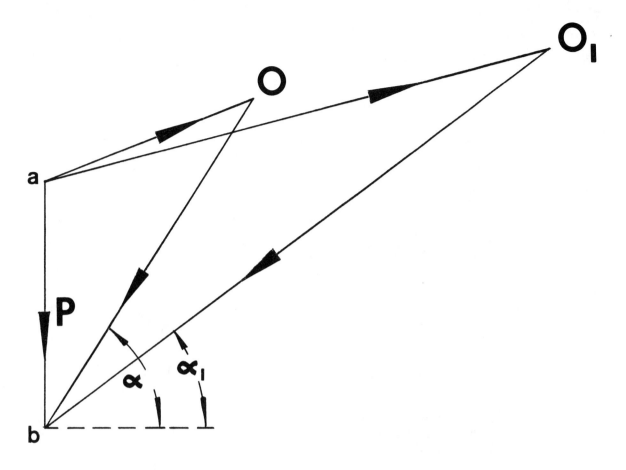

DOUBLE TOGGLE ACTION CLAMP:

It is a clamp actuated by two sets of toggle action
mechanisms to get an inner horizontal force when the
handle of the clamp is in the regular vertical position, and,
at the same time, a higher mechanical advantage.

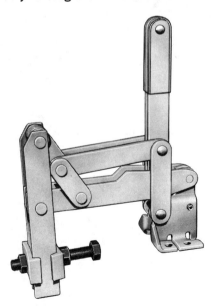

THE CAM ACTION CLAMP:

The ordinary toggle action clamp does not permit much variation at its locking position.

Therefore, if clamping is needed for a variation of thicknesses, a cam action clamp using the toggle action could be used.

It has a cam and follower and it locks at several positions.

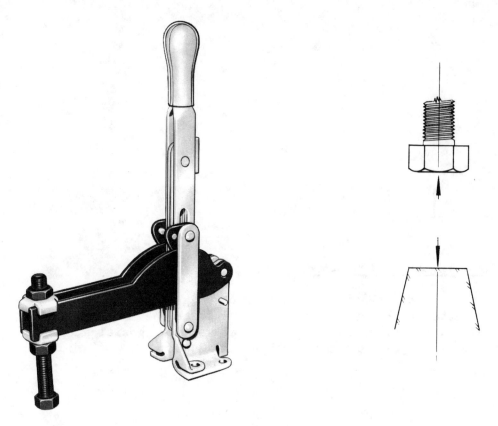

In short, we can summarize the most important features of toggle action clamps as follows:

(1) Very high ratio of leverage not existing in cam-lever, screws, wrench, etc.

(2) It is very fast.

(3) It has a very good and secure method of locking which may hold as long as it is needed without additional human effort or pressure.

THE VARIOUS SHAPES & FORMS OF TOGGLE ACTION CLAMPS

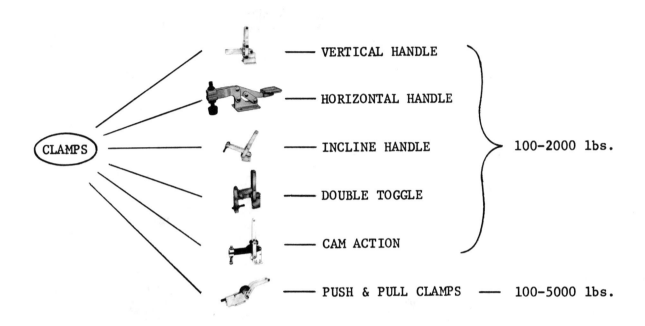

CLAMPS
— VERTICAL HANDLE
— HORIZONTAL HANDLE
— INCLINE HANDLE
— DOUBLE TOGGLE
— CAM ACTION
} 100-2000 lbs.

— PUSH & PULL CLAMPS — 100-5000 lbs.

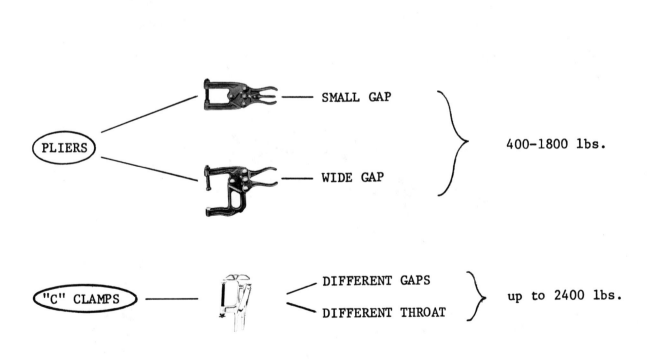

PLIERS
— SMALL GAP
— WIDE GAP
} 400-1800 lbs.

"C" CLAMPS —
— DIFFERENT GAPS
— DIFFERENT THROAT
} up to 2400 lbs.

AIR-OPERATED (OR OIL) FOR AUTOMATION

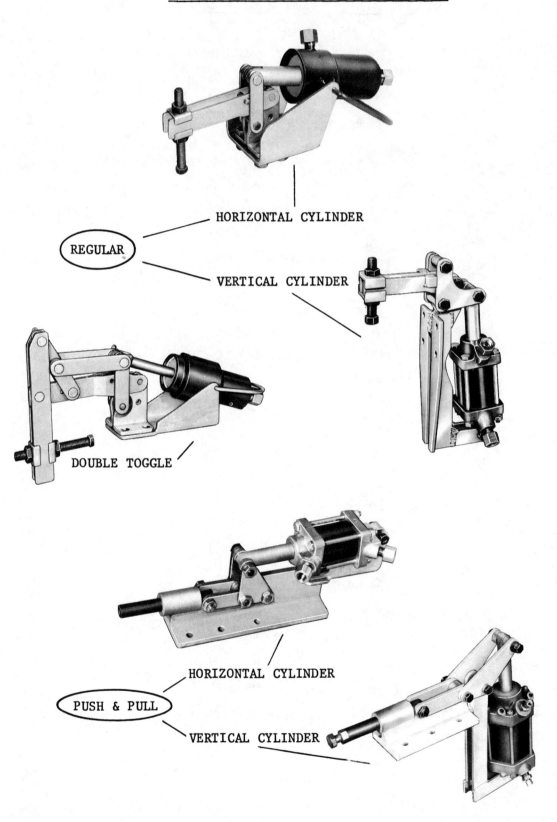

HORIZONTAL CYLINDER

REGULAR

VERTICAL CYLINDER

DOUBLE TOGGLE

HORIZONTAL CYLINDER

PUSH & PULL

VERTICAL CYLINDER

(4) It is simple, easy to get, and to maintain.

(5) As it is made in volume, it is inexpensive.

(6) It has its handle, mounting base and holes incorporated into it.

(7) It has an adjustable spindle to take care of wear and can be adjusted accordingly.

(8) The designer has a very wide selection to fit any profile or to fit any application.

(9) Air Clamps are essential for automation and for mass production.

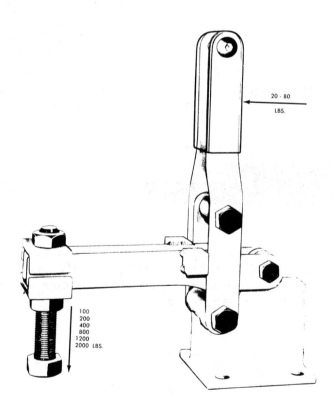

WELDING FIXTURE

A welding jig or fixture is a piece of equipment containing several parts that form a weldment in such a way, that it keeps a fixed dimensional relationship among the different parts.

It has to fulfill the following requirements:

1. To <u>accommodate</u> each part in its proper place.

2. It has to <u>hold</u> it in its place with minimum deviation, allowing for reasonable expansion without warping or twisting.

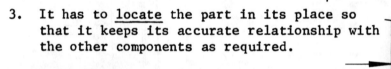

3. It has to <u>locate</u> the part in its place so that it keeps its accurate relationship with the other components as required.

4. It has to be capable of performing some straining effort to bring the different components together, specifically when welding more than one sub-assembly.

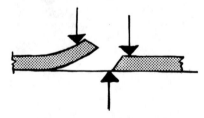

5. It has to give effective support and backing to all parts in order to prevent any deviation.

6. It has to be simple in shape and light in weight
    without any overdesign.

7. It has to permit easy access to welding with mini-
    mum parts exposed to sputters and heat, especially
    screws, pins, hinges, etc.

8. It has to provide the maximum flat or horizontal
    welding position and has to avoid overhead and
    vertical welding positions.

9. The welded part must be easy to remove from the jig
    or fixture with the least effort. And, in cases
    where effort is needed, some push clamps or cam-
    levers can do the work of the ejections.

10. The fixture has to be accurate to prevent any need
    for additional machining, grinding or any other
    preparation.

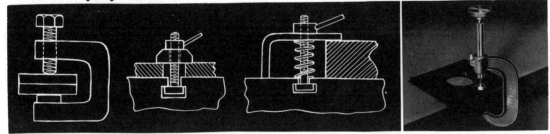

11. It must be rigid enough to compensate for any strain
    exerted by the expansion of welded parts.

12. It has to be robust enough not to deflect under its
    weight due to improper support.

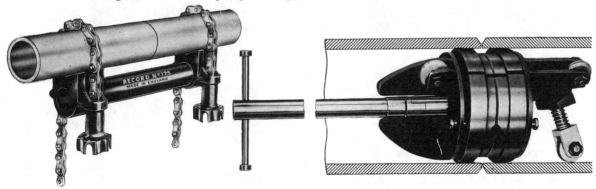

41

13. Each part must fit into its own proper place without the possibility of having to use any other part.

14. A welding jig or fixture has to be designed in light of its accuracy, durability, number of weldments to be made, etc.

15. Parts in a jig which are subjected to excessive wear and abrasion must have hard shoes or a hard face welded layer.

16. A fixture has to be a good conductor of electricity and provide good grounding.

17. An important welding fixture has to be easy to maintain stored in a proper place and kept in good condition.

18. A fixture must have enough tolerance to take care of different deviations and dimensions of rolled steel parts or steel castings which usually do not come with the same tolerance.

19. A fixture has to be a unit with all the parts, such as clamps, pins, levers, an integral part of it.

20. Where components are jig drilled or machined for other purposes, advantages should be taken of the holes and faces for positioning in the welding jig. Fitted pegs and machined stops can often take the place of more complex fitting.

21. Fixtures which need to be turned or tilted during welding must have an easy means of support as legs or curved surfaces; we have to reduce this turning to a minimum to prevent "welder fatigue" from taking place.

22. Proper bed-plates are helpful when used. They may be heavy iron with T-slots or may be made of rolled steel of I-beams and channels as grillage with or without a sheet steel top of a proper thickness.

    An important feature of bed-plates is the drilling, cutting or welding of parts to it. Those parts could be restored by grinding. Consequently, the bed-plate will then be ready for re-use.

23. All locking means as clamps, wedges, pins, etc. must give a firm and definite locking capability.

24. A welding fixture has to be designed so that it could be used again for other similar purposes, sizes, shapes, etc. (standarization through families).

25. Machining and tool room work are highly expensive and time consuming. Therefore, they should be avoided as much as possible and replaced by other means of adjusting, such as ready made components of jigs and fixtures.

26. Accuracy means cost. The greater the accuracy, the more we pay. Accuracy should correspond to the type of work required. Therefore, accuracy should be practiced, bearing in mind the function and purpose of the welded part and where it is going to fit.

POSITIONERS

A positioner is a handling device or machine that enables
a workpiece to be put into a variety of positions.

Positioners may vary in complexity from a simple turntable,
axle, to a highly mechanized and controlled machine.

*Special purpose rotating frame jig with
hydraulically-operated lever clamps. Quick release
C clamps used for positioning and holding bulb
angle*

The main purpose of a welding positioner is to allow welding to be done in a flat position. A properly designed
positioner has many advantages:

     1. <u>Increased production</u> by using <u>larger</u> guage
electrodes, higher currents and welding machines and,
therefore, higher deposit rates, fewer runs and less
descaling.

     2. <u>Improved quality</u>, because there is less
possibility of slag inclusions, undercut and operator
fatigue.

     3. <u>Less skilled operators</u>, because welding
is easier and there is greater control over welding
proceedure.

     4. <u>Reduced handling time</u> for cranes and other
shop equipment while welding is being done.

Rocker-type positioner of 8 ton capacity with power driven variable speed table.

5. <u>Require</u> <u>less</u> <u>floor</u> <u>space</u> than cranes and
are removed entirely when they are not in use.

6. <u>Greater</u> <u>safety</u> because the assembly is
firmly clamped down and there is no chance for turning
over or slipping accidentally.

Variable speed range positioner with vertical adjustment of table

Special-purpose rotating frame jig for welding of mineral waggons

Relieve your pins for better fixturing

Have you overlooked your fixtures as a possible source of some of your machining problems? More than a snug fit is required for good location control. Here are some new angles that can help you get the most out of your pin fixtures.

By JACK LAUDICK
Lyons, Kans.

Don't be misled by the simple form of locator pins. There are right ways and wrong ways of using them in jigs and fixtures. They can either help or hinder in providing for accurate and easy work location, as well as rapid work loading and consistent quality machining.

The classical simple pin fixture has two pins. They are located on the fixture so as to mate with two specific existing holes in a part, such as in a bolt circle, thus consistently locating parts in a desired fashion. This location system is simple, positive, and relatively foolproof compared to edge location. It is so superior that it is often advisable to add holes in a part solely for fixturing. These locator holes might be added to portions of the part that will be removed when the part is finished.

However, a compromise is required at the very outset of using a two-pin fixture. For close location control of the part, each pin diameter should be virtually the same as the hole it is to fit into. But easy part loading and unloading requires a loose fit. The compromise is a snug slip fit.

Then there is the problem of machining consecutive parts. If all the parts in a lot are to fit the same two fixed pins in the fixture, the

spacing between the two locating holes must be held constant. A snug pin fit allows no spacing leeway. Maintaining such a close control of hole spacing in workpieces is costly and impractical if they're only needed for fixturing.

Using smaller pins is not an effective answer. Some leeway is gained but location control is sacrificed. For example, parts with holes spaced closer or father than nominal would fit snug and seem to locate well. But parts with nominally spaced holes could locate loosely in a variety of positions. The possible angular error in location would magnify dimensional errors in the workpiece extremities and the machined part

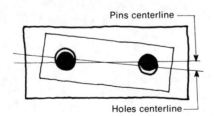

Pins centerline

Holes centerline

geometry would be erratic. Quality demands a snug pin fit at both holes.

Returning to their drawing board, tooling designers very early recognized the basic error in attempting to use two full round pins

as locators. That approach attempted to provide full location at two positions simultaneously. Analysis showed that a part fixtured on one full round pin is constrained in all respects except that it can rotate about that pin. The second pin is needed only to pre-

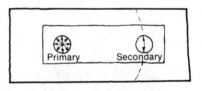

Primary   Secondary

vent that rotation and, to do that, it requires a full diameter on only a limited part of its surface. The rest of the full diameter surface contributes nothing to part location except interference. Thus, shops soon learned that to provide full constraint, only one of the pins needed to be full round and, to avoid interference, one must be relieved. The easily produced diamond pin came into common use to provide that relief.

As an example of the relief possible, consider that chords 0.030 in. on either side of a 1.000-in. diameter hole are 0.002 in. shorter than the diameter itself. Thus, a pin with a full diameter portion

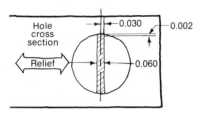

0.002 in. smaller than the hole diameter and relieved in all other areas would essentially be free to move laterally within a zone 0.060 in. wide. The 0.002 in. difference between hole and pin diameters is translated into 0.060 in. relief.

It is apparent from these figures that the larger the proportion of full diameter area on the pin, the less it will function as a relieved locator and the more it will act like a simple round pin. Various problems in using pin fixtures—hole damage, pin breakage, loading/unloading difficulties—can be traced to pin designs that do not provide sufficient relief. The patented oval locator was conceived to provide more relief than

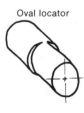

Oval locator

Diamond locator

the commonly used diamond pin locator.

Pin orientation is another source of problems in using a relieved pin in fixtures. Various

users seem to believe that the full round portion of the pin should be oriented perpendicular to the surface of the part being cut. However, the relieved locator pin should prevent rotation of the part about the primary round locator

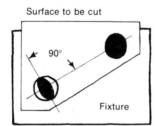

Right

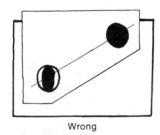

Wrong

pin. As such, it must be installed with its major axis oriented 90 degrees to a centerline joining both pins. Any other alignment results in some loss of relief.

There is another adverse effect to improper pin orientation. The secondary hole located by a relieved pin tends to position along the minor axis of that pin. If this

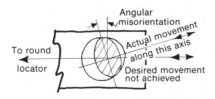

axis is not in line with the centerline joining the primary and secondary pins, angular misorientation can occur. With this improper orientation, slight, otherwise unexplained variations in machined part geometry can result even with snug fitting pins.

Here are some ways you can use relieved locators.

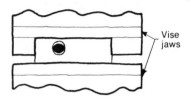

Relieved pin allows edges of vise jaws to establish location in one direction.

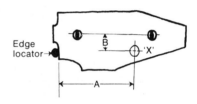

Hole at X is machined to close tolerances relative to two planes—one established by simple edge locator.

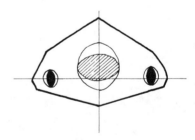

Three relieved locators provide full constraint without interfering with loading and unloading.

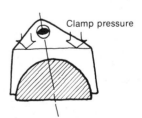

Clamp pressure holds part against large, half-round locator. Addition of secondary relieved locator solved problems caused by cocking of parts and clamping distortion.

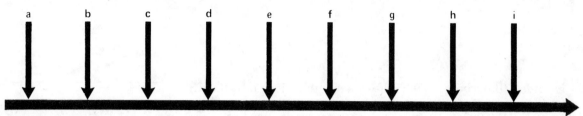

VALUE ADDED ASSEMBLY SEQUENCE (SERIES FUNCTION)
(SNYCHRONOUS OR NONSYNCHRONOUS)

a   b   c   d   e   f   g   h   i

LOWEST COST VALUE   HIGHEST COST VALUE

SCHEMATIC 1

A typical value-added series-function sequence. Shown here as an in-line sequence, it can also be designed in a rotary or closed-loop sequence. The primary defect in assembly via a series function is the rapid fall-off in efficiency, since each station contributes its own inefficiency to the overall inefficiency of the system. As an example, if each station in this sequence has an efficiency factor of 99.1%—which is high for any given station—the efficiency of the system is 0.991^9 or 92.1%, which is unacceptably low. Given a large enough number of stations, this type of system is theoretically unable to produce finished assemblies.

Reprinted from Manufacturing Engineering, October 1979

Systems Engineering in Advanced Assembly

SUBASSEMBLY MACHINE modules which are part of an automatic Demand system designed for the automatic assembly of hydraulic control valves. Rate of production: 1500 per hour.

The growing complexity of modern manufacturing is bringing systems engineering to the fore in parts assembly. Here's a look at the systems approach as it has been developed by Swanson-Erie

DANIEL B. DALLAS
Editorial Director

VALUE ADDED SEQUENCE (SERIES PARALLEL)

SUPPLY MODULE

a b c

50/MIN

BUFFER STORAGE

REJECT FUNCTION

SUPPLY MODULE

d e f

50/MIN

REJECT FUNCTION

SUPPLY MODULE

g h i

50/MIN

REJECT FUNCTION

47/MIN

40/MIN

DEMAND MODULE

SCHEMATIC II

A typical value-added series-parallel assembly sequence. Sometimes called the Supply and Demand System—and more often simply the Demand System—its design is predicated on the ability of the supply modules to provide perfect sub-assemblies to the demand module. To accomplish this, the supply modules operate at a pace faster than the cadence of the demand module (which may be of in-line or closed-loop design). The supply modules will periodically produce defective or incomplete subassemblies, which are rejected. However, the increased speed of the supply modules means that their buffer storages will always be full and waiting for part pickup by the demand modules.

HIGH CYCLICAL SPEEDS, OPTIMUM yield, minimal parts shrinkage, and much greater productivity... These are the criteria characterizing advanced assembly systems in American industry. Making the business of automatic assembly even more difficult is the fact that tolerances are growing tighter and individual assemblies are becoming increasingly complex. Time was when the Model T contained some 5000 individual components; today there are more than 5000 parts in an automatic transmission. The import of this is that if assembly had remained unautomated, product design would have stagnated and total part costs would have gone through the roof. To provide the necessary automation, builders of automated assembly equipment — and their customers — are breaking new ground in the field of systems engineering. An outstanding example is seen in the work of Swanson-Erie, an Erie, PA builder of automated machine tools and assembly equipment.

Synchronous v. Nonsynchronous. Swanson-Erie's approach to assembly machine design involves using the best of the synchronous line concepts, while eliminating certain of its deficiencies.

Normally, nonsynchronous lines present problems in machine dynamics — synchronous lines do not. However, nonsynchronous lines give the individual operator a greater degree of flexibility in performing work assignments at any given station. More specifically, the operator is not geared to the line's cadence — only to its average throughput rate. In addition, some production

managers favor the options open to them with a nonsynchronous line in the event of downtime.

Thus, the nonsynchronous line suffers the twin drawbacks of relatively crude machine dynamics and resulting slow operations — usually in the neighborhood of 900 pallets or parts per hour maximum.

A principal kinematic drawback to the nonsynchronous line is the sequencing of machine functions at an operating station. As an example, if a nonsynchronous line is designed to have a gross throughput of 900 parts or pallets per 60-minute operating hour, its cycle time is a mere four seconds. A portion of this four-second interval is utilized in moving the palletized workpiece into and out of the station. Still another segment is used for locking and unlocking the pallet in position. The remainder of the four-second interval is used to actually perform the work and to verify its successful performance through probing, which is also sequentially performed.

The significance of this is that while the nonsynchronous line is more operator-flexible, its cadence must be slow enough to provide for both sequential work transfer and work performance. As a result, it is not capable of performing higher speed assembly operations without loss of dynamic control.

Swanson's response to high output assembly requirements has been the development and production proving of two assembly systems utilizing high-speed synchronous machine elements in conjunction with buffer banks and

manual assembly loops. One system development is called the Demand System; the other is the Swanson SynchroBank Assembly System.

THE DEMAND SYSTEM

In addition to the potential of high-speed operation, the Demand System has still another advantage — it produces perfect parts or subassemblies to the downstream assembly modules. Faulty and incomplete subassemblies can be produced, of course, but they are eliminated in the modules before they reach the intermodule connecting tracks. As a result, only acceptable products come off the end of the line.

Calculating Efficiency. This feature of the Demand System can best be appreciated by examining a conventional in-line system involving nine discrete operations. (See Schematic Drawing I.) If an efficiency factor of 99.1% is assumed for each value-added operation performed in the line, the overall efficiency of this system will be 0.991^9 or 92.1%. Assuming a continuation of this sequence, the dropoff in efficiency continues catastrophically so that at the 15th station it would amount to 87.3%. At the 20th station it would be 83.45%, and at the 30th station efficiency would be down to 76.2%.

The ideal goal of 100% efficiency is attained in the Demand System, however. The system accomplishes this goal by forcing — artificially if necessary — the assembly of a product by subassemblies. It is often called a satellite system of *supply to demand* modules.

PALLET TRANSFER INTO POSITION
READ MEMORY PIN
ASSEMBLY OPERATION
VERIFICATION
UNLOCK PALLET & TRANSFER PART OUT

MODULE I

ASSEMBLY OPERATION
VERIFICATION & READ MEMORY PIN
ASSEMBLY OPERATION
VERIFICATION
REJECT
UNLOCK PALLET & TRANSFER PART OUT

PALLET TRANSFER
BUFFER STORAGE

READ MEMORY PIN
ASSEMBLY OPERATION
VERIFICATION
ASSEMBLY OPERATION
VERIFICATION & READ MEMORY PIN

MODULE II

ASSEMBLY OPERATION
VERIFICATION & READ PIN
ASSEMBLY OPERATION
VERIFICATION
REJECT
UNLOCK PALLET & TRANSFER PART OUT

PALLET TRANSFER
BUFFER STORAGE

READ MEMORY PIN
ASSEMBLY OPERATION
VERIFICATION & READ MEMORY PIN
ASSEMBLY OPERATION
VERIFICATION & READ MEMORY PIN

MODULE III

ASSEMBLY OPERATION
VERIFICATION & READ PIN
ASSEMBLY OPERATION
VERIFICATION
REJECT
UNLOAD

EMPTY PALLET TRACK FOR CLEANING AND RETURN

TYPICAL SYNCHROBANK SYSTEM

SCHEMATIC III

A three-module SynchroBank System—the next step up in assembly line sophistication. The assembly operations indicated in this drawing are not necessarily produced in the individual modules, although they may be. In some instances, they are shunted out of the modules and into dial index machines not shown for multiple assembly operations. Of special importance are the buffer banks shown between the modules. Since Module I operates at a faster rate of speed than Module II—and II is faster than III—the flow of parts is continuous, with the excess loads being contained in the buffer storages.

To illustrate its efficiency, let it be assumed that the demand objective (i.e., desired output) is 40 completed assemblies per minute. (See Schematic II.) To meet this demand, the supply function of the upstream or satellite supply modules are designed to provide, say, 50 subassemblies per minute. Like the series function shown in Schematic I, the supply (subassembly) modules do not work at 100% efficiency. But if a 99% efficiency factor per operation is assumed, the overall efficiency of each three-part subassembly supply module is 0.99^3 or 97%.

Accordingly, each subassembly module can produce in excess of 48½ perfect subassemblies (average) per minute — more than enough to feed a line moving at a cadence of say 46 cycles per minute. Thus the demand line has a reasonable expectation of producing well in excess of 40 completed, product-verified assemblies per minute.

Excess Flow. Because the supply modules are feeding subassemblies to the line faster than they can be consumed, they must be periodically deactivated. This is accomplished by maximum and minimum limit switches installed in the intermodule connecting tracks. When the minimum switch is energized, the supply module's cycle is initiated; when the maximum switch is energized, the supply module stops its operation.

Today, the entire process is most frequently programmably controlled, with any number of monitoring means. An advantage this provides is shift register memory, i.e., the ability of a signal at one station to cancel out operations at subsequent stations, and to have a discretionary (selective) discharge function (good, faulty, incomplete).

Wraparound Memory. If for one rea-

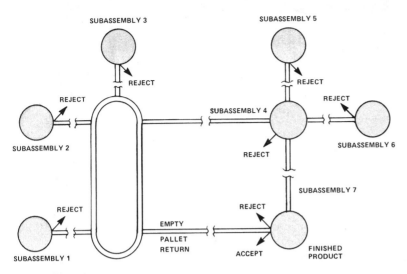

SUBASSEMBLY 3

SUBASSEMBLY 5

REJECT

REJECT

REJECT

REJECT

SUBASSEMBLY 4

REJECT

SUBASSEMBLY 2

SUBASSEMBLY 6

REJECT

SUBASSEMBLY 7

REJECT

REJECT

EMPTY

PALLET

RETURN

SUBASSEMBLY 1

ACCEPT

FINISHED PRODUCT

COMBINATION OF DEMAND AND SYNCHROBANK SYSTEMS

SCHEMATIC IV

A SynchroBank and Demand system in combination. Perhaps the most significant aspect of this design is the versatility it displays. The complexity of modern industrial parts poses correspondingly complex problems in assembly. But no matter how complex the parts may be, there is a system or combination of systems to effect the assembly. Importantly, systems of the type discussed in this article were impractical— and in most cases impossible—prior to the advent of the programmable controller.

son or another an assembly is not properly completed at a given station, a probe signals an incomplete or a faulty assembly operation. This causes deactivation of all further work on the subassembly. With Swanson-developed Wraparound memory, as in the case of a reusable but incomplete assembly, the memory can be programmed to lock out discharge (complete or incomplete). As a result, the subassembly continues to advance through the machine a second time. This time the assembly operation at the station which failed in the previous cycle may be completed. But if the assembly is still not effected, the controller opens the reject gate, thus removing the incomplete (apparently defective) subassembly from the system. The advantage of this is that a great deal of operator inspection of partially completed assemblies is eliminated, and parts shrinkage is reduced.

Flexibility. Thus the essence of the Demand System is that numerous operations are performed off-line in such a way that only perfect parts and subassemblies in the required amounts are presented to the final assembly process. Imperfect subassemblies are eliminated upstream, and incomplete subassemblies may be given a second chance through the Wraparound memory feature of the programmable controller.

Several other advantages of the Demand System should be noted. One is that it permits staggered installation. It need not be laid out in the straight-line design shown in Schematic II. This means that the entire complex need not be laid out on the user's floor at one time — or that the system need be ordered in its entirety at one time.

Still another advantage is flexibility. Through the conveyor shunts and re-

turns, it is possible to have either short or prolonged periods of downtime in individual modules without losing the benefits of the overall system.

THE SYNCHROBANK SYSTEM

The next step in advanced state-of-the-art assembly systems engineering is the SynchroBank. An example is shown in Schematic III. Three assembly modules are used in this system. Each module is a synchronous carousel assembly system. Module I operates independently and at a faster cyclical rate than Module II. And, Module II operates independently at a different cycle rate than Module III.

Work Sequence. Work begins with the entry of pallets from the first buffer to the carousel (Module I). The pallets, manually loaded with matrix parts, now begin their movement through the system. Various components are added to the palletized matrix part. Continuing down the line, the matrix part in its palletized fixture picks up additional components which are each individually probed for acceptance or rejection. Should there be an incomplete or faulty assembly, it is removed from the system by way of a rejection gate on each of the three modules. (Wraparound memory was not used in this instance.)

But, if the part — now a subassembly — is acceptable, its pallet is shunted into an elevator that carries it to an overhead track. The track moves it into the buffers (a series of spiral storage conveyors) which carries it to the downstream module.

Continuing the flow throughout the system, still more components are added as the palletized matrix part advances through Module II. In several instances additional parts or subassemblies are added by off-line dial indexers. Again, each part must pass inspec-

tion or leave the system through the reject position. If it passes, it is again entered into the buffer bank.

Finished Assembly. After completing the series of value-added operations performed on the third module, the part exists as a finished assembly. The empty pallets are unloaded into a line that carries them through a wash conveyor and then back to the first storage tower for eventual reloading and reentry into the system at Module I. Thus, the cycle is completed.

Combination System. Schematic IV illustrates an eight-module machine system in which the Demand System and the SynchroBank System are combined into a single system.

It can be seen from this schematic that subassemblies are supplied through intermodule buffer banks to the main demand loop — this being in itself a closed-loop SynchroBank System. The benefits of subassembly series/parallel machine arrangements which compose the Demand System are also integrated into a SynchroBank with its marriage of synchronous machine modules and individual buffer banks.

Principal Benefit — Systems Engineering. The availability of standard machines of all types and configurations plus the availability of dependable controls have greatly benefited Swanson and its ability to develop integrated machine systems which can be individually tailored to the mechanized assembly requirements of high production. These systems challenge the routine — and often oversimplified — machines which, in the past, were often the only equipment available to manufacturing engineers charged with responsibility for effecting complex assemblies at the lowest possible cost. As such, they do much to reverse the continuing decline of manufacturing productivity. ∎

Reprinted from Control Engineering, November 1980

Object Detection Techniques Range from Limit Switches to Lasers

Photoelectric detectors (visible and infrared), proximity detectors, air streams, lasers, limit switches, fiber optic light pipes, ultrasonic ranging, rf-based sensors, and capacitive sensors are all used in object detection for providing input signals for parts counting on manufacturing lines. How they work, and some typical applications is the subject of this article. No attempt has been made to round up all manufacturers in the business as they are too numerous.

HENRY M. MORRIS, CONTROL ENGINEERING

One of the better examples of where a variety of object detection techniques are employed is found in the bottling or canning plant. The presence or absence of the bottle and its orientation (upside down or rightside up) must be sensed before the filling operation can be initiated. Then the system has to detect whether sufficient product has been loaded into the can or bottle. Presence or absence sensing is again performed at the bottle-capper. Afterwards, the bottles have to be checked to ensure that all have been capped. The process sequence then activates a sensor to ensure that there is a carton or case in position for filling. Finally, the control system must check that no bottles have been omitted from the carton or case. Canning offers similar problems, further complicated by the fact that the cans are cardboard, aluminum, or steel, which are opaque, necessitating different techniques.

In other industrial applications, moving parts of machines must be sensed in their proper position before the next operational sequence can be initiated by the controller.

Detection techniques

The detection of the presence or absence of an object can be performed in two major ways: contact or non-contact. Contact sensors depend on physical contact in order to detect the presence of absence of the object, while non-contact methods sense the presence or absence in several differing ways.

The simplest method of non-contact sensing is where the sensor emits a signal (light, air, and others) which passes across the area being sensed and is picked up by the receiver. If an object passes through the area, the signal path is altered and the presence is detected. This technique is called the "through-beam" technique.

A modification of the through-beam technique uses a reflector to return the signal to the receiver. In this technique, the separate receiver and transmitter do not have to be directly across from each other, or can be combined in a

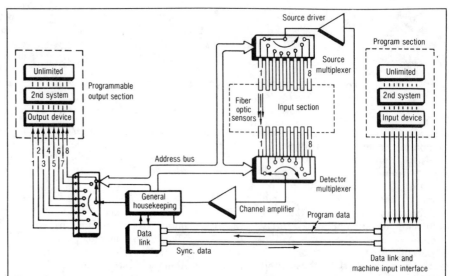

This programmable fiber optic multipoint sensing system is used in applications where several sensing inputs and one or more outputs are needed along with operation of a given channel for a predetermined time. The system operates randomly and without fixed rates or sequences, as needed. Dolan-Jenner Industries.

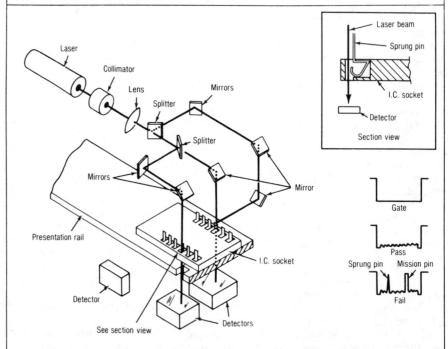

A single laser source has its beams focused and then split into several beams which are directed onto an IC socket to detect missing pins or those which are over-sprung. Two holes are inspected at a time as the part moves past the inspector "on-the-fly." Automation Systems, Inc.

single housing, as required. This technique is commonly known as retro-reflective sensing.

A third technique of non-contact sensing relies on the object to reflect the signal back to the receiver. A shiny object reflecting the optical signal back to an optical receiver is being subjected to "specular scanning" while a matte-finished object would be subjected to "diffuse scanning." Some optical sensor manufacturers call the technique which uses the reflection from an object back to the sensor as proximity sensing, although it is better termed reflective sensing.

By contrast, in the proximity method, the presence of an object within the zone of detection causes a disruption of detector operation, either by reflecting, absorbing, or altering the sensing field state from the state where no object is present. This alteration of the state of sensing field varies with the type of proximity sensor being used.

Limit switches

There are several types of electronic and mechanical devices marketed as "limit switches." For our purposes, here, we shall use the term to refer to the mechanical type. Most electronic types are marketed as proximity limit switches.

Mechanical limit switch. A mechanical limit switch is a contact type of sensor. The actual "touching" of the object being sensed may, in practice, be performed by a mechanical linkage which actuates the limit switch when an object is present or absent.

Electrical connections can be single- or double-pole and parallel-serial wiring combinations can be implemented, virtually without limit. Some types of proximity switches cannot be wired either in series or parallel with each other or with mechanicals.

An advantage of mechanical switches is that they can be mounted side-by-side while most proximity types will interfere with each other if mounted too closely.

Mechanicals can be as accurate as some proximity switches in sensing application to within 0.001 in. However, mechanicals do suffer from drifting. After several million operations the trip point may have to be changed a little, but the plus and minus variation about the trip point will be about the same as when the switch was new. It is often the case that the mechanical part which actuates the limit switch will wear out faster than the switch so that the switch, in essence, will be far more accurate than the device it is mounted on.

Modularity is a key factor in today's limit switches. Most are broken down into four modules. The actuator, the operating head; the switch module; and the wiring (receptacle) module. Several companies offer proximity modules which can plug into the wiring (receptacle) module from a mechanical limit switch. This saves money and allows a greater versatility of sensing options to the end-user.

Hall effect. The switching element itself has undergone some changes from the day where it was simply a snap-action switch. One example from MicroSwitch Div. of Honeywell, Inc. (Freeport, IL) is the solid-state version of their LS series, which uses a Hall effect sensor as the switch. In operation, a magnet is brought close to the Hall effect chip by the action of the mechanical lever-arm. The solid-state digital output interfaces directly with most electronic circuits: discrete transistor circuits, integrated logic circuits (RTL. DTL, TTL, and HTL), and silicon controlled rectifiers (SCRs). Naturally, application of this product is limited to areas where there are no high level magnetic fields present.

Other areas of concern with mechanical limit switches include the unavoidable contact bounce and an operational frequency limited to two or three operations per second. These are eliminated by the solid state Hall-effect device described above.

Sources for mechanical limit switches include: Allen-Bradley Co. (Milwaukee, WI), Square D Co. (Milwaukee, WI), The Cutler-Hammer Div. of Eaton Corp. (Milwaukee, WI), Omron Electronics, Inc. (Schaumburg, IL), General Electric Co.; General Purpose Control Department (Bloomington, IL), and Micro Switch Division of Honeywell, Inc.

Pneumatic limit switches. Up to now we have examined only the electrical-output mechanical limit switches. There are also pneumatic-output mechanical limit switches on the market.

In operation, these devices permit air to flow, or not to flow, like their electrical counterparts allow electricity to flow or not to flow. Most have actuators not too dissimilar to many of those found on the electrical-output types of mechanical limit switches, making it possible to install either to serve in the same capacity.

Sources for pneumatic limit switches include: C. A. Norgren Co. (Littleton, CO), Miller Fluid Power Corp. (Bensenville, IL), The ARO Corp.(Bryan, OH), and Numatics, Inc. (Highland, MI), to name just a few.

Back-pressure switches

Anyone who has driven into a gas station and rung the bell, which alerts the attendant to their presence, has activated a back-pressure switch. The air in the hose attached to the air switch, located in the gas station structure, has a certain volume of air within it. Depending on the manufacturer, the air stream can be continuous or static within the hose. Back-pressure from the car's compressing the hose with its tires triggers a back-pressure sensitive switch which activates the bell.

This same concept can be applied to many industrial applications where the object being sensed is heavy enough to compress the hose, yet able to pass over without damaging it. One source for air switches used in driveways is Milton Industries, Inc. (Chicago, IL).

Proximity detectors

Proximity detectors rely on the presence of an object within a certain distance of the sensor. The presence of this object changes the properties of the sensing field in some fashion. It is this change in the sensing field that is detected by the circuitry, not the actual object, itself. There are several methods employed for non-contact "proximity" sensing.

Eddy currents. Probably the most-popular method of sensing the presence of an object uses eddy currents. Since eddy currents cannot be created in non-metallic materials, this method is limited to sensing metallic objects.

In operation, an electromagnetic field is generated by an oscillator circuit. Whenever a metallic object enters the field generated by the sensor, a counter current (eddy current) is set up in the metal object. This changes the inductance of the sensing field which detunes the oscillator circuit. An internal circuit, such as a Schmitt trigger, senses the increase in current required to keep the oscillator circuit functioning and actuates the solid-state switched output as desired.

Companies which produce eddy current sensors include: Allen-Bradley Co., Omron Electronics, Cutler-Hammer Division of Eaton Corp., Micro Switch Div. of Honeywell, Inc., General Electric Co., General Purpose Control Department, and Square D Co.

Magnetic sensors. The second most-often used proximity sensing technique relies on magnetic properties of the materials used in the construction of the object being sensed. Where eddy current devices can sense any type of metal, magnetic devices are limited to ferrous materials.

Magnetic sensors are activated whenever a ferrous-based material is present within the zone of detection. The most common application is where the sensor is used as an input for a tachometer. Every time a tooth on a

The FYHD Series of solid-state non-contact proximity sensors can sense a piston's cushion or collar through its own metallic protective face.—Micro Switch Div., Honeywell, Inc.

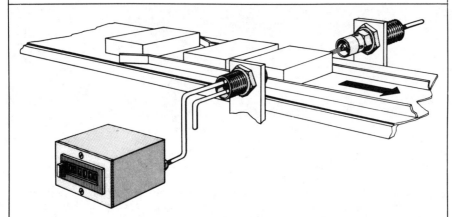

As parts move past the sensors, they interrupt the air flow and signal the system so that they can be counted.

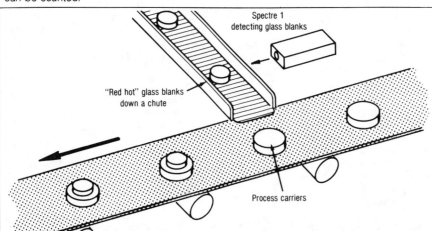

Spectre 1 detecting glass blanks

"Red hot" glass blanks down a chute

Process carriers

The Spectre 1 is a radar-type proximity sensor which can be used next to a chute delivering red hot glass blanks.—Automation Concepts, Inc.

gear wheel or magnet mounted to a wheel passes through the field generated by this type of sensor, the change in inductance is sensed.

There are basically four different types of magnetic proximity sensors.

■ One, described above, generates an ac field which is interrupted by the sensed object. The frequency of the interruptions can be calibrated to rpm, fpm, or other engineering-units. Sources include AIRPAX, (Ft. Lauderdale, FL), Electro Corp. (Sarasota, FL), and Transducer Systems Inc. (Willow Grove, PA).

■ A second type of magnetic proximity sensor uses an external magnet to close a reed switch in the sensor. This is most often used in sensing that the position of a mechanical part of a machine is correct prior to allowing the machine to continue on in its sequence. One source for this type of magnetic proximity switch is General Electric Co., General Purpose Control Department.

■ The third type of magnetic proximity sensor uses the Hall effect. In operation, whenever a magnet is brought near a thin strip of semiconductor material (Hall generator) through which a constant control current is passed, such that the magnet's field is directed at right angles to the face of the semiconductor, a small voltage appears at the contacts placed across the narrow dimension of the semiconductor strip. As the magnet is removed, the Hall voltage reduces to zero.

Micro Switch offers a plastic mounted Hall effect position sensor. The sensor is mounted inside of a threaded plastic bushing. The device features a single, digital current sinking output which directly interfaces with microprocessors, digital logic, linear components, discrete transistors, and SCRS.

■ The fourth magnetic technique employs the Weigand effect. A Weigand module consists of a short piece of Weigand wire with a copper sensing coil wound around it.

A Weigand wire is a small diameter wire that has been selectively work-hardened so that there is a differing magnetic permeability between the surface and core of the wire. When subjected to a magnetic field, the wire emits a well-defined pulse that requires little in the way of shaping. This pulse induces a voltage in the surrounding sensing coil. The wire is insensitive to polarity and will emit a pulse whether the magnetic field is flowing from north to south or from south to north.

A Weigand proximity sensor will sense the presence or absence of ferromagnetic material, causing the sensor to emit a pulse of ±1 V. Any ferromagnetic protrusion will cause the sensor to emit the sensing pulse. Similarly, the removal of a ferromagnetic material will also induce the pulse output.

Weigand effect devices are available from Sensor Engineering Co. (Branford, CT).

Ultrasonic ranging. Ultrasonics have long been used for detection of levels of materials in containment vessels. The latest effort to bring ultrasonic ranging into the industrial scene is being conducted by Polariod Corp. (Cambridge, MA). The transducer developed for their SX-70 automatic focusing camera is now available for use in industrial applications.

The ultrasonic ranging system uses an electrostatic transducer to generate brief pulses of sound which reflect off objects and return echoes to their source. By measuring the time interval between the sending pulse and received echo, the distance from the source can be measured. Likewise, no object, no echo.

In operation, four brief pulses are sent out at different frequencies: 60 kHz, 57 kHz, 53 kHz, and 50 kHz. The 50 kHz signal is the last and longest pulse. A filter allows all four returning wavelengths to pass equally immediately after the pulses have been sent, but then progressively narrows to the 50 kHz frequency as it waits. Since that frequency is the least attenuated by air of the four, the transducer is concentrated on the frequency most likely to return from long distances.

A designers kit is available for product engineers so that they can familiarize themselves with the capabilities of the sonar transducer. The kit includes two instrument-grade transducers, a modified ultrasonic circuit board, two Polapulse batteries (6 V), a battery holder, and a technical manual.

RF absorption. RF absorbtion is not the same as radar. In use, the rf signal is absorbed by the mass of an object. This technique is primarily used in level sensing, but, can be applied to object detection where the rf emission will not effect operating personnel. One source

for this type of proximity detector is Delavan Electronics, Inc. (Scottsdale, AZ).

RF reflection. RF reflection is radar. In use, the rf signal reflects off of the object being sensed and is then received. Where radar can tell if an object is present, it cannot tell if the object is properly full. RF absorbtion can distinguish between full and empty containers. One source for rf reflection detection devices is Automation Concepts (Syracuse, NY).

Radiation techniques. Beta or gamma radiation is employed by several companies to sense level, mass flow, and moisture. The same equipment may be modified to tell if an object is present or not, and whether it is full, partially filled, or empty.

In operation, a Beta source projects radiation to a reciever. The amount of radiation absorbed can determine whether the contents of the object being sensed is within desired limits. Too little radiation, too full a container. Too much radiation, too empty a container.

Some sources for radiation emissive equipment include: Kay-Ray, Inc. (Arlington Hts., IL), Texas Nuclear (Austin, TX), Ohmart Corp. (Cincinnati, OH), and AccuRay (Columbus, OH).

Capacitive sensing. Primarily used for level sensing, the capacitive concept can be applied to assembly lines where the objects to be sensed pass between two plates that are capacitively coupled to each other. The change in capacitance induced by the presence of an object can be sensed and used in control systems. One source for capacitive sensors would be Electromatic Components Ltd. (Arlington Heights, IL).

Fluidics and fluid logic

Fluidics was heralded in the early 60's as a replacement for many cumbersome electronics controls of the day. It offered small size logic elements, no moving parts, and was inherently non-explosive. All these features are now found in electronic integrated circuits and fluidics has faded from the limelight. However, don't write it off just yet. Several companies are still vitally con-

cerned with the proliferation of the technology and feel that it can hold its own in the marketplace.

Fluidics differs from fluid logic in the fact that it requires no moving parts to perform the logic. Combinations of wall attachment (Coanda effect), stream interaction, and vortices are used to create the various logic elements. These logic elements perform exactly the same functions as their electronic counterparts, only at the speed of sound instead of the speed of light. Linear, proportional, and digital circuitry can be created to suit the application.

Any electronic circuit can be duplicated in the fluidics format. Interfacing fluidics and electronics is accomplished via I/P and P/I devices so that the best of each world can be connected together into a system that neither technology could duplicate alone.

C. A. Norgren Co. points out that in today's electronics-oriented world fluidics still offers the possibility of providing control, computation, or timing circuitry that would not be vunerable to: heat, cold, nuclear radiation, acceleration, vibration, shock, magnetic thrust, or electronic jamming.

Fluid logic requires moving parts in the system to perform logic functions. Logic elements are usually small air valves that duplicate a logic function by moving the air from one port to another port in a predetermined sequence. Fluid logic operates at standard plant air pressures of 80 psi whereas fluidics systems use less than 1 psi.

Miller Fluid Power offers fluid logic modules that are ready to use and which can be mainfolded into whatever control system is required by the end user.

Both fluidics and fluid logic systems use similar sensors to sense objects. Fixed gap sensors are used for smaller parts and adjustable gap sensors are intended for larger parts. In operation both send an ''air barrier'' which is a stream of moving air which travels from a transmitter nozzle to a receiver port. Anything which interrupts this air stream activates the system.

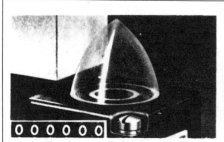

When a metal object enters the electromagnetic field, a counter current (eddy current) is set up in the metal object. This changes the inductance of the sensing field which detunes the sensor's oscillator thereby actuating an output signal.–Square D Company

Either an electrical or pneumatic limit switch can be used to sense the presence or absence of an object. The electric switch, on the left, is from General Purpose Controls Department of General Electric Co., and the pneumatic limit switch is from ARO Corp.

Channels of pulsed IR, each with its own emitter and receiver form a curtain of light. Any penetration of that curtain is sensed and the appropriate control action is initiated by the circuitry in the sensor.

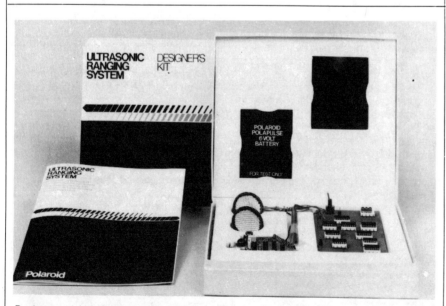

Designers wishing to create their own ultrasonic proximity or ranging system can take advantage of this kit put out by Polaroid. All the parts necessary and detailed technical data are included.

Fluid logic and fluidics offer proximity sensors, too. In operation, these nozzles expel a circular stream of air which is quiescent in the center. An object which enters the operational field of this sensor reflects eddies back to a centrally located return port. The returning pressure in this port activates the system.

The ARO Corp. and the Festo Corp. (Port Washington, NY) are two additional companies offering lines of fluid logic and fluidic control systems.

Photoelectric switches

Photoelectric switches are quite often used for non-contact sensing of objects. Through-beam, retroreflective, and reflective techniques are used. Both infrared and visible light devices are on the market. The visible light sources may be incandescent or LED while the infrared sources are invariably LEDS.

Light emitting diodes offer the advantage that they can be modulated directly whereas incandescent lamps necessitate a mechanical chopper of some sort when modulation is required.

Modulating the light offers several advantages. One advantage is that the system becomes oblivious to ambient lighting conditions. Another is that the modulation can allow greater intensities of light than steady-state operation thus making it possible to project a light beam over a substantially longer distance; or to penetrate haze, smoke, or other suspended particulate matter.

Photoelectric switches can be set to actuate in the presence or absence of received light. This opens the door to applying logic functions optically. Switches can be wired so that AND and NAND or OR and NOR functions can be implemented.

Photoelectric switches come either in separate transmitter and receiver pairs or in configurations where the transmitter and receiver are mounted in a single housing.

Companies that offer photoelectric controls include: Micro Switch Div. of Honeywell, Inc., General Electric Co., General Purpose Control Department, Scientific Technology, Inc. (Mountain View, CA), Banner Engineering Corp. (Minneapolis, MN), and Omron Electronics, Inc.

Fiber optics

Fiber optic bundles can be added to existing photoelectric switches to form object sensors and can be combined to implement logic functions. Several applications are described in the August 1980 CE Pp. 61 and 62, entitled Applying Fiber Optics to Photoelectric Switches by Alton H. Krueger, Dolan-

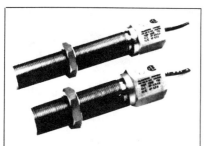

These magnetic sensors are used to count the number of teeth on a gear or to sense the presence of an object between the probes. —Airpax

This air stream sensor by C.A. Norgren Company is available with a variety of sensing nozzles. Changes can be made in the field without disrupting the installation.

This Weigand effect sensor can be purchased in a hermetically-sealed package for use in dirty environments. —Sensor Engineering Company

Jenner Industries, Inc. Woburn, MA.

Dolan-Jenner has also developed a multipoint programmable sensing system using fiber optics for use in those applications that require several sensing inputs and one or more outputs to interface with microcomputers or minicomputers.

The program selection allows the use of "light operate" or "dark operate" modes of operation and permits operation of any channel for a predetermined period of time; thereby avoiding sequential channel operation at fixed time frames. It is intended for those applications where a programmable controller is not warranted because of cost or complexity.

Input can be from a relay or switch contacts, transducers, memory devices, CMOS. or TTL. The output section provides a channel signature for each emitter and detector pair, resulting in the capacity of actuating one or more. output devices. The actual architecture can be configured to suit the application.

Lasers

Lasers have found many places in industry. Automation Systems, Inc., (Brookfield, CT) markets systems using lasers for three basic types of sensing applications.

■ The first is based on reflectivity. In this case, the laser is most often used to sense surface flaws or other characteristics such as threads. The depth of holes can be easily calculated using reflection.

■ A second technique is "shadowing." The shadowing method is used in the same fashion as that of optical comparators. The exception being that with a scanning type laser, very accurate dimensional checks can be made while the part is moving. Dimemsions can be sensed in the 0.0001 in. range.

■ The third technique is transmission. In some cases the material of a particular part is translucent or nearly clear; such as a glass bottle. In these instances, the transmission of light energy through the product will be used to detect flaws or suspended contaminants in a liquid product. The properties of light diffraction are also used in conjunction with this approach.

Laser beam inspection has been used in wire drawing, centerless grinding, and to detect assembly defects on automatic lines.

Hewlett-Packard Co. (Palo Alto, CA) offers the HP 3850A industrial distance meter which is a laser-based device that can be used to detect the presence of an object that might otherwise be inaccessible to other techniques. It can also measure the distance from the

sensor to that object from as far as eight kilometers away.

Hot billets in street mills are an excellent application for this device. It can sense the presence and size of the billet from a distance far enough away to be safe for the electronics. It can also supply velocity and acceleration data to the process controller.

Temperature sensing

Non-contact, infrared-based temperature sensors can be used for object detection when the object will be significantly different in temperature than the surroundings. One application could be where the sensor monitors the presence or absence of frozen foods as they pass on a conveyor on their way to be boxed in master cartons. Steel and other metal fabrication centers can use these devices to sense the presence or absence of hot ingots, billets, etc.

Some suppliers of infrared temperature sensing equipment include: Ircon, Inc. (Skokie, IL), Raytek, Inc. (Mountain View, CA) and Vanzetti Infrared and Computer Systems, Inc. (Canton, MA).

Load cells

Load cells have long been applied for sensing the passing of an object on a conveyor line. Most of this sensing has been for weighing purposes, however, the applications are not limited to weight alone. The simple fact that something is pressing down on the cell indicates that an object is present. Electrical load cells are the more popular form and can be obtained from several manufacturers. Some of these companies include: Interface, Inc. (Scottsdale, AZ), ASEA Electronics Corp. (White Plains, NY), Transducers, Inc. (Whittier, CA), Acurex/Autodata (Mountain View, CA), and Datametrics, Inc. (Wilmington, MA).

Vibrating wire A unique load sensor which could be used for object detection is made by K-Tron Corp. (Glassboro, NJ). For a detailed description of this device refer to CE, November, 1979, p. 45, Microcomputer Ups Resolution in Mass-Sensitive Scale by S. J. Bailey.

Pneumatic load cells Moore Products Co. Spring House, PA produces a pneumatic force transmitter which utilizes the force-balance design. It transmits a pneumatic output signal that is proportional to the force applied to the input.

Hydraulic load cells Two companies which offer hydraulic load cells are the Enerpac Division of Applied Power, Inc. (Butler, WI) and A. H. Emery Co. (New Canaan, CT). These devices are found most often in weighing platform scale applications. □

CHAPTER TWO

WORKHOLDING AND CLAMPING

Reprinted from Stanki I Instrument, Vol. 49, Number 6, 1978

INCREASING THE ACCURACY AND STIFFNESS OF COLLET CHUCKS

STANKI I INSTRUMENT, Vol. 49, Issue 6, 1978, pp. 21—23

A. I. Ivolgin   U.D.C. 621.941.234.3-229.323.4

Experimental investigations conducted at the Khabarovsk Machine Tool Factory aimed at raising the accuracy and stiffness of collet chucks in single spindle automatic lathes are reviewed. Basic requirements to be met by collet and mating parts design are formulated. A new technology for collet manufacture is proposed. A special methodology for identifying collet errors in the total inaccuracy of the clamping mechanism serves to determine the accuracy of collets manufactured with the new technological process.

To clamp cold finished bars of the 3rd to 4th accuracy class on single spindle bar automatics a clamping mechanism[1] is employed in which the clamping element is a collet of the third type[2] that possesses lower radial accuracy and stiffness because of the presence of additional clamping elements, namely a compression bushing and thrust nut (Fig. 1). In connexion with growing demands for the accuracy and stiffness of collets, experimental investigations aimed at improving these parameters were organized at the Khabarovsk Machine Tool Factory.

It is known[2,3] that the stiffness of any collet chuck depends to large degree upon the technological clearance angle α and technological diameter clearance (interference) $\Delta d = d_c - d_w$, where α is half the difference between the taper angle of the compression bushing and the collet taper before the collet segments are separated; d_c is the technological diameter of the collet bore (before segment separation); d_w is the bar (workpiece blank) diameter.

The methodology of the investigations was developed so as to reveal optimal (in terms of clamping stiffness and accuracy) values of the technological angle and diameter clearances and was based on the theoretical assumptions cited below in which for simplicity it is assumed that the collet segments have not been opened up (expanded).

Depending upon the sign of the diameter clearance Δd

two variants of clamping schemes are possible. In the first variant a positive technological clearance $\Delta d = 0,05-0,15$mm is provided between the collet and blank. At first we take $\alpha = 0$. If one assumes that the webs between segments have not been cut, then only the tapered surfaces of the compression bushing and the collet come in contact in clamping and clearance Δd remains between the blank and collet bore. In this case the diameters of the collet and compression bushing tapers are equal in all their cross sections and clamping does not take place.

In a collet with separated segments the latter contact the blank when the collet taper enters the compression bushing taper; here in any transverse section the collet taper diameter is greater than the compression bushing taper diameter by an amount close to Δd.

Contact between the tapers may take place only at the end points of cross-sections (Fig. 2(a)). Since the nominal value of the taper diameters in end sections AA and BB is noticeably different, clearances h_1 and h_2 in the central longitudinal collet jaw section also differ.

In the end sections the collet jaw is skewed by the amount $\Delta h = h_1 - h_2$. Work clamping begins on the front collet edge at three points (for a three-jaw collet). As the clamping force increases the collet jaw rotates about the contact point with the work. Contact between the collet and the compression bushing is impaired in all cross-sections (except AA), starting with section BB. Contact with the work in those sections is not guaranteed. The accuracy and stiffness of such clamping is not high.

Giving a certain positive value to angle α (in practice α = 10′−30′) in principle does not change the described clamping scheme (see Fig. 1).

In the second variant the size of the collet bore is such that a negative technological clearance (interference) Δd (Fig. 2(b)) exists between the collet and the work. When α = 0 the smaller diameters in all collet taper cross-sections are under the larger diameters of the compression bushing taper. The segments contact the bushing along the central taper generatrix, and if one considers jaw deformation − in areas symmetrically located with respect to that generatrix. The collet and the clamped bar come in contact as the edge generatrices of cylindrical surface in places where the segments are separated. Clearly such clamping is stiffer and more accurate. The closer the bar diameter is to the technological collet size ($\Delta d \to 0$), the greater is the collet contact area with the compression bushing and the workpiece, and clamping accuracy and stiffness are accordingly higher.

The following limitations and assumptions are made in both variants: the angle α must not have a negative value because in this case one has the least stiff clamping scheme[2]; the effect of the segment jamming force in the collet band joining them on the position of these segments in the clamping mechanism is negligible compared with mechanism clamping forces.

To experimentally verify the above theoretical assumptions in real life operating conditions (deformation of contacting parts and attainable manufacturing quality of collet chuck working surfaces), three-segment collets for round bars were made by the same process from 9KhS steel and were tested on a Model 1B136 lathe (collet dimensions were taken from GOST 2876−70). The collets differ only in the taper angle (α = 0°; 0°10′; 0°20′; 0°30′) and the clamping bore diameter (d_c = 18; 24; 30 and 36mm).

The accuracy parameters of the collet chuck were tested following the methodology of GOST 18100−72, test 1.9 (Fig. 3), according to which the measuring end of the dial indicator touches the inspection mandrel surface at a distance of 80mm from lathe spindle end and is directed perpendicular to the mandrel generatrix. The spindle is slowly rotated one complete revolution and in this way the radial run-out of the mandrel held in the collet is determined as being the algebraic difference of the indicator readings. Numerous experiments have shown that when the indicator is placed closer to the spindle end the mandrel run-out is diminished, therefore, only the extreme indicator position was employed.

In the present experiments as an accuracy assessment criterion, the relative accuracy characteristic, namely the scatter ΔK of run-out of a mandrel held in the collet was employed, as obtained by eight measurements conducted at eight different spindle angle positions at the instant the mandrel was clamped, six of which were oriented with respect to the collet (the vertical plane passes between the segments or through their centre) and two with respect to the clamping mechanism pawls (plane of the pawl pair is horizontal or vertical).

As the evaluation criterion for compliance, which is the reciprocal to clamping stiffness, served the residual displacement ΔC of a collet held mandrel produced by the application and removal of a force P = 500kgf (see Fig. 3). This displacement was the result of collet skewness in the clamping mechanism and skewness of the inspection mandrel in the collet. It was determined

as the difference between the run-out of the inspection mandrel before and after force application. The diametral technological clearance (interference) was produced by selecting the size of the inspection mandrel.

Figure 4 shows the change in the relative accuracy characteristic ΔK and the compliance characteristic ΔC depending upon the angle and diametral technological clearances. No clear influence of collet bore size on both characteristics and of the spindle position at the instant of mandrel clamping on ΔK was observed. The plots show average values of readings for all collet sizes and for all spindle positions. Angular collet orientation with respect to the compression bushing (see Fig. 1(b)) was chosen for the collet with the best absolute accuracy.

Analysis of the plots shows that the accuracy and stiffness characteristics of clamping mechanisms with collets of the third type in a large measure depend upon the angle and diametral technological clearance. The best accuracy and stiffness was found in collets having zero clearance angle and diametral technological interference of 0,05−0,15mm.

Based on the recommendations in References 2 to 4 and the results of the present investigations the following requirements were formulated for the design of collets and compression bushings for lathes Model 1B124, 1B136, 1B125, 1B140, 1E125, 11F25, 11F40, and 1E165, which are aimed at increasing clamping accuracy and stiffness.

1. The nominal taper angles of collets and compression bushings should be equal. The tolerances for these angles should be minimal and placed with respect to the nominal value so as to eliminate the possibility of the development of a negative clearance angle α.

2. The collet bore size should ensure an interference of 0,1−0,15mm between collet and bar, taking into account the bar size tolerance.

3. Form deviations of the collet bore in longitudinal and transverse sections should not exceed the tolerances for these parameters as stated for 1st accuracy class bores.

4. The slots for angular orientation of the collet with respect to the compression bushing should be made displaced from the central taper generatrix (see Fig. 1(b)).

5. The maximum value of the jaw contact angle β (see Fig. 1(b)) should be in the range of 60−80°.

6. Scatter in the chord H length (see Fig. 1(b)), for all collet segments should not exceed 0,2mm.

To realize the previously described design requirements a production process for the manufacture of collets was developed and tried[5], which ensured high, absolute accuracy. Its distinguishing features are as follows.

1. When manufacturing collet blanks, the metal fibres should be directed along the collet axis.

2. After rough turning the collet is subjected to preliminary heat treatment (quenching and tempering).

3. The slots between the collet segments are milled by means of a dividing attachment such that the wall between the slot and the bore is not less than 3mm (see Fig. 1(b) and 5). The stepped slot profile helps to maintain angular accuracy when the segments are separated.

4. All external and internal cylindrical and tapered surfaces as well as fillets are rough ground which ensures the necessary equality of the segment walls (within 0,1mm) and reduces the possibility of jaw fracture as a result of stress concentration in places which were undercut in machining.

5. The collet is finish ground before the segments are

separated. First the bore and front end are ground in one setup. Then using the ground bore as a location the opposite end is ground keeping it parallel within 0,003–0,005mm. Finish grinding of external cylindrical and tapered surfaces (and rough grinding of guiding band A) is done on a special mandrel (see Fig. 5) on whose guiding surface the collet 4 is mounted with a clearance not exceeding 0,005mm which is provided by size selection or grinding to size in place. The collet clamping force is adjusted by nut 7 and elastic washer 6 between the nut and thrust bushing 5 which prevents bending of the mandrel axis. The main part of the grinding torque is taken up by driving pin 2 which is pressed into mandrel 3 and enters the slot between the collet segments.

6. Segment expansion (0,6–0,8mm on diameter) is done in sand in a special fixture (Fig. 6). Heating time depends upon the heating temperature and collet size. The position of the jaws uniformly displaced with respect to the mandrel axis by the tapered surface B is maintained stable with the aid of a fixture. The segments are heated through windows in the holding cylinder which at the same time prevents overheating the collet guide band A.

7. Finish grinding of the guiding band A is done after segment expansion using a mandrel (see Fig. 5) and hollow nut 1.

To measure the technological accuracy of the collet a method of separating collet errors from total clamping mechanism errors was employed and is presented below. The radial run-out index Δ_m of an inspection mandrel clamped in the collet can be divided into two parts: collet run-out Δ_c caused by collet manufacturing errors (technological collet error), and mandrel run-out Δ_r resulting from manufacturing and assembly errors of remaining parts of the clamping mechanism (manufacturing error of the clamping mechanism).

The relative position of the clamping mechanism elements remains unchanged and therefore the angular direction of the Δ_r component with respect to the spindle end face is also constant. The angular position of the collet may only be variable during its orientation with respect to the compression bushing and accordingly the Δ_c component. The collet position is fixed by a pin pressed in a radial hole in the compression bushing whose projecting part enters one of the collet slots (see Fig. 1(b)).

When the collet rotates through 360° run-outs Δ_r and Δ_c mutually cancel or mutually strengthen each other, as a result of which one can obtain not only a better accuracy reading Δ_m min, but also a worse Δ_m max. If $\Delta_c < \Delta_r$, then Δ_m max $= \Delta_r + \Delta_c$; Δ_m min $= \Delta_r - \Delta_c$. Hence

$$\Delta_c = \frac{\Delta_m \text{max} - \Delta_m \text{min}}{2}; \quad \Delta_r = \frac{\Delta_m \text{max} + \Delta_m \text{min}}{2}. \tag{1}$$

Similarly when $\Delta_c > \Delta_r$

$$\Delta_c = \frac{\Delta_m \text{max} + \Delta_m \text{min}}{2}; \quad \Delta_r = \frac{\Delta_m \text{max} - \Delta_m \text{min}}{2}. \tag{2}$$

Having checked the accuracy of several collets in a particular clamping mechanism and thus determined for each collet half-the-sum and half-the-difference of

maximum and minimum mandrel run-outs, one can identify the Δ_r of this mechanism from the total number of paired indices Δ_r and Δ_c, because Δ_r indices are approximately equal, and Δ_c indices differ, and as a rule, significantly. After elimination of Δ_r we have in the 'remainder' Δ_c values of the respective collets. It is enough to have six–eight collets for the average value of Δ_r to be taken as an objective quantity for quick and accurate determination of the technological accuracy of these and remaining collets. As a result of many years of experience in using the above described methodology, accuracy of the described technological process was determined and the manufacturing error Δ_r of clamping mechanisms in single spindle automatic lathes was identified. Fig. 7 shows the scatter field of the respective statistical data. On the abscissa axis is the index $K = d_c/L = 0,3–1,2$ (relates only to collet accuracy), where L is the collet jaw length (see Fig. 1). Jaw length for all collets in a specific lathe model is the same. Then when size d_c is reduced the quantity K diminishes and as a result the technological accuracy of the collet is increased.

Analysis of the plots (see Fig. 7) shows that when the above described technological process is employed, one should consider as the optimum manufacturing and assembly error for clamping mechanisms $\Delta_{r.opt} = 0,03$mm. For such a value of Δ_r, inspection mandrel run-out Δ_m min $= 0–0,03$mm, which ensures that the requirements of test 1.9 GOST 18100–72 are met.

Conclusions

1. By observing the design and technological measures described in the article one may obtain a high manufacturing accuracy in clamping collets with a significantly reduced scatter field of the accuracy indices of the collet clamping mechanism and also a higher clamping stiffness.

2. Further increase in the clamping accuracy in single-spindle automatic lathes can be obtained by improving the manufacturing and assembly accuracy of clamping mechanisms.

REFERENCES

1. IVOLGIN, A. I. *Clamping mechanism for single spindle automatics.* Machines & Tooling, No. 9, 1971.

2. DALSKII, A. M. *Collet clamping mechanisms.* Mashinostroenie, M., 1966.

3. ORLIKOV, M. L., KUZNETSOV, Yu. P. *Selection of collet parameters.* Machines & Tooling, No. 9, 1971.

4. KUZNETSOV, Yu. N. *Calculating the contact angle of collet and spindle.* Machines & Tooling, No. 8, 1970.

5. IVOLGIN, A. I. *New technology for manufacturing collets for single spindle automatic lathes.* 'Experience of production leaders and innovators. Introduction of NOT' Express-information. NIIMASh, Issue 12, Abstract 110, 1973.

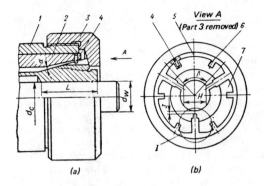

Fig. 1 Clamping elements (a) and angular orientation scheme (b) of type III collet: 1 — spindle; 2 — compression bushing; 3 — thrust nut; 4 — pin; 5 — central taper generatrix; 6 — slot for angular collet orientation; 7 — collet; I — slot prior to segment separation.

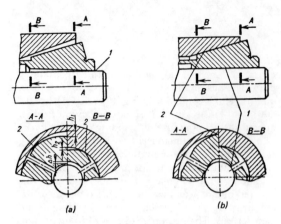

Fig. 2 Variants of clamping schemes: (a) $\Delta d > 0$; (b) $\Delta d \leqslant 0$; 1 — contact with workpiece; 2 — contact with compression bushing.

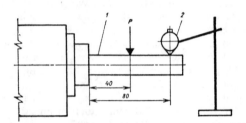

Fig. 3 Inspection method for collet clamping accuracy and stiffness: 1 — inspection mandrel; 2 — dial indicator.

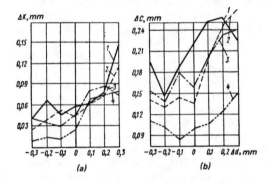

Fig. 4 Dependence of accuracy (a) and stiffness (b) characteristics upon angle and diametral technological clearance: for $\alpha = 30'$ (curve 1), $20'$ (curve 2), $10'$ (curve 3) and $0'$ (curve 4).

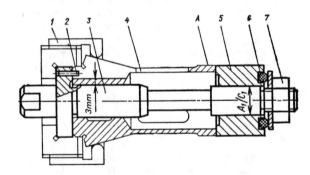

Fig. 5 Mandrel for external finish grinding of collets.

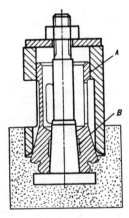

Fig. 6 Fixture for segment expansion.

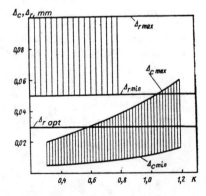

Fig. 7 Accuracy characteristics Δ_c min — Δ_c max of the technological process and manufacturing error Δ_r min — Δ_r max of the clamping mechanism.

Reprinted from Tooling & Production, April 1977

Counter-centrifugal chucks for high speeds

The need for a chuck able to handle the trend toward higher spindle speeds and horsepowers on today's turning machines has been recognized by machine builders, tool suppliers and chuck manufacturers alike. In late 1975, The S-P Mfg Corp, Cleveland, OH, released its first commercially available counter-centrifugal chucks.

The first unit, Serial No 1, is a 10″ chuck which became part of a Model 1225 universal bar chucker from The Lodge & Shipley Co, Cincinnati, OH. The machine was delivered to Duquesne Mine Supply Co, Pittsburgh, PA, in December 1975. Duquesne manufactures special industrial fasteners, including special bolts for off-road vehicle manufacturers. Located in coal mining country as it is, Duquesne also supplies mine hardware and is a leading producer of mine electrical power line goods.

"We analyzed our future tooling requirements about five years back," relates Plant Manager Bill Woods, "and high on our list was the need for an NC bar chucker with some refinements that weren't available then.

"Duquesne has a very large product mix. We deal in specials, short runs, a wide variety of metal stock and an even wider variety of finish requirements. Our chucker had to have a smooth, constant surface speed control, spindle speeds in the high range and rapid changeover capability.

"That produced two major requirements, a constant-pressure chuck that would be safe up to 3000 rpm and a better numerical control than was then on the market."

Woods' notes from those days show that he had actually built a specification for a counter-centrifugal chuck, but no one was building one. It wasn't until the SME show in Detroit that Bill saw a Lodge & Shipley demonstrator with a GE 550 TX series control from the General Electric Co, Salem, VA, and a 10″ S-P counter-centrifugal chuck.

How they work

Counter-centrifugal chucks are designed with a trunnion-mounted counterweight-lever for each master jaw. As rpm increases, the centrifugal force developed by the integral counterweights directly offsets centrifugal force developed by the chuck top jaws, **Figure 1**.

A tooling package is available for the S-P chucks which includes a set of hard serrated inserts with bases, hard serrated master jaw pads, soft blank master jaw pads, soft blank top jaws (steel and aluminum), jaw boring fixture and jaw force gage.

The serrated inserts are also a new design, superseding the traditional hard stepped top jaws. They provide longer gripping length, improved gripping range, easier changeover and better adjacent tool clearance. The inserts have a jaw "bite" design which further enhances gripping capability of the chucks.

Another innovation is the master jaw pads which are mounted on the ends of master jaws in a manner similar to conventional collet pads, with gripping action characteristics of a collet. Higher driving force is developed because the serrations penetrate the workpiece rather than holding it by friction alone. This is particularly advantageous on first operations where heavy cuts are most apt to be made and where resulting serration marks on the workpiece are acceptable. Where the marks are not permissible, the alternative is soft pads bored-in-place (friction gripping).

The combination of constant surface speed available from the SCR control and the constant workholding forces has given Duquesne better surface finishes, **Figure 2**, with longer tool life and better chip-loading characteristics, according to Woods. Considering the many changeovers required in a normal day, which means a big investment in tools, this has turned out to be a real money saver.

Duquesne has used the counter-centrifugal chuck in many kinds of setups — master jaw pads, hard top jaws and a Madison-Kosta face driver mounted in soft top jaws for work between centers. For any given job, workholding force remains constant throughout the whole job.

The GE numerical control is the 550 TX model with memory, editing and tape punch capabilities. The usual routine is to write a program "upstairs" (in the programming office) and bring it to the machine for debugging. Woods reports that exactly twice in the history of the machine has a tape been sent back because it could not be debugged using the memory and editing features of the control.

Frequently, the better surface characteristics obtainable with higher surface speed and gripping force control allow the elimination of a cut during the editing process.

"We would never have bought a 3000-rpm machine if the counter-centrifugal chuck had not been included," says Woods. "It just isn't safe to run those speeds without it. This is a densely populated shop — 65 welders, blacksmiths, forgers and machinists. If a part lets go, someone is bound to get hurt — or worse."

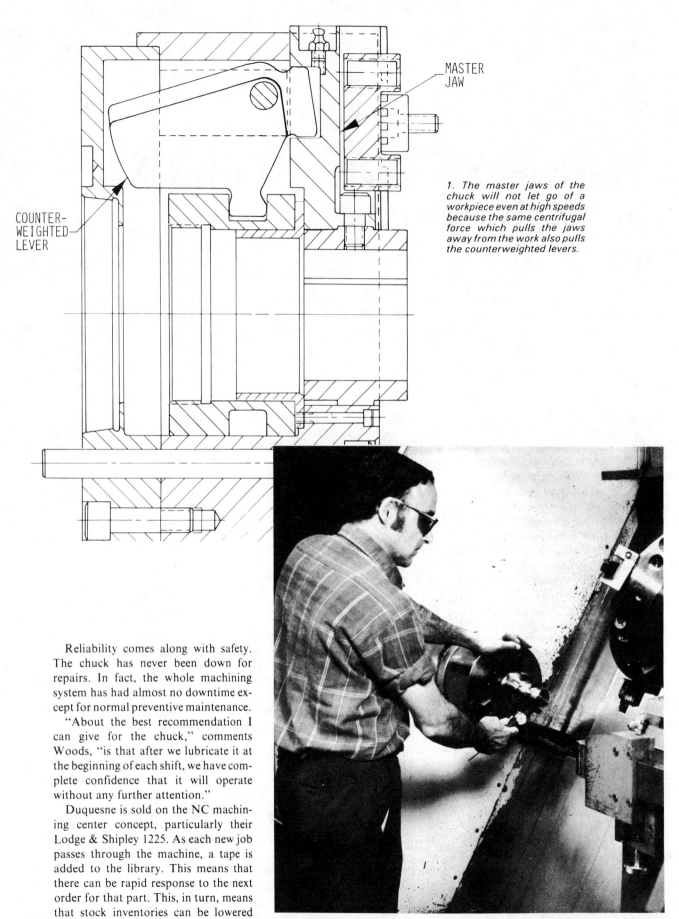

COUNTER-
WEIGHTED
LEVER

MASTER
JAW

1. The master jaws of the chuck will not let go of a workpiece even at high speeds because the same centrifugal force which pulls the jaws away from the work also pulls the counterweighted levers.

Reliability comes along with safety. The chuck has never been down for repairs. In fact, the whole machining system has had almost no downtime except for normal preventive maintenance.

"About the best recommendation I can give for the chuck," comments Woods, "is that after we lubricate it at the beginning of each shift, we have complete confidence that it will operate without any further attention."

Duquesne is sold on the NC machining center concept, particularly their Lodge & Shipley 1225. As each new job passes through the machine, a tape is added to the library. This means that there can be rapid response to the next order for that part. This, in turn, means that stock inventories can be lowered without hurting the ability to service customers.

2. The chuck is mounted on a Lodge & Shipley lathe. Duquesne is pleased with the accuracy and surface finish made possible by this installation.

Reprinted from Modern Machine Shop, February 1980

Chuck Pressure Gages: What Do They Say?

A chuck's gripping pressure decreases as spindle speed increases. For safety purposes, use a chuck pressure gage to determine the gripping force of power chucks out in the shop.

By RICHARD C. SPOONER, President
PowerHold Products, Inc.
Rockfall, Connecticut

With today's emphasis on high speed machining, it's more important than ever to know what the actual capabilities of the lathe chucks are out in the shop. Lathe chucks represent a potential danger in terms of personnel injury, machine damage, and loss of tooling due to an accident. These dangers can be minimized with the use of a chuck pressure gage.

For years, many manufacturers established the maximum operating speed of their chucks at a point where 50 percent of the initial or static gripping force was lost due to the centrifugal force. This speed was based on using standard soft blank top jaws and gripping the workpiece at the ideal position (the point where the chuck shows the maximum holding force). The fact of the matter is, however, that when these rules or unspoken standards were being used, there were no lathes or other machine tools that could really challenge the holding power of the chuck. Standard turning equipment and cutting tools were not ready for high speeds during the 1960's.

Then in the 1970's, lathe builders increased the horsepower and speed, and cutting tool people started emphasizing ceramics. Out of necessity, counter-centrifugal chucks started appearing on the scene. These chucks are designed to overcome the effects of centrifugal force as the spindle speed increases (see Figure 1). For example, before counter-centrifugal chucks were introduced, a 12-inch-diameter chuck might have been rated at a top speed of 2200 rpm. Again, this is where 50 percent of its initial grip is lost. Today, a counter-centrifugal chuck may be run at 3000 rpm and hold up to 80 percent of its initial gripping force.

This new technology means increased emphasis on the chucking of workpieces to be machined. Previously, designing special top jaws was fairly routine. All that was required was to know the configuration of the workpiece and how the top jaw mounts to the chuck. Between those two very defined parameters, the draftsman could come up with a design in a short period of time.

But with high speed turning, all that has changed. Part tolerance, balance, grip position, height, and weight of the top jaw all are extremely important considerations. With a power chuck, unlike a manual chuck, the operator has no feel or sense of gripping force; this function is provided automatically with a pneumatic or hydraulic cylinder.

The manufacturing engineer knows that he must take full advantage of all available horsepower and speed to remain competitive in today's market. Operation at higher speeds means it is necessary to accurately monitor the gripping force of the chuck at the speed it will be run and with the jaws that will be used.

Determining Safe Conditions

Until recently, it was difficult for a machine operator to recognize that a gradual decline in gripping pressure might mean losing a part. In fact, it might take two or three days to find

Fig. 1—The gripping pressure of a chuck decreases as the spindle speed increases.

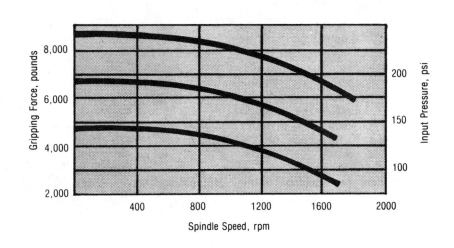

out why a chuck had let a workpiece slip. Such factors as tool geometry, top jaw design, feeds, speeds, material, and chuck maintenance would be considered. But hours or perhaps days of downtime would be required for an investigation of the chuck's integrity, including disassembly, cleaning, reassembly, retooling, and testing. And if the chuck was found to be in good operating condition, the operator would have to go back to the assembly to check for cylinder leakage, drawbar obstruction, or other interferences—meaning more downtime and loss of production.

But now, the use of a chuck pressure gage can indicate in minutes whether the chuck is producing the gripping force it was designed for. This diagnostic ability is not only important for chuck users but also for the machine tool builder's field service engineer.

The frequency of application of the chuck pressure gage depends upon the type of work being done and the interval between routine chuck maintenance. In general, power chucks that are used two shifts daily should be checked no less than once a month. When a machine is first set up in the shop, the gage should be used to check the actual gripping force of the chuck for future comparison. Thereafter, periodic checks will indicate if there is a gradual decline in gripping force, which may be caused by lack of proper lubrication, a build-up of chips in the chuck mechanism or, perhaps, a cylinder leak.

Description

The chuck pressure gage is really a matter of preventive maintenance and should be considered a very modest investment in employee safety and turning equipment up-time.

As shown in Figure 2, two types of chuck pressure gages are offered: one for static measurement and one for dynamic measurement. If care is used in top jaw design, a static measurement of grip force is usually sufficient if the chuck is rated at a higher speed than the lathe is capable of producing. A decline in static gripping pressure may indicate the chuck is in need of lubrication and/or cleaning. On the other hand, the 4,000 to 6,000 rpm machines on the market today more nearly equal the chuck ratings and require a dynamic measurement of gripping force at top speed.

It's important to remember that there is a big difference in the clamping force of a chuck measured statically and dynamically. Centrifugal force tends to try to open the chuck the faster it turns, allowing noticeable grip losses at higher speeds. In a standard chuck, the top rated speed is usually the point at which the chuck has lost 50 pecent of its initial grip. This point can be reached very quickly in high speed turning.

As soon as the high speed turning machines are employed to their fullest capacities, chuck pressure gages will become more commonplace. Most major machine tool builders now gage and certify chuck gripping force before they ship new machines. Users in the field are beginning to follow suit, and the ease of use of the chuck pressure gage is helping the process along.

Fig. 2—The pressure gage above is used for static testing of power chucks while the gage on the right is used for dynamic testing of chucks.

Using The Gage

The chuck pressure gage weighs about 12 pounds and is easily carried from one machine to another for testing. Equipped with a carrying case, service engineers can take the gage on service calls.

Technically, there is no limit as to the size jaws that can be used with the chuck pressure gage. In actual practice, however, the gage should be used on those chucks that are 18 inches in diameter or less.

The chuck pressure gage measures the total gripping force of a power chuck to indicate whether or not it is performing according to the manufacturer's specifications. To determine the static gripping force, the pressure gage is inserted in the chuck jaws and power is applied to the chuck. A self-contained hydraulic system with three plungers takes the load from the chuck jaws and transmits it through a reducer piston to indicate the gripping pressure.

When a chuck is being run at high speed, the gage could be thrown from the chuck jaws just as though it were a workpiece. Therefore, to determine the dynamic gripping force, the pressure gage is bolted to the face of the chuck. This procedure not only insures safety, but also provides a true running, concentric gage installation. For dynamic testing, a rod is inserted in a coupling and extends to a stationary member of the lathe. This prevents the gage from rotating so that the pressure can be read through any safety enclosure.

For the large user of power chucks, the gage can be an extremely valuable comparative tool. For instance, assume that a 12-inch chuck generates 20,000 pounds of grip at a certain drawbar pull. The same drawbar pull on a different brand of chuck may only generate half the gripping power. This testing procedure could indicate which chucks should be used for specific jobs.

Another use for the gage involves gripping fragile workpieces which might collapse under high gripping pressures. The chuck pressure gage can be used to determine the ideal gripping force; this documented force can then be used each time a similar job is set up. **MMS**

ABOUT THE AUTHOR

Richard C. Spooner began his career in workholding as a Field Service Engineer for PowerHold Products, Inc., in 1966 following his graduation from Porter School of Engineering Design. He has continued specializing in workholding devices and was a design consultant for a period of two years. His expertise in workholding has increased his responsibilities and has helped him attain his present position of Vice President and General Manager. Mr. Spooner, a recognized expert on workholding, has been a principal speaker at many technical meetings throughout the country.

Reprinted from Manufacturing Engineering, March 1978

Power Workholding

. . . can help improve machining accuracy

TODAY'S POWER WORKHOLDING SYSTEMS promote productivity by increasing the speed and dependability of the clamping function. In machining operations, the resulting improvement in workpiece accuracy can also help simplify processing and cut down on equipment requirements.

At the North Terminal Co., Hingham, MA, these benefits are particularly evident in the manufacture of receivers used in building the well-known Winchester rifle. North Terminal is a division of Winchester Western which, in turn, is a division of Olin Corp.

Single Power Sources. In designing and building special equipment for machining receivers, the company applied the concept of clamping parts at multiple work stations while using only one power source. A good example of the concept is a basic rotary indexing fixture like that shown in *Figure* 1. This is a demonstration fixture incorporating components manufactured by Vlier Engineering, Burbank, CA, Div. Barry Wright Corp.

Mounted around an air-operated in-tensifier are four clamping stations, each consisting of a four-way pneumatic control valve, a holding valve, and a swing clamp. The system also includes a rapid exhaust valve and a rotary union that connects the intensifier to the rotary fixture.

A pneumatic valve is used to introduce shop air to the intensifier which, in turn, forces oil through the holding valve into the clamps. When actuated in the opposite direction, the control valve sends air to a pilot rod that releases the holding valve and clamps and permits oil to flow back into the intensifier reservoir. An advantage of the holding valve is that it can be relieved with low-pressure air of only 50 to 70 psi (344.7 to 482.6 kPa).

Special Machines. North Terminal has applied these and other Vlier components in the design of special rotary indexing machines for processing the rifle receivers. A typical machine consists of a standard rotary table on which the company mounts its own fixtures, slides, and spindles. Receivers for the Winchester Model 9422 are run on the eight-station machine shown in *Figure* 2. A 10 in.^{3} (164 cm^3) intensifier in the center of the table powers the clamping system consisting of a four-way air valve, a holding valve, and two threaded cylinders at each station. One of the cylinders locates the receiver in the fixture; a second, larger cylinder provides final, positive clamping.

Operations performed at the seven

2. EIGHT-STATION ROTARY INDEXING MACHINE incorporates automatic power clamping at each working station. Positive clamping helps assure machining accuracy.

working stations include drilling, reaming, boring, and counterboring. Previously, these parts were run on four different drill presses, requiring excessive handling and refixturing. This created a problem in maintaining centerline distances between holes. Surface finish on the holes posed an additional problem.

Both difficulties are eliminated in the eight-station machine. Greater machining accuracy is credited for reducing the number of operations on one hole from four down to three. Loading each part just once into a single fixture equipped with positive clamping is an important factor contributing to accuracy.

A similar eight-station machine was built to handle 22 caliber rim fire receivers. A Vlier hydraulic work sup-

1. FOUR-STATION INDEXING FIXTURE featuring a holding valve that provides independent control of each clamping station and requires only one power source.

port at each station assures correct part positioning before two swing clamps are actuated to apply final clamping pressure. The clamps swing through 90° to facilitate part loading and unloading. Each of the eight fixtures holds two parts.

The clamping function is controlled by means of pneumatic logic in which small air valves are used to control the main hydraulic valves. Efficiency of this control arrangement enables the operator to also run a broaching machine.

Operations performed on these receivers include drilling, reaming, milling, tapping, counterboring, and deburring. The rotary indexing machine and broaching machine now do the work that previously required 15 individual machines.

In a third application of power workholding, North Terminal runs a different model of Winchester receiver on a six-station machine performing drilling and milling operations. A double fixture at each station holds two parts. Sequence valves are used to assure correct workpiece positioning. Two pressure limiting valves adjusted separately provide reduced clamping pressure on those areas of each receiver that are subject to deflection. Each fixture is equipped with four 90° swing clamps and five threaded cylinders.

Versatile Units. Vlier sequence valves and pressure limiting valves are versatile components. With the sequence valves, for example, parts can first be located with a line pressure between 400 and 800 psi (2758 and 5516 kPa), then clamped securely at a force between 800 and 1200 psi (5516 to 8274 kPa), and finally supported at a line pressure between 1200 and 1600 psi (8274 to 11 032 kPa). After the sequencing operation, the entire clamping system will lock in at the full intensifier hydraulic pressure of approximately 2000 psi (13 790 kPa).

The pressure limiting valve lowers line pressures to all clamps plumbed to the downstream lines. It is adjustable between 400 and 1300 psi (2758 and 8963 kPa) of hydraulic pressure. For example, if using several threaded cylinders rated at 500 lb (2224 N) of clamping force and it is necessary to reduce the force to 100 lb (445 N) to prevent part distortion or damage, a pressure limiting valve can be installed ahead of the clamping cylinders requiring the lower line pressures, which results in lower clamping forces.

North Terminal's use of power workholding is aimed at improving machining accuracy. The company has also been able to combine operations and reduce machine requirements while increasing its production capacity. ∎

Presented at SME's Chicago Area Conference, October 1977

Evaluation of Chucking System Needs for High Speed Turning

By Gerald E. Mueller
Houdaille Industries, Inc.

The trend toward high speed machining has brought about an influx of changes among machine builders, chuck manufacturers and tooling suppliers. These changes have brought about new concepts with relatively unknown technical areas. To the end user these areas can be problematic. The most troublesome topics dealing with workholding, are the loss of gripping force and the distortion of thin walled objects at elevated speeds. Employing methods of graphical solution and applying formulas, help to reduce trial-and-error engineering. Different jaw counter-balancing systems are discussed, pointing out methods of approach. Machining methods are discussed to reduce manufacturing problems and improve performance.

INTRODUCTION

The trend of new high speed NC lathes is toward increased production, improved accuracy, surface finish, and more rapid changeover capabilities. These are typical demands industry has faced over the years. However, now more emphasis is placed on their significance.

New machines rapidly face obsolescence when concepts improve. The end user must make a precise analysis prior to his investment, to determine the best choice of capital equipment for present and future needs.

Increased production, means more returns per investment dollar. A variety of approaches to increased production are available. These approaches vary in significance and also in their applicability to a broad range of machining circumstances.

A production increase is accomplished by three major means; achieving higher cutting speeds and increased metal removal rates while improving uptime. Stress hasn't been so widely placed on improved uptime, except at the manufacturing engineering level.

Quick-change devices are available as options at the O.E.M. level to drastically reduce changeover times. Since NC machines provide low to medium production, changeover is an important factor of the efficiency spectrum.

Increased feed rates and higher speeds are the major topics of discussion among machine builders, chuck manufacturers, tooling suppliers and end users.

Machine builders are supplying machines which can perform not only at higher horsepower peaks, but also at elevated rpm. Higher rpm often require improved machine base structures to resist vibrations, new bearing designs to withstand increased peripherial speeds and refined bearing lubrication systems to withstand resultant heat gains. Improved bar-feed designs allow for increased rotational speeds while reducing vibrations. Constant surface speed rates have progressed to meet advanced tooling necessities.

Cutting tool manufacturers are continually improving existing carbides, coated carbides and ceramic tools. Tooling is more rigid and sturdy to overcome vibration problems, interrupted cuts and increased removal rates. Speeds are steadily rising to provide optimum metal removal.

Tooling using carbide and carbide bases allow for heavy depth of cuts and an extension of rpm limits. This approach uses brute horsepower to boost removal rates. It proves most effective for large amounts of metal removal but causes excessive power consumption and in extreme cases, causes system fatigue from exorbitant loads. Machines using this type of tooling are generally not classified as high speed machines, except when used in conjunction with ceramic tools for finishing operations.

Initially ceramic tooling concepts with high cutting speeds, reduced depths of cut, and increased numbers of machining passes met much opposition. Its benefits have finally been realized and accepted on a broad scale. Advancements with respect to strength and durability have made increased chip loads and longer life possible.

Ceramic tools and the high speed concept, lend themselves to certain types of applications. Smooth surface finishes are readily achieved, along with accurately held dimensional tolerances. Parts manufactured from forgings and castings or slugged bar stock, are most effectively machined with the high speed approach.

Less stock removal is required on pre-formed work materials. Generally a single roughing and one finish pass is possible to complete the part surface. Carbide tooling requires the same number of passes for this type of application, but is accomplished at much lower speeds.

Slugged bar stock parts can be readily machined with the high speed concept for several reasons. Traditionally, bar cut-off operations have been very slow due to tool configuration and chatter. This time consumption can be very expensive due to NC machine and operator costs. Recently, the concept of slugging parts has received a great deal of analysis.

A steel handler can load a bar on the cut-off, facing and centering machine set on automatic cycle, while he performs other functions. This can be done at low machine costs and virtually no operator expense. Besides the reduction of cut-off costs, the most important savings can be attained from machining.

Slugs are easier to handle and can be machined at higher speeds. Rotation of bar stock at high rpm is not practical due to vibrations. Bar feed mechanism designs do not perform well at elevated speeds. Slugged workpieces are easily supported by a center, for stable working conditions.

Although this technique does not satisfy everyone's needs, it often means improved performance, greater safety, and less maintenance.

THE WORKHOLDING ELEMENT

The machine, workholding device and cutting tool are considered separate entities, but their performance relies on the ability to function in harmony. The workholding element requires particular attention since its performance is critical.

An improper chucking system can be hazardous, restrict performance or limit the machine's use. Its performance becomes particularly important in high speed applications where gripping force losses could mean tool breakage due to part slippage, or an accident from the workpiece escaping the jaw's grasp.

The chuck's rating is important to insure satisfactory operation, while maintenance is a key factor to preserve that rating. Since no concrete formula is available to determine a chuck's rating and standardization among chuck manufacturers doesn't exist, a significant difference may result from one chuck to another, although the manufacturers have similar ratings.

Parameters which play an important role in determining the rating are the maximum draw-force, clamping factor, lubrication and anti-friction constituents, jaw weight and position of its center of mass, and the percentage jaw force loss acceptable by a manufacturer.

The most common method of chuck rating is to specify the maximum clamping force. At low rotational speeds, the dynamic gripping force approximates the static force. However, at higher speeds this no longer holds true and the effects of centrifugal force must be considered. At elevated speeds it is possible to completely loose the jaw force with a conventional chuck. To achieve the maximum static clamping force, the chuck's input, mechanical advantage and frictional losses must be taken into account.

THE MAXIMUM DRAWFORCE

The chuck's input only varies slightly since rotating air and hydraulic cylinder forces are standardized to a degree. Cylinder designs and safety features differ significantly and deserve special attention at high speed operation. A cylinder should have features which maintain cylinder force even when the supply of the force media is lost.

CLAMPING FACTOR OR MECHANICAL ADVANTAGE

The clamping factor or mechanical advantage can vary from less than 1:1 to a ratio of 4:1 or more. A high input to output ratio can be advantageous if frictional losses don't deteriorate the gains. Low ratios require larger and more cumbersome cylinders which add weight to the revolving components. Constant speed changes on NC machines are required for optimum performance. This added weight can be detrimental to the drive system and braking elements. Weight (inertia) increases power consumption and accelerates drive system wear and reduces brake life.

LUBRICATION AND FRICTION

Costly frictional losses can be a result of inadequate lubrication or poor chuck design. To maintain a chuck's factory performance, it is necessary to grease the chuck at regular intervals. In severe cases, disassembly may be required, when the mechanism becomes overly contaminated with grit and chips. Poor lubrication canals restrict the proper flow of grease. New greases should force the old broken down lubricants out. New greases should not ooze out of the chuck, except in cases of over lubrication. Poor lubrication systems have a tendency to fling grease. Lack of grease enhances premature wear and can reduce jaw force up to as much as 50%.

In the presence of coolant environments, lubricants tend to wash away. Grease compounds containing molybdenum disulfide provide good lubricative and adhesive characteristics, and errosion resistance.

A good chuck design encloses the chuck mechanism completely. The base jaws and jaw guides should be shielded as much as possible to eliminate coolant and chips from entering, and grease from escaping.

The difference in performance between laboratory and actual working conditions is determined by the consequences of these variables. They affect the chuck's function of achieving maximum performance.

JAW WEIGHT AND POSITION

The jaw performance is an important parameter in determining the efficiency rating of the chuck, since it directly relates to clamping force at elevated speeds. Jaw weight, geometry and radial location affect the force at high speeds. The jaw weight, geometry and radial location of the center of gravity are a product in determining the moment of inertia for rotating elements. This inertia factor (P_c), is a function of the square of the rpm. The jaw force loss can be easily calculated by the following equation:

$$\text{(jaw force loss)} \quad P_c = (.112)(G)(r)\frac{n^2}{100} \quad \text{equation \#1}$$

Where P_c (Kp) equals jaw force loss, G (Kg) the weight of base and top jaws plus T-nuts and bolts, r (cm) the radius of center of gravity of the jaw unit and n (rpm).

The clamping force of a rotating three jaw chuck at n (rpm) is given as the sum of all three jaws:

$$\text{(jaw force @ n)} \quad P_{sp} = P_o - 3P_c \quad \text{equation \#2}$$

P_{sp} (Kp) represents the total jaw force at n spindle speed, P_o (Kp) indicates the total static gripping force and P_c (Kp) the jaw force loss due to centrifugal effects at n (rpm). P_o can be determined by using a static gripping force gauge or by calculating the drawforce input and multiplying this quantity by

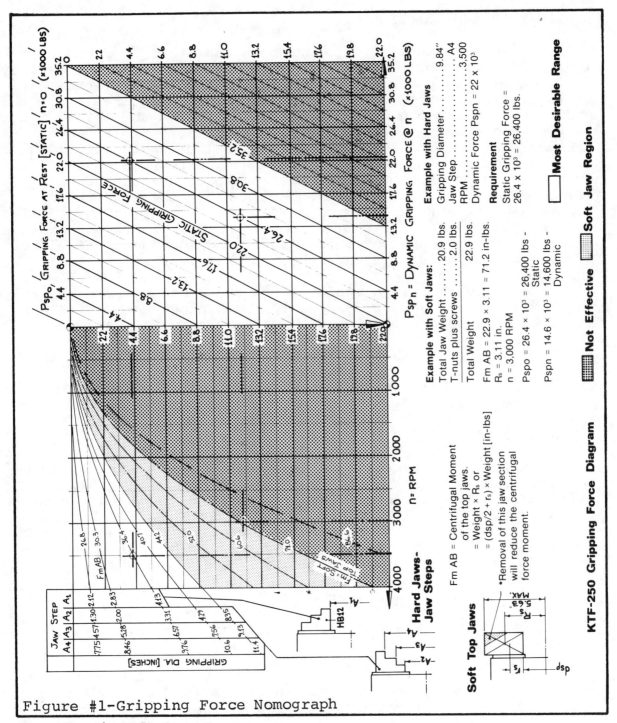

Figure #1-Gripping Force Nomograph

the clamping factor.

Jaw force losses become critical at high speeds since losses increase with the square of the speed. These calculations can become overbearing when constant chuck use is involved and different speeds and clamping diameters are desired. To alleviate this problem, nomographs are available from certain chuck manufacturers to aid in the calculation of dynamic jaw force. Figure #1 shows such a graph. This sample shows the Universal-Forkardt model 3KTGF-250 (10") countercentrifugal power chuck.

This nomograph is particularly helpful for several reasons. It dispenses with lengthy exponential calculations. It gives you the security of knowing how your chuck should ideally operate throughout the entire chucking range and with different top tooling. Maximum force values insure proper limits for workpiece rotation with solid or sturdy geometry. For workpieces of thin walled construction, it gives force limits which help to eliminate distortion problems. For soft jaw calculations, an easy to use formula is given to indicate safety areas of chuck performance.

It is easily understood that for optimum operation, the lightest possible jaw assembly is required, along with a jaw position which is located close to the chuck's center of rotation. Figure #1 shows this principle with jaw steps A4, A3, A2 and A1. These hard stepped top jaws can be positioned or reversed to achieve better dynamic gripping characteristics. Gripping a 4" diameter could be accomplished by either of two means, jaw step A2 or A1.

When looking at the schematic of jaw geometry showing steps A2, A3, and A4 it can be seen that the jaw's center of gravity is between steps A3 and A4, and nearest to A4. Gripping a 4" diameter on step A2 would make the jaw weight at the outer most position making r of equation #1 the greatest.

Gripping on step A1 in this instance would reduce the (r) dimension and bring the center of mass of the jaw close to the center of rotation. This makes gripping in this position most efficient because jaw losses are less at elevated speeds.

Gripping on step A4 would be most effective in the case of gripping an O.D. of 7.75 inches through 9.13 inches. The center of mass would be brought nearer to center. However, it is not always possible to grip on the most advantageous positions, due to workpiece geometry.

Since these options exist, it is imperative that the programmer indicate the proper jaw position to the operator. High speed applications become critical and failure to notify the operator could result in an accident.

Also note that the removal of a section of the soft top jaw as shown, would not affect jaw strength but reduces weight and moves the center of gravity (CG) closer to the center of rotation. These techniques can be helpful in achieving greater performance and insure a higher gripping force safety factor.

The example shown in Figure #1 with hard jaws at 3,500 rpm maintains 83% gripping efficiency. The variables have not been so critical, to allow for excellent efficiency. The example with soft jaws is not quite so effective. An efficiency of 55% at 3,000 rpm is realized since the jaw weight and (CG) have not been so favorable.

Various chuck manufacturers advertise concepts using jaw surfaces with serrations, machined directly onto the base jaws. Another technique allows for the mounting of collet pads onto the base jaws. These concepts attempt to reduce the inertia effects while reducing the bending moment of typically designed top jaws. These approaches are helpful to some degree, with bar stock workpieces which fit within the chuck bore.

They are attempting to approximate the concept of a collet
type bar chuck. These approaches vary in applicability and significance of achievement. These chucks generally have large
bore sizes with respect to outer diameter. Collet type chucks
do not require such large O.D.'s since the method of guiding
the collet is not done radially.

Collet type chucks are not affected by centrifugal force,
since the pads are contained by the chuck body. Most bar machines are designed for use with collet chucks. They offer
large bar sizes with small outer diameter restrictions. The diameter restrictions are a result of turret design. For optimum
performance jaw chucks can not compete with collet chucks having small O.D. sizes and large bores. Jaw chucks do however
have more versatile characteristics than collet chucks.

Chucks with large bores and small outer diameters generally have friction problems, due to jaw guide length reductions.
The lack of sufficient jaw guide length causes the jaws to bind
during movement. The above mentioned chucking approaches usually leave the jaw guides unprotected, and invite chips and
grit to enter from lack of proper shielding. Earlier during
the discussion of friction, we mentioned friction can consume
up to 50% gripping efficiency.

The common fault of most jaw chuck collet pad approaches is
the lack of peripheral support. Too often, this lack of support
allows the insert to spring back at the outer edges. In practice, no finite accomplishments are realized. Rough stock is
not perfectly round. A pad with surrounding circular geometry
will not make even continuous contact. This can be seen by
looking at the gripping pattern of the serrations on a rough
workpiece. At the center, near the support, the serrations are
embedded deep within the material's surface. The marks fade as
the edges are approached. A collet type chuck has 360° peripheral support.

The effective driving force of both concepts shown in
Figure #2 may be identical.   Force is a product of stress and
area. If the area is decreased, the stress or pressure is increased. Due to the decrease in area, the cobblestone serrations of Figure #2a will bury deeper into the material. The
shear stress of a material may be greater than the coefficient
of friction. Most of the contact area shown in Figure #2b will
be driven by friction rather than by interlocked material to
provide positive drive.

Parts having premachined gripping
surfaces must be analized in a different manner. The premachined surfaces
are almost perfectly round and can be
considered as such. When pad forms or
soft jaws are used, the gripping nature
assumes a different pattern. The gripping characteristics are six contact
surfaces instead of three. This can
be seen in Figure #2 when assuming a
three jaw chuck.

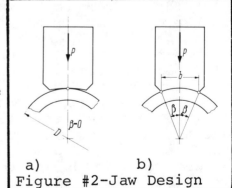

a)   b)

Figure #2-Jaw Design

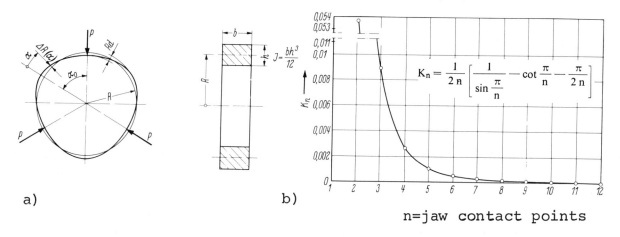

$$K_n = \frac{1}{2n}\left[\frac{1}{\sin\frac{\pi}{n}} - \cot\frac{\pi}{n} - \frac{\pi}{2n}\right]$$

a)   b)

n=jaw contact points

Figure #3-Distortion of Thin Walled Objects

Gripping marks on premachined surfaces are generally not acceptable. Therefore, smooth jaw surfaces are used and friction is the only driving force. The increased contact area in this case drastically increases the drive, while giving a wrap around effect on the workpiece. This is especially helpful in eliminating distortion of thin walled parts.

DISTORTION OF THIN WALLED OBJECTS

The distortion of thin walled parts is a very complex subject. To the manufacturing engineer, this can be a troublesome topic as well. Machining of such parts requires precise analysis due to their fragile nature. Gripping these components at high speeds can also be troublesome.

A high gripping force will distort the part, while too low of a force invites accidents. Forkardt has undertaken in depth studies to determine traits of commonality among thin walled objects and the results of gripping with numerous jaw arrangements. Figure #3 shows a segment of their research findings.

Figure #3a shows the distortion which occurs with a three jaw chuck and how it applies to the neutral axis of a ring shaped object. Figure #3b shows the effects of increased jaw contact as it relates to the K_n factor. K_n is a constant used to determine the total distortion (R_d) of the neutral ring axis. R_d is easily calculated from the following formula:

$$\left[\begin{array}{c}\text{distortion of neutral} \\ \text{ring axis}\end{array}\right] \quad R_d = (K_n)(P_{sp})\frac{R^3}{(J)(E)} \quad \text{equation \#3}$$

R_d(cm) is the total distortion of the neutral axis of the ring material. K_n is a dimensionless factor which is determined by the number of jaw contact points, P_{sp}(Kp) represents the sum of the jaw force, R(cm) is the neutral axis of the ring material, J (cm^4) is the polar moment of inertia, and E(Kp/cm^2) is the modulus of elasticity of the ring material.

K_n is given graphically and is determined by the equation given in Figure #3b. Additional information pertaining to Δ R and $\propto$ are available from the chuck manufacturer. This topic

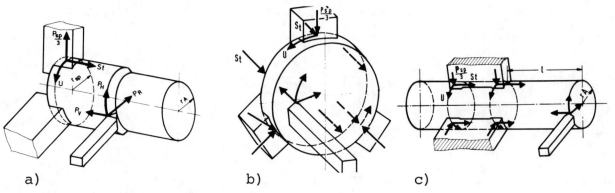

a)   b)   c)

Figure #4-Various Machining Exercises

becomes so complex that it cannot be covered in this text in any
further detail. Equation #3 is often referred to as the Kohlhage
equation after the man who did extensive research on this topic.

To eliminate distortion, the maximum permissible clamping
force on a thin walled workpiece can be determined. We must
also know the minimum possible clamping force to insure proper
driving of the component. In critical cases, this is required
so a force can be chosen within the extremes for optimum performance.

Figure #4 shows typical types of chucking examples. Figure
#4a is a vector force diagram illustration. Figure #4b is of a
flat workpiece where a bending moment from the cutting tool does
not exist. Figure #4c is such a case where the moments (P_R x l)
and (Pv x rA) exist.

Three main cutting tool forces are involved in the force
diagram: P_H the main cutting force, P_V feed force and P_R pas-
sive force. P_V and P_R have counter reactions at the jaw by
force St or by having the part placed against a tool stop.
Stops such as a jaw step or the chuck body are common.

Drive is created by friction between the jaws and workpiece. This counter reaction must over come the main cutting
force moment (P_H x r).

[1] This subject becomes complex due to the variety of different possible chucking procedures. A simplified equation giving
the clamping force requirements for a specific situation is
shown below. Equation #4 applies only to the machining of flat
and tailstock supported components. Provisions for the effects
of tilting and pullout are not included.

$$\begin{bmatrix} \text{Force} \\ \text{Requirements} \end{bmatrix} \quad P_{sp} = S \cdot \frac{P_H \cdot dA}{\mu_{sp} \cdot d_{sp}} = S \cdot \frac{N \cdot 716 \cdot 2}{n \cdot \frac{d_{sp}}{2} \cdot \mu_{sp}} \qquad \text{equation \#4}$$

1) portions of the given information were extracted from the
   Forkardt Workholding-Chucking techniques manual.
   Paul Forkardt KG, Duesseldorf, West Germany, 1974.

KEY TO SYMBOLS

P_{sp}=clamping force (sum)
U =circumferencial force
St =supporting force
P_H =main cutting force
P_V =feed force
P_R =passive force

S =safety factor
dA=cutting diameter=$2r$
d_{sp}=clamping diameter=$2r_{sp}$
N =power rating of machine
n = spindle speed
μ_{sp}=friction value

VARIOUS FRICTION VALUES
WITH DIFFERENT JAW FORMS

μ_{sp}=.15 with smooth jaw surfaces
   =.25 with cobblestone serra-
        tions on jaws
   =.35 with sharp jaw serrations

VARIOUS FRICTION VALUES
WITH DIFFERENT WORK SURFACES

μ_{sp}=.26 on unmachined com-
        ponent
   =.17 on rough machined
        component
   =.10 on finish turned
        component

The main cutting force can be calculated by the simplified formula below:

(main cutting force) $\qquad P_H = a \cdot s \cdot K_m \qquad\qquad$ equation #5

(a) indicates depth of cut in mm, (s) feed in mm/rev, and (K_m) a dimensionless material constant with dull tool factor included. This factor can be found in almost any technical journal where recommended cutting tool speeds and feeds are indicated. These formulae were derived by H.J. Warnecke for his dissertation at the Braunschweig Technical college in 1962.

ACCEPTABLE JAW FORCE LOSSES AND RATINGS

The reason for a lack of standardization among chuck manufacturers' ratings is quite obvious. The numerous variables coupled with the multitude of machining exercises make standardization almost impossible.

Different manufacturers accept different jaw force loss percentages. Some manufacturers rate the maximum rotational speed when only one fourth of the maximum static jaw force remains, while others accept one half of the static force. These percentages are typical for the rating of conventional chucks.

One half the static rating should give an adequate safety factor. However, little may be accomplished at so low a speed. One fourth the static rating leaves little regard for safety. Using a principle of one third the maximum static force for the rating of standard chucks is practical.

Chuck manufacturers attempt to supply devices which will function adequately for general purpose use. Exotic machining applications deserve additional attention. All high speed applications should be given critical scrutiny, since some designs do provide greater performance and safety features than others.

CHUCK DESIGNS

A multitude of chuck designs are available to industry. This magnitude is quite clear, when it is understood that each chuck manufacturer offers several models. This variety can be broken down into three major catagories for mode of operation. Of these three types, the lever and wedge-hook models are most popular and operate in similar fashion. The wedge-block version, functions under a completely different principle.

Lever and wedge-hook type power chucks have served industry for many years. Development of high speed chucks, made the adaptation of these two principles with counter-centrifugal balancing methods the most obvious approach.

Lever Chucks

A typical lever chuck is shown in Figure #5. The diagram depicts the mode of operation. Push and pull forces created by axial movement of a hydraulic or air operated cylinder, induce an "in and out" radial jaw movement. Generally the cylinder is located at the opposite end of the spindle. The chuck and cylinder are connected by means of a drawtube or drawbar. This transfer of motion is accomplished by rotation of levers, located by cross-pins at each jaw. A 90° transformation of motion is accomplished by rotating the lever to change axial movement into radial motion. The difference in link length of the lever accomplishes different clamping factors or mechanical advantages which were discussed earlier.

Wedge-hook Chucks

The wedge-hook chuck shown in Figure #6 also uses a similar principle to change the axial movement of the cylinder into "in and out" radial jaw motion. Wedges are used instead of levers. The wedge has a Tee configuration and is located at a 10°-20° incline. The change in incline produces a different clamping factor. Action of the cylinder causes the Tee section to slide on the puller ramp, causing "in and out" radial jaw motion. The term wedge applies, since the small angle acts to provide a wedging action which serves as a semi-sticking surface. This sticking helps to a degree, to insure against jaw force loss even if the cylinder force is diminished.

Both of these approaches have a limited jaw stroke. To achieve various clamping diameter extremes, the top tooling must be relocated. This is a time consuming effort.

The inner components require precision machining to insure equal jaw movement of all three jaws. Figure #6 also shows that the piston rod can be ground and guided within the spindle bore. This prevents the puller from tilting. The puller is the part connected to the wedgehook or lever. Tilting of this element causes unequal jaw movement and loss of accuracy. Guiding this element reduces run out and increases repetitive accuracy.

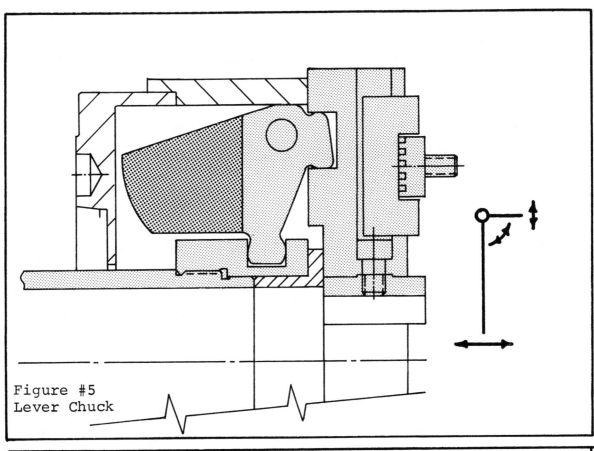

Figure #5
Lever Chuck

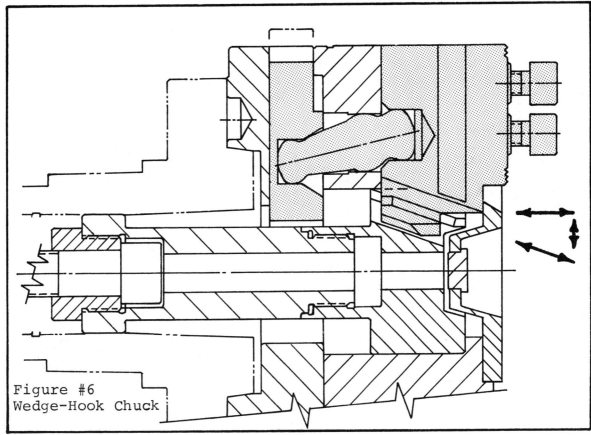

Figure #6
Wedge-Hook Chuck

COUNTER-CENTRIFUGAL BALANCING

The application of counter-centrifugal balancing methods (flyweight compensation) is done in a variety of ways. Each manufacturer applies a slightly different technique, which they feel provides the best solution.

Figures #5 and #6 show different techniques of applying this flyweight action. The most common method used on lever chucks is shown in Figure #5.

The shaded portion located at the rear of the lever shown in Figure #5, is the mass which counter-reacts the combined mass of the base jaw, bolts, Tee-nuts and top jaw. This shaded portion is a solid connection between the cylinder, drawrod and the base jaw.

Figure #6 shows a slightly different method where the flyweight mass is located independently from the drawrod. Only a solid connection between the counterweight and the base jaw is offered. This was done since the centrifugal force directly relates to the jaw and counterweight, not the cylinder drawrod. Connection to the drawrod was felt to function as a dampener.

Normally the counterweight mass is designed for a "mean" jaw position with standard top tooling. This counterweight is of a specific weight. The inertia of the top jaw can vary drastically, due to the range of positions and reversal features. The base jaws are generally used in a normal mean position of the limited jaw stroke. The base jaw therefore causes no problems for the calculation of centrifugal force. The difference of top jaw position and inertia makes the counterbalancing difficult.

Nomographs such as shown in Figure #1 show the effects. Small gripping diameters can be overly compensated for and thus, an increase in gripping force can be accomplished at higher speeds. Large gripping diameters or rather large inertia factors are under compensated for, and greater force losses are encountered.

Some manufacturers offer interchangeable weight packages which can be mounted on the counterbalancing system. This works fine for extreme cases. An operator must be very careful not to forget to change the balance weight, if the next set-up operation calls for an increase in inertia. This could very easily become a safety hazard.

Wedge-block Chucks

Figure #7 shows the wedge-block type power chuck. It uses the principle of a rotary actuator instead of a cylinder with axial movement. The actuator is connected to the chuck by means of a torque tube which is splined on both ends. The mating spline can be seen on the chuck's back face, shown as the innermost portion of Figure #7a. Rotation of the torque tube moves the helical gear. The gear is responsible for moving the three wedge-blocks simultaneously. The wedge-blocks have inclined rack teeth on two sides. One side mates with the helical gear. The other rack tooth side mates with the base jaw. A

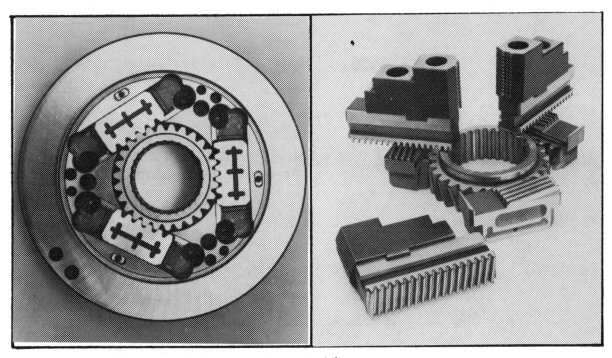

a)   b)

Figure #7-Wedge-Block Chuck

portion of this rack on the base jaw side has been removed as
seen in Figure #7b. Normal rotational movement of the actuator
operates the jaws in a fully engaged rack tooth stroke. Overriding of the actuator allows for disengagement of the mating
rack teeth between wedge-block and base jaw. This allows for
rapid disconnect and changeover of jaws or jaw position. This
feature provides a substantial time savings over conventional
power chucks, while also offering a larger jaw stroke. The
mechanical advantage of the wedge-block chuck is so great that
jaw forces in the range of three times the magnitude of conventional chucks are realized.

     This chuck does not have centrifugal counter-balancing
since, even if two-thirds the jaw force is lost, its force is
greater than static flyweight chuck maximum readings. This
chuck can be used for high speed operation where solid or
sturdy workpiece geometries are to be machined. Pressure reduction allows for reduced clamping force on thin walled objects. A variation of this wedge-block design has been made,
which allows for ultra-high speed operation, of a 6" diameter
chuck at 10,000 rpm.

REFERENCES

Chucks for High-speed Turning, R.N. Stauffer, Manufacturing Engineering, March 1977

Counter-Centrifugal Chucks for High Speeds, Tooling & Production, April 1977

Einfluss des Futters auf die Verspannung verformungs-empfind-licher Werkstuecke, H. Antoni, Werkstatt und Betrieb, Heft 12, 1976

Faster Chucks with N/C, J. Thornton, AMM/MN-Tooling Section, May 16, 1977

Forkardt Workholding--Chucking Techniques, Paul Forkardt KG, 1974

High-Speed Chuck & Cylinder Combos, Metlfax, July 1977

New Chucks Meet High-speed Needs, B.A. White, American Machinist, McGraw-Hill, June 1977

Power Chucking for Profit in NC Applications, W.A. Mossner, Tooling & Production, September 1976

The Quick-Jaw-Change Chuck...A New Way to Cut NC Downtime, W. A. Mossner, Modern Machine Shop, December 1976

Time to Come to Grips with NC Chucking Problems, W.J. Reed, Machine and Tool Blue Book, July 1975

Reprinted from Manufacturing Engineering, May 1977

The OD Collet Fixture

– tooling for speedy setup and ultimate precision in the gung-ho world of mass production

DANIEL B. DALLAS
Editorial Director

WHEN USED IN THEIR CONVENTIONAL range of applications, collets exert a constrictive force. This force may be used to hold a cutting tool, as in the spindle of a vertical milling machine.

In other applications, they're used to hold parts to be machined. Drill rod, for example, is finished to tight enough tolerances to be held in head-stock collets. Important advantages of collets in both types of work include high holding power, attributable to the large amount of surface area brought into contact with tool or workpiece, and an absence of the clamping marks normally left by chucks.

In either application, the force exerted by collets is *constrictive*. It's possible, however, to design and manufacture collets that exert an *expansionary* force. This is the design approach taken by Drewco Corp., Franksville, WI, in the design and manufacture of a broad range of special holding devices. An example is seen in *Figure* 1.

Design Principles. The mandrel shown in *Figure* 2 illustrates the extremely simple design principles of an expanding collet — or an OD collet as it is normally referred to. Basic components are (1) a body, (2) OD collet, (3) slider, (4) sleeve, and (5) locknut. To expand this collet, an operator simply turns the locknut. This exerts force on the sleeve, which in turn exerts force on the slider. Advance of the conical surface of the slider — matched by a correspondingly conical surface on the mandrel body — forces the collet to expand outward.

While this is an extremely basic design, it can be utilized in more sophisticated forms to hold a variety of otherwise hard-to-hold parts. An outstanding example is seen in *Figure* 3.

An OD and Two Ends. The workpiece in this instance is a cylindrical casting measuring approximately 5 1/8″ (130-mm) OD x 10 5/16″ (262-mm) long. Its ID is finish bored prior to machining of the OD. In addition to the OD and shoulder operation, the part must be finished on both ends — a mixture of machining assignments that normally requires at least two setups. In this instance, however, all machining is done in one setup on a Seneca Falls lathe.

Technically, this is a mandrel. It can be more accurately described as a lathe fixture, however, since its design permits pinpoint positioning of the work prior to machining. As a fixture, it is held on the lathe axis by engaging the spindle nose. At the other end, the lathe tailstock engages an insert in the end cap. Thus, the fixture is accurately held on the machine axis, and it is fully supported throughout the machining operation.

Two collets are used in this design. Both collets are keyed to the mandrel body to assure continuing precision throughout the life of the tool. In other words, keying the collets precludes the possibility of even minor errors in con-

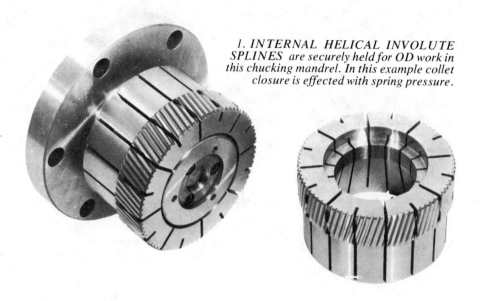

1. INTERNAL HELICAL INVOLUTE SPLINES are securely held for OD work in this chucking mandrel. In this example collet closure is effected with spring pressure.

gear housings. The housings are part of a tractor transmission system. The job to be done is the finish reaming of six holes on an 8″ (203-mm) bolt circle. Accuracy must be held to 0.0008″ (0.02-mm) TIR.

It will be noted that there is a slight difference in size between the upper and lower views of the slider/collet combination. It will also be noted that a small spacer ring is used to back up the slider in the upper view. These differences are attributable to differences in the two parts to be machined.

As in the preceding example, this system utilizes a pneumatically actuated drawbar to expand the collet. The drawbar is threaded to a drawpin. Retraction of the drawpin forces the slider back, and this in turn forces the collet to expand. Collet expansion locks the workpiece securely in place against the rest buttons shown.

Because the accuracy requirements of this job are especially tight, pinpoint adjustment of the collet assembly is mandatory. Adjustment is accomplished with four centering screws, which provide for movement equal to approximately 1/2 the clearance in the adjacent socket head capscrew holes. The procedure followed is to snug the capscrews, then put a workpiece on the collet assembly and determine its runout with an indicator. Careful adjustment of the centering screws brings the assembly to within-tolerance alignment. The workpiece is then removed and the cap-

centricity through tolerance accumulations.

Actuation of this system is through the action of an expander cap threaded to the drawbar. As the drawbar retracts, it forces the collets and the slider to the left. When the left-hand collet contacts the conical surface of the mandrel, both collets begin their expansion. This expansion — no more than several thousandths — continues until both collets are securely holding the workpiece through the exertion of force on the workpiece ID.

The action of the locating pin is an especially interesting aspect of this design. A rack is machined on both the drawbar and the pin. A pinion gear engages both. As the drawbar retracts to expand the collets, the rotating pinion withdraws the pin from contact with the workpiece. Thus the way is cleared for machining of the locating surface.

Heald Bore Matic Fixture. Still another outstanding example of the use of OD collets in fixturing is seen in *Figure* 4. This unit is designed to hold two different types of planetary

2. THE BASIC PRINCIPLE of OD collet action is illustrated in this drawing. Locknut force on the slider cams the collet outward.

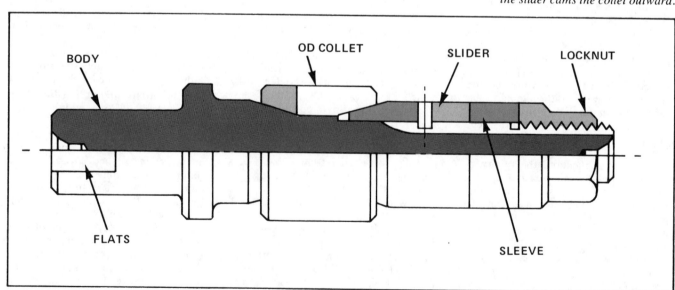

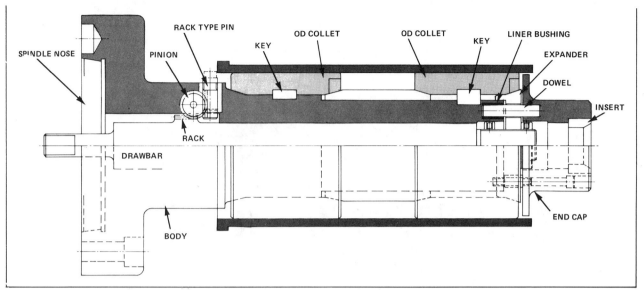

3. RACKS AND PINION are incorporated in this design to retract the workpiece locating pin. This permits machining of all exterior surfaces.

screws tightened to the maximum torque required.

Collet Manufacture. In producing its collets, Drewco generally follows the following procedure: First, the collets are turned and bored, then milled as required, and partially slotted. Partial slotting means that a small webbing is left at the normally open ends of the slots. At this point they're heat treated, then rough and finish ground. The webbing is then broken with a rubber wheel.

Up to this point, the processing is fairly routine. Drewco has added still another step, however — the sealing of the collet slots. The thinking behind this design innovation is that collet slots filled with an elastic sealant will not collect chips and smaller particles carried in the coolant. Without this protection, some wear might eventually be detected on the slider and on the interior surface of the collet.

The sealer used for this application resembles an RTV (Room Temperature Vulcanizing) rubber. It is, however, a thermoset rather than a thermoplastic. While no information is available on the precise formulation, the mixture preferred by Drewco is sold under the trade name of Plast-O-Meric®.

Application of this sealer starts with a thorough degreasing of the collet followed by application of a primer, which serves as an adhesive. The primer is allowed to dry for 60 minutes. Next, the collet is mounted on a fixture — designed to protect the ID —

and sealant is poured into the surrounding container. The entire fixture is then baked at temperatures ranging between 370-400°F (188-204°C). The bake time is a variable, and it generally is a function of collet size. Following the baking operation, the exterior sealer is peeled off in a lathe operation, and the collet is finished.

The Applications. The question arises: Who needs this type of fixturing? The answer, according to Drewco's management, is the major auto manufacturers, off-the-road vehicle manufacturers, and numerous others involved in relatively high production. The production requirements do have to be relatively high to justify this type of holding device — or if monthly production requirements are relatively low, the duration of production must be indefinitely long.

But the buyers of this highly specialized tooling, according to Drewco, are generally interested in reduced setup time. Fixtures based on the expanding collet principle are an excellent way to get it. ∎

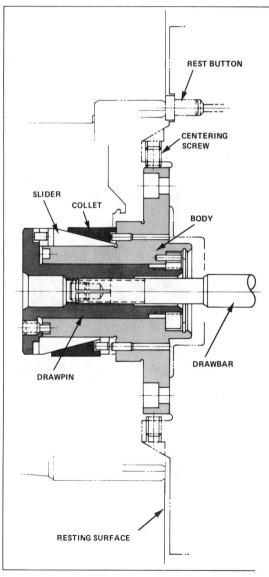

4. TWO DIFFERENT WORKPIECES can be held in this Bore Matic fixture simply by changing sliders and collets. Parts machined are planetary gear housings.

Reprinted from Tooling & Production, June 1981

CLAMPING IDEAS

More metalworking companies are realizing too much time is lost in handling workpieces and materials while performing loading, positioning, clamping, unclamping, and unloading operations. Automatic hydraulic clamping devices and systems reduce the nonproductive time and motion involved even when modern methods and techniques of machining, assembling and fabricating are used.

1. *Compact cylinders work in limited areas and tight spaces.*

A number of metalworking industry studies have produced some revealing statistics on productivity. They show that even the most modern machine tools, which have dramatically increased machining speeds in recent years, still only work to about one-third their capacity—or less than 3 hr in an 8-hr day. Specifically, they are producing only 17 to 38 percent of the time. The remaining downtime is taken up with the workpiece handling operations mentioned above. It's the same tedious operations over and over again. In fact, the more a part is handled, the more time is lost.

Depending on the amount of time it takes for machining, the timesaving advantages offered by hydraulic clamping vary considerably. Obviously, if a part is loaded and clamped to be milled for 4 hr, the time spent in nonproductive workhandling is insignificant. But if machining time is 2 min and it takes 4 min to manually load, clamp, unclamp and unload the part, 10 finished parts per hour are produced or 80 parts in an 8-hr shift. Using hydraulic clamping, the load/unload cycle time might be reduced to 1 min, and with machining time remaining the same, 160 parts in an 8-hr shift could be produced. This is a 100-percent increase in production on one machine alone. The significant point is that the shorter the handling time compared to the machining time, the greater the production.

An automatic hydraulic clamping system offers many inherent advantages whether milling, drilling, boring, cutting, tapping, welding or performing other accurate machining jobs. Here are a few:

• A single throw of a lever delivers positive clamping with constant, repeatable holding forces.

• With higher clamping flexibility and with higher accuracy and precision, most wasteful rejects due to clamping errors are eliminated, reducing material costs and wasted machining time.

• Parts with bulging curved surfaces or narrowing fine points can be secured.

• Unlike mechanical clamping, which can work loose when subjected to vibration, hydraulic clamping exerts continuous force against the part being machined to better resist machining torque.

• It helps reduce tool chatter, which can shorten tool life and possibly cause severe tool damage.

• Setup times using fixtures incorporating built-in hydraulic clamping are reduced; once the fixture is built to handle a particular part, it can be stored until needed.

Basic cylinder/clamp styles

Compact, threaded cylinders. The entire external body of these hydraulic cylinders is threaded so they can be screwed directly into a fixture. Simply drilling and tapping a single hole is all that is required. Threaded brackets are also available. They feature a spring return for

2. On this application, a sequencing valve is used in conjunction with swing clamps to delay clamp action until the precise moment the piece is located. The intensifiers are mounted on a mobile cart for use on other machine tools.

3. Eight threaded cylinders are used in this double fixture to secure eight parts.

rapid work release and threaded holes in the end of the piston rods for attaching various clamping-component accessories, including conical points, spherical buttons and toggle pads. The cylinders are easily adjustable by screwing in or out.

Standard-cylinders. Standard type cylinders offer up to two full inches of travel as well as heavy clamping forces. Mounting can be accomplished with brackets or clamps which grasp the body. A wide variety of component accessories can be screwed into the threaded holes in the ends for holding curved parts, angular shapes and other odd shapes.

Hollow cylinders (pistons). A through-bore provided the full length of the piston rod allows a mounting bolt to pass completely through the cylinder and permits the fixture designer to apply either a pushing or pulling force. For pushing applications, the top end of the piston can push against any flat surface. For pulling or squeezing applications, a bolt or shaft passes through the hollow piston, with a nut or shoulder resting on the top of the piston, while the other end is mounted through the fixture.

Fully automatic swing clamps. Swing clamps apply a rotary swinging motion and vertical clamping action for positive clamping and rapid horizontal clearing upon release. The clamp arm normally swings clockwise into position and upon unclamping swings back counterclockwise from this zero position. However, by adjusting the clamp, it can be made to swing in the other direction.

Self-adjusting work supports. Used to support parts that are being machined, these work supports offer resistance to clamping forces and/or the weight of the workpiece, acting counter to the direction of the heavy-duty support plunger. They automatically adjust to work loads and are ideal for support needed at hard-to-reach places, while reducing the possibility of misalignment and mismachined parts.

Retracting clamps. Special clamps for particular applications can be developed in minimum time and cost through the use of retracting clamps. By welding a clamping member to the movable arm, the member moves forward and down to clamp the workpiece. The retracting clamp can be mounted in any position, vertically, or horizontally. Typical applications are shown in the box at the end of this article.

Following are a few examples of how productivity increased using innovative hydraulic clamping methods.

Case History 1

ETM Corp, Monrovia, CA, uses standard and threaded cylinders to clamp difficult-to-hold, odd-shaped parts during machining operations. The company has become dedicated to the use of hydraulic clamping since they tried and measured the increased speeds and efficiencies of their first hydraulic system a few years ago. Today, they use hydraulic clamping at more than 90 percent of the machining stations throughout their plant. They turn out over 50 shapes and configurations of high-precision dental pliers used by orthodontists around the world.

Specifically, the fixture shown in **Figure 1** uses two threaded cylinders to automatically clamp and unclamp two parts on a milling machine. Compared to manual clamping methods, the time saved is more than 80 percent.

Other fixtures used at ETM perform automatic locating and positioning, multiple sequencing and leverage clamping. With the use of fixtures incorporating automatic hydraulic clamping, they have increased their total plant production by 32 percent.   **...continued**

...continued

4. An innovative manufacturer found a way to clamp six parts by stacking them on locating pins and securing them with a single hollow cylinder.

Case History 2

The old way of doing things—drilling one piece at a time and using a different fixture for each different size—is now obsolete at the Barry Controls Div, Barry Wright Corp, Watertown, MA. They use a back-to-back, double intensifier system and universal fixtures that handle three different sizes of US Navy rudder-assembly, shock-mounting plates. Speed is substantially increased because time-consuming changes are eliminated. The setup is seen in **Figure 2**.

The increase in production from 16 to 60 pcs/hr is also the result of a number of other important features designed into the system. Locating is performed faster, with more accuracy and automatically with threaded cylinders. Clamping is achieved with swing-type clamps. A sequencing valve is used for delaying swing-clamp action until the precise moment the piece is located. Flex lines and quick disconnects are used to speed changeovers. Also, the intensifiers are mounted on a mobile cart for use on other machine tools. Additionally, while one piece is being machined, the operator is loading another piece on the other side for even greater timesavings.

Case History 3

Vlier Engineering, Burbank, CA, the company that provided the information for this article, uses their own hydraulic-clamping products to manufacture their hydraulic clamps, **Figure 3**. They have more than tripled production from 40 pcs/hr to more than 135 pcs/hr when producing swing-clamp arm parts. Before the special fixture was built, they used to mill each part separately and averaged 1½ min/part in loading and unloading time.

The old way took 18 min to load and unload eight parts; the new way takes only 3½ min. Four parts on each side are clamped simultaneously, milled and unclamped on a Sundstrand milling machine. One side is completed, then the fixture is simply rotated for milling the other side.

Case History 4

One company, using a single hollow cylinder, clamps six parts simultaneously and significantly speeds their manufacturing operation. The setup, seen in **Figure 4**, replaces the tedious process of one-at-a-time loading, locating, clamping, machining and unloading. Milling time for the six parts on a Cincinnati end mill is 3½ min and the total loading/unloading time for the parts is 3 min.

Using hand-clamping methods required an operator's full time attention. Now with more free time, the operator's responsibilities have been expanded to include the operation of a machine adjacent to the mill. The company has estimated that their overall production increase has exceeded 100 percent since incorporating the new clamping techniques.

Building your system

In addition to the specific clamping components needed to meet specific work demands, each Vlier system consists of two other important parts. First is the power source, and then the associated accessories such as valves and tubing.

5. Hydraulic intensifiers can produce up to 2750 psi of clamping force using standard shop air. A see-through reservoir makes it easy to check the fluid level.

The most widely used power source is the hydropneumatic intensifier or "booster." This device simply concentrates and converts shop air pressure to a high hydraulic pressure. A typical installation is seen in **Figure 5**. The basic law of physics used to compute the amount of final hydraulic pressure is:

$$\text{Pressure x Area} = \text{Force}$$

For example, when 100 psi of shop air is applied to an intensifier's air chamber piston area of 22 sq in and reduced to a hydraulic piston area of 1 sq in (for a final ratio of 22:1), it produces 2200 psi of hydraulic fluid pressure: 100 x 22 = 2200 lb.

Each intensifier includes a reservoir made of see-through plastic for ease of checking fluid, and may be mounted at any angle, in any position, allowing various placement possibilities. When mounted on a mobile cart and fitted with quick-disconnect couplings, it may be used to power hydraulic clamping at various locations throughout the plant.

Also available are two types of manual pump-type power sources for applications where cycle time is not important or where air is not available.

When an air-powered intensifier is used as the power source, the system has two distinct sides: An air (low-pressure) side and a hydraulic (high-pressure) side. There are inherent advantages to this type of arrangement.

First, most of the controls are located on the air side, thus less expensive low pressure controls may be used. These include a filter-lubricator-regulator which controls the amount of air entering the system and the amount of high pressure clamping force at the fixture. This device also provides essential lubrication which keeps the system running smoothly. In addition, there are manual and electric 3-way valves and a manual 4-way valve for controlling the clamping/unclamping cycle, and a check valve to ensure that the workpiece is held securely even if air pressure is lost.

Other advantages of this air/hydraulic system are that the high-pressure hose or tubing is a single line. As both the output and return flow through the same line, there is very little heat buildup in the fluid due to friction.

General rules for the low-pressure side of the system are simple:
• Use the largest-size air line possible.
• Stay as close to the air compressor as possible.
• Keep the number of lines that come

out of the air-feeder line to as few as possible.
• Install a water trap on the line if water is a problem.

On the high-pressure side, when multiple functions such as locating, clamping and support are all done on a common fixture, they must take place in a logical, controlled order. The control (sequencing) of these functions is accomplished by using a sequencing valve. With multiple valves, each adjusted to sequence at a different pressure, the various functions occur in a controlled order.

Very often a large production run of relatively small parts can be expedited by using multiple work stations in conjunction with automated drilling, reaming, tapping etc. A holding valve provides the capability of selectively unloading one station in a multiple-station fixture.

Generally, on the high-pressure side of the system, use ¼" tubing wherever possible. This is because the entire side is a straight out-and-back, single-line circuit. If flexible tubing must be used, keep it to a minimum. It has an expansion factor of as much as ⅛ cu in/linear ft, which requires additional capacity.

Application examples

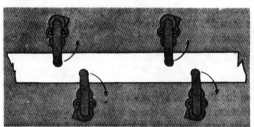

Swing clamps holding long piece of bar stock swing away for easy removal.

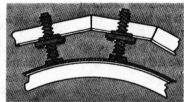

Cylindrical part is being securely held by threaded cylinders.

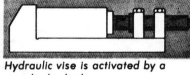

Hydraulic vise is activated by a standard cylinder.

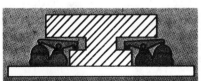

Clamping into a groove with swing clamps permits easy removal of parts.

Where side space is limited, standard cylinders activate rocker arms.

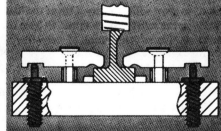

Part is firmly held by strap clamps automated with threaded cylinders.

Reprinted from Manufacturing Engineering, March 1978

Make it Mandrels

... to simplify setups for cylindrical grinding

1. TYPICAL CYLINDRICAL GRINDING setup with the part mounted on a precision mandrel.

THE TIME REQUIRED for setup and changeover is of obvious concern in operations involving short run precision parts. In cylindrical grinding, mandrels have a well-earned reputation for keeping these nonproductive time elements to a minimum.

The Ort Tool Co., Erie, MI, a producer of special machines, fixtures, dies, and tooling, provides a good case in point. The company has solved setup and changeover problems in its cylindrical grinding department through the use of Stieber Glidebush mandrels. These units are included in the Stieber line of clamping tools sold and serviced in the U. S. by Starcut Sales, Inc., Farmington, MI.

Advantages Noted. Angelo Milano, president of Ort Tool, says that most of the company's cylindrical grinding work involves parts that are typically high precision, in a variety of odd sizes, and in small lots — sometimes only one or two of a kind. This means that fast, accurate changeover in setups is a must, if the operation is to be economical.

Figure 1 shows a typical application with the part loaded on a mandrel and ready for grinding. The hand-operated mandrel is held between centers. Prior to using Stieber mandrels, Ort Tool had to chuck the parts or, in some cases, make up special arbors.

Milano points to what he considers the most appealing feature of these devices. "With only five basic sizes of clamping tools we can easily cover a work ID range in excess of 3″ (76.2 mm) with consistently accurate concentricity." *Figure* 2 shows an assortment of typical parts mounted on various sizes of mandrels.

How They Work. The basic construction of Stieber Type GDS Glidebush mandrels consists of a tapered shaft assembly, threaded on one end, over which various sizes of basic sleeves with a corresponding taper can be fitted and held with a retaining ring. The dif-

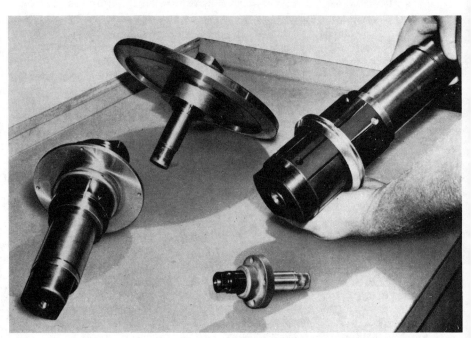

2. ASSORTMENT OF TYPICAL PARTS mounted on various sizes of mandrels. Axial positioning of the part is an inherent feature of the mandrel design.

ferent interchangeable sizes of glidebushes are screwed to the threaded end of the shaft.

In operation, manual clockwise rotation of the clamping bolt in the end of the shaft forces the glidebush up the taper (or base sleeve taper when sleeves are used) and expands the glidebush OD to hold the workpiece. An additional retaining ring groove is provided at one end of the basic arbor for an axial stop. Positive axial positioning by both radial and axial clamp-

ing provides maximum torque transmission. When the clamping bolt is released, the glidebush diameter contracts, and the part can be removed.

Type GDS mandrels are available in five basic diameter sizes and in 57 various intermediate sizes. These range in diameter from 0.591″ (15 mm) to 3.543″ (90 mm). Each of the 57 sizes has an expansion range of 0.047″ (1.2 mm) for those used with the three smaller basic mandrel sizes, and 0.095″ (2.4 mm) for those used with the two larger sizes. ∎

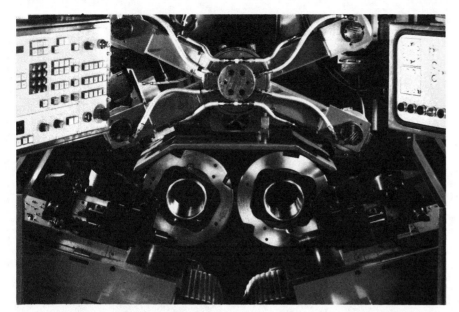

High-performance developments, like countercentrifugal and indexing chucks, allow high-production machines to achieve even higher productive capability. The photo shows indexing chucks set up on a Warner & Swasey 2-MSC to machine both sides of two cylindrical pipe couplings in a single chucking.

Reprinted from Tooling & Production, November 1980

Trends in
chucking

by Richard G Green
Managing Editor

Of all the workholding devices used in metalworking manufacturing the most universal is the spindle-mounted chuck used to grip essentially round workpieces in rotating-type metalcutting machines, such as lathes, chuckers and boring machines. Although several types of workholding devices fit this broad description—such as collets and magnetic

chucks—we have chosen to limit this article to that category of chucks that grip workpieces by means of movable jaws, actuated either manually or by means of a pneumatic or hydraulic power source.

Within this category, the number of jaws mounted to a chuck can vary from two to three, four or more. One of the most commonly used is the three-jaw, self-centering geared scroll chuck, which may be operated by a power chuck wrench on larger machines. Another is the four-jaw chuck, with the jaws operated independently—useful for holding odd-shaped parts or on jobs where a casting or forging needs truing up before machining. Combination chucks, with both self-centering and independent jaw action, give the user a compromise between the other two. The self-centering movement of the jaws on combination chucks can be either geared scroll or power operated.

Power operation of chucks is achieved either through lever action or wedge action. Jaw movement on the lever type is by means of a lever crank working between the drawbar and the jaws. The wedge-type chuck achieves jaw movement by the action of the end of the master jaw in a tapered wedge, which is moved longitudinally by the power cylinder.

Within these broad, generic classifications of chucks there are many variations and proprietary versions. In gathering information for this article, we contacted the majority of chuck manu-

Erickson Tool Co, Solon, OH. The company produces two types of counter-centrifugal chucks—standard and front-actuated—in 10″-, 12″- and 15″-dia sizes. The standard countercentrifugal type (see photo) can be used for high-speed turning operations and features a patented internal jaw lock that maintains gripping pressure in the event of hydraulic failure. The company's front-actuated countercentrifugal chucks are designed for machines capable of delivering hydraulic fluid to the spindle nose. This chuck's self-contained internal hydraulic actuator eliminates the need for external plumbing. In addition,

the chuck enables the spindle to be open for bar work. The company's countercentrifugal chucks are designed with wedge-type jaws for increased accuracy. The company also produces a line of All American power chucks that are of large through-hole design for chucking and bar work up to 7″ dia. An 8:1 lever ratio insures secure gripping power, and the top jaws are available in either hardened, serrated, reversible style for normal chucking applications or soft blank style that can be bored to suit part requirements. These chucks are available in 8″, 10″, 12″, 15″ and 18″ dia, with mountings to match all

machine spindle-nose configurations.

facturers in order to be able to present a reasonable cross section of the types of chucks available on the market and specific features offered by each. The products of those chuck manufacturers who responded to our survey are described in the showcase sections throughout this article. Space limitations do not permit a comprehensive listing of the complete product lines of each company, but inquiry numbers assigned to each will allow the reader to request additional information from any of the companies.

Machine builders' points of view

A chuck is a specially designed device for gripping a round workpiece in order to machine it. However, it is not useful in performing its function unless it is also attached to the spindle of a turning machine. For this reason, it would seem that a good source of information on the proper use of a chuck would be the turning machine builders. Therefore, we contacted a number of them in order to gain some insight into their recommendations, and guidelines for the proper use of chucks on their machines.

Safety is the prime concern of turning machine manufacturers, and, with respect to chucks, safety relates to spindle operating speed. Safe operation of a given chuck in a given application involves many factors, including workpiece rigidity; chuck type and condition; clamping force achieved; jaw type; workpiece size, shape and finish; and type of cut and tooling used in the cut.

A spokesman for one of the leading companies in the turning machine field—Warner & Swasey Co, Cleveland, OH—shared some of the company's views on the subject of the proper selection, care and use of chucks: "The three-jaw, geared scroll has traditionally held the well-deserved position of being the universal chuck for general applications. When properly serviced and lubricated, it provides a strong grip within a wide range of gripping diameters. There are some limits to the versatility of a geared scroll, however, in today's higher-production applications.

"Warner & Swasey produces some geared scroll chucks for applications involving turret lathes, and occasionally the SC column-type NC machines. Most of the chucks required on larger Warner

A

PowerHold Products, Inc, Rockfall, CT. This company specializes in workholding devices, including air and hydraulic power chucks, counterbalanced chucks, index chucks, compensating chucks and chuck actuators, in addition to other types of workholding devices. Chucks range in size from 4″ to 18″, with related air or hydraulic actuators.

PowerHold chuck and collet combinations are designed to reduce the time required to change turning machines from bar to chucking work. Shown in photo A is Model 380, 2/3-jaw power chuck which can be used as a two-jaw chuck, three-jaw chuck, collet chuck and faceplate for mounting special chucking fixtures. It is set up for chucking work in an automatic turret lathe with special top jaws for handling this irregular part.

PowerHold Products recently introduced a new line of 6″-, 8″- and 10″-dia six-jaw precision chucks (photo B) for secondary-operation lathe and boring machines. Designed for safety in high-speed applications, these six-jaw chucks can accommodate any common top jaws and mount directly to A-type spindle noses.

The company's chuck pressure gages may be used for the dynamic testing of rotating and stationary chucks or for the static testing of stationary chucks only (photo C). It is always recommended that the gage be bolted directly to the

B

face of the chuck, especially in high-speed applications. The company recommends that, for safety purposes, a chuck pressure gage should be used to determine the gripping force of all power chucks in the shop, and to help insure operator safety and prevent the loss of parts in process.

C

...continued

& Swasey slant-bed and 3-SC column machines are geared scroll chucks supplied by outside manufacturers.

"We take great care to assure that a compatible combination of top-jaw tooling, chuck and actuating equipment is assembled to form an effective workholding system. An NC turning lathe is a highly versatile machine, requiring fewer setup changes to produce a wide variety of turning jobs. A versatile chucking set-up, therefore, is basic to maximizing the use of machine features. This need for chuck versatility has encouraged a trend toward through-hole chucks for general use, chucks designed for both chucking work and bar work with collet pad inserts. Most of the standard chucks supplied on Warner & Swasey SC column-type machines are air or hydraulically operated, and drawbar or draw tube actuated. The brand is determined by customer specification and the field engineer's recommendations, based upon the requirements of a particular machine.

"Different chucks vary widely in performance, from brand to brand, or even between different chucks of the same model and size. Chucks used on our machines are expected to grip accurately within 0.001″ for chucks 15″ and smaller, and within 0.002″ or 0.003″ for chucks 15″ and larger. Representative of expected gripping force, a 10″ chuck should provide from 5000 to 8500 lb per jaw, and a 24″ from a minimum of 10,000 lb to a maximum of 22,000 lb per jaw. Countercentrifugal chucks are expected to maintain 75 percent of their static gripping force at their maximum rated speed. Consideration is given to the suitability of a chuck and actuator combination in working with the high-low pressure control available on many Warner & Swasey machines. The effectiveness of safety check valves used is also considered when choosing from available workholding equipment.

"As in other areas of machine tool development, the industry as a whole has benefited greatly from the results of specialized workholding research and development efforts. Innovation in internal chuck designs, such as wedge or wedge-bar, cam and lever jaw chucks provides operating characteristics that can be applied to specific holding problems. Countercentrifugal chucks allow the maximum benefit to be gained from high-speed spindles and improved cutting materials. High-performance developments like countercentrifugals and indexing chucks allow high-production machines to achieve even higher productive capability."

South Bend Lathe, Inc, South Bend, IN, emphasizes not only the safety aspects of chucks used in their lathes, but also their accuracy. They use chucks made by a chuck manufacturer, because they consider them a specialized business.

The company's Maurie Gallaher says, "Our requirements have always been for a chuck of precise quality that will not detract from the performance of our lathe. Recently, we introduced a CNC turning center which runs at 6000 rpm.

SMW Systems, Inc, Cincinnati, OH. Company's line of self-contained, large through-hole power chucks includes Models BB, SP and LP, all of which share common operating principles and features. A 25″-dia Model BB, shown in the photo, has a 12″-dia through-hole. These chucks use self-contained cylinders and actuators, eliminating the need for draw tubes and rear-mounted cylinders. The internal air cylinder actuates master jaws through 10-degree wedges. A nonreturn safety valve provides added assurance against air-line-failure accidents. Chucks in this series are available in sizes from 5½″ to 31½″, with through-holes from 1″ to 15″.

The company's Model KNCS is a power chuck designed for short- to medium-production NC turning, and allows a complete set of top jaw tooling to be changed in less than one minute. The chuck's wedge bar design allows it to run at relatively high speeds without significant loss of grip force. This model is available in sizes from 5½″ to 25″. The newest chuck, Model KDHV, is a self-contained, front-mounted hydraulic type built to accommodate higher speeds, maximum spindle through-hole work space and variable-grip force. A valve system has been developed to maintain grip force if the hydraulic system should fail for any reason.

Hydra-Lock Corp, Warren, MI. This company (formerly known as A & C Engineering Co) pioneered the concept of full-surface holding. In the Hydra-Lock concept, hydraulic pressure is exerted against the clamping portion of a sleeve on arbors and chucks. The pressure may be produced by a self-contained system within the tool, or generated externally with pumps or power cylinders. As the fluid surrounds the steel sleeve, it uniformly contracts (or expands) the sleeve since the hydraulic pressure is equal in all directions. This causes the sleeve to expand and completely fill the entire area between the part and the arbor or chuck. Expansion or contraction of sleeve diameters allows ample clearance for loading and unloading of parts. Releasing hydraulic pressure returns the expanded sleeve to its original size. The tooling is designed to work within the elastic limits of the various materials used.

Since the applied pressure is absolutely uniform and is applied over the entire area surrounded by the sleeve, there are no localized pressure points as with mechanical tooling. This results in very high accuracies. Even though the part is out-of-round, Hydra-Lock can pick up and establish a theoretical centerline for machining concentric parts. Holding power is maximized, since a large por-

tion of the surface is used for holding. In effect, Hydra-Lock tooling permits the operator to have a variable, controllable, interference fit without marring finished surfaces. Because of the large gripping area, serrations are not required. Hydra-Locks are available in any size from ⅜". They are used for precise inspection requirements (0.000 030″ tir) as well as for heavy-

duty, high-torque machining jobs (0.002″ tir), and can be designed with a half-inch movement. Shown in the photo is a reversible Hydra-Lock chuck for grinding tapered roller bearing races concentric with involute tooth forms. Part is reversed without releasing initial grip. Precision alignment is obtained by built-in Hydra-Lock expanding dowels. Circle E120.

Rohm Tool Corp, Miami, FL. This company is the exclusive representative for and importer of Gamet power chucks and actuators from France. These high-speed chucks incorporate countercentrifugal weights built into the jaw-actuating mechanism. The cam-

lever design is self-locking under load and will not release the workpiece if power fails. In addition to these standard high-speed chucks, Rohm Tool also features the Gamet high-speed chucks with integral actuating cylinder and check valves to eliminate the need for a rear-end actuator and draw tube. Countercentrifugal jaw actuation is incorporated and the design retains the safety features in common with the new range of standard Gamet chucks described above.

The chuck is powered by hydraulic oil pressure and the oil supply enters the chuck via holes which are drilled into the spindle wall and spindle nose. The entry of oil to the spindle connections is via a hydraulic distributor bearing located within the headstock of the machine. Quick-change jaws (less than 30 sec) permit maximum production utilization of NC machines.

Pratt Burnerd America, Inc, Springfield, MI. Offers Constant Grip high-speed chuck in five sizes—6¼″, 8″, 10″, 12″ and 15″ dia. These three-jaw power chucks harness centrifugal force by means of a full-wedge counterbalance system that maintains any selected static grip at any speed within their specified range—up to 6000 rpm for the 6¼″ dia. The range of each is limited by the size of the chuck and by the strength of its component parts. Gripping force can be set over a wide range—up to 10,480 lb per jaw at maximum drawbar pull in the largest size. The repetitive gripping capacity of the CG chuck is within 0.001″. The company also makes a line of self-contained power chucks and three-jaw, four-jaw and six-jaw manual scroll chucks, plus other workholding devices. Circle E125.

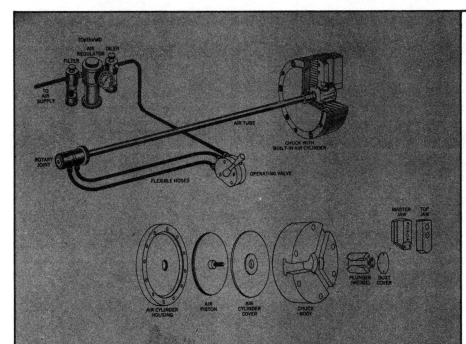

Northfield Precision Instrument Corp, Island Park, Long Island, NY. Company's line of air-operated chucks—in sizes from 3″ to 12″, two- and three-jaw versions—emphasizes high accuracy, with the ability to automatically maintain 0.0001″ concentricity and end lengths without the necessity of indicating each piece. The chucks are air activated through an air rotary joint and ¼″-dia hose (see drawing). By increase of pressure through a standard regulator, they may be used for semi-rough machining as well as for fine-finish machining. Chucks incorporate a safety design which, in the event of air pressure failure, will not release the workpiece until air pressure is reestablished. Circle E126.

Metrology Systems Corp, Plainview, NY. Company produces a line of air-actuated chucks in sizes from 3″ to 10″ dia, in two- and three-jaw configurations. Through-hole chucks are available with up to 2½″-dia unobstructed hole through the center. The company's new E series front-mounted, self-contained through-hole air chucks (see photo) feature a new seal design which assures reliable operation at speeds up to 3150 rpm. Accuracy attained to 0.000 050″. These chucks are double acting for both ID and OD workholding and can be activated while rotating. Model E-15 has a 1½″ through hole and 1500-lb jaw force at 80 psi.

Machine Tool Div, Logansport Machine Co, Inc, Logansport, IN. Features of Logansport chucks (no relationship to Logan lathes) include recessed master jaws to facilitate mounting of tooling plates and prevent damage to master jaws in the event the cutting-tool path is programmed too close to the face of the chuck, lever block assembly design which allows full block contact for minimum wear, and chuck bodies machined from one-piece steel castings.

Company's line of Tach III high-speed power chucks range in size from 10″ to 24″ dia, with through-hole sizes from 2⁹⁄₁₆″ to 6″ (15″, Model 388-15, shown in photo). This line features heavy gripping forces, high-low capability, close repeatability and force-offset counterbalancing.

The Logansport Galaxy series of chucks range in size from 10″ to 18″, with through-hole sizes from 2.56″ to 5.75″. These are front-actuated hydraulically and, in the event of pressure failure or drop, dual pilot-operated check valves maintain operating pressure until interlocks stop the machine. The company also makes standard power chucks in two-, three- and 2 in 1 four-jaw styles.

One of our biggest problems has been to obtain a chuck that will operate safely at this speed. To date, we found one 4"-dia chuck that is rated at this speed, and are testing a 6"-dia chuck that is recommended for this speed.

"We would not attempt to recommend a chuck design, but do feel that with the improvement in tooling—i.e., ceramics, coated carbides etc—the requirements are for chucks that will run at substantially higher speeds than have been used in the past. I believe it is obvious that with high spindle speeds and spindle runout of 0.000 025" a chuck has to be a precise component so as not to detract from these accuracies and capabilities."

According to Joseph F Maurus, manager, NC Product Line, Automated Machine Div, Babcock & Wilcox, Bloomfield Hills, MI, the company uses a great variety of chuck styles and sizes on their NC lathes. He says, "Due to the versatility of NC turning equipment, many things must be considered, starting with the customer's part and required tolerances. On our particular equipment, chuck sizes can range from as small as 6" to 32" dia.

"Part processing is our next consideration. On our Detroit 1100 CNC vertical lathe, we can machine a part at a turning speed of up to 1500 rpm in a 32"-dia chuck. When using a 32"-dia chuck turning a part at 1500 rpm, workholding becomes critical. At this speed, we are generating 13,000 sfm at the outside diameter of the 32"-dia chuck. With normal chucking equipment, the available jaw force is almost negligible. It becomes obligatory to look at this type situation very early in the proposal stage.

"Our guideline with normal chucking equipment is 'No chuck is to be turned at a speed that will exceed 6000 sfm measured at the largest diameter in the chucking setup *unless* an anticentrifugal or counterbalanced chuck is used. An unbalanced or unsymmetrical part could further limit the rpm and should be considered.' In the proposal stage, we look at this aspect of the job very closely and specify the proper chuck.

"Because the machine user could put any type of chuck on the machine, we mount a red warning plaque on the front guard of every machine. The warning is also stressed in the machine quotation, service manuals and programming manuals. This is basically all we can do to protect our customers, chuck suppliers and ourselves. In some cases, with the

Cushman Industries, Inc, Hartford, CT. The company offers a complete line of manually operated chucks ranging from 5" to 48" dia.

Types available include independent, self-centering, Accra-Set and combination. Their manual chucks are designed for use on American Standard Spindle types A, D and L plus Straight Recess. Cushman offers power chucks in a wide variety of types and sizes from 6" to 48" dia. The most recent addition to the company's power chuck line is a front-mounted, self-contained unit designed for NC turning machines and available in 10", 12" and 15" sizes. Designated the 430 series, these chucks reduce space requirements by eliminating the rear-mounted cylinder; the hydraulic cylinder is self-contained in the chuck body. The hydraulics are fed directly into the chuck actuating mechanism through ports in the spindle nose. The 430 series features a built-in pressure compensator which adjusts to volume increases or decreases caused by oil temperature variations.

Counterbalanced for high-speed operations up to 3500 rpm, the 430 series chucks allow low- or high-speed operations. For added versatility, high/low gripping capabilities permit chucking a part at high pressure for roughing; after relaxing, it can then be gripped at a lower force without losing concentricity or causing deformation.

The company's line of manual chucks has been expanded with the addition of the new Series 50 three-jaw, self-centering chucks (top right). They are available in sizes from 5" to 12".

Designed to give a positive digital readout of 40,000 lb per jaw with an accuracy of ±100 lb, the Cushman Jaw Force Analyzer (right) can monitor loss of jaw force through periodic readout

due to contamination or improper lubrication, and verify that a workholding system meets setup requirements. It can also assess maintenance needs, qualify jaw force requirements for specific setups, determine the exact capability of a chuck's holding power over a period of time, ascertain when a system is being overstressed, and provide an accurate method of evaluating tool design in relation to chuck performance succumbing to centrifugal force.

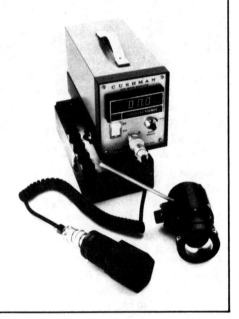

customer's knowledge, we have limited main-drive-motor horsepower should it be determined that the chuck actuating equipment requested or demanded is not adequate for proper workholding and safety."

Reprinted from Manufacturing Engineering, August 1979

Quick-change Chuck for Short-run Production

*Job shop work means small lots —
and plenty of them. Here's how a highly
versatile chuck is keeping parts going
out the door at one small company*

GARY S. VASILASH
Assistant Editor

RAPID-FIRE CHANGEOVER is essential at the MCG Machine Co., Roseville, MI. The reason: it's a job shop that runs parts in lots ranging from 10 to a few thousand. As Bill Peifer, MCG's president, puts it, "Jobs can change once in four days or twice a day."

An Answer. According to Peifer, one of the main things that helps MCG complete its orders on time is a 12" (actually 315-mm) Forkardt KTNC power-operated chuck. The chuck is utilized on a Lodge & Shipley 12/25U turning machine, as shown in *Figure* 1. The German-made product is available in the U.S. from Universal Engineering Div., Houdaille Industries, Frankenmuth, MI.

Bill Peifer says that the chuck is used exclusively on the turning machine. The job shop does approximately 60% military work and 40% commercial, so its parts mix varies widely. Some of the items held by the chuck for heavy duty machining are spindles, gears, hubs, housings, pinions, and retainers. A shock absorber piston made of SAE 1144 steel is being turned in *Figure* 1.

Speed and Versatility. Prior to the purchase of the KTNC, Peifer says that a great deal of time was spent at the welding machine making up jaws to handle the various parts. "In six months," he notes, "we've never had to make up special jaws for a job. We use this chuck for everything."

1. *CHUCK holding shock absorber piston
in a turning machine. Other jobs run
include spindles, gears, hubs, and pinions.*

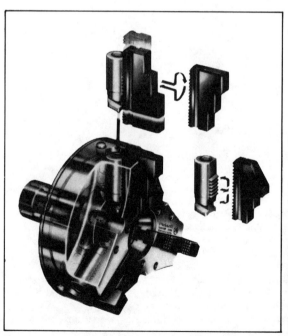

2. *JAW REMOVAL
AND REVERSAL
is easily effected in less than
◄ two minutes. Driving jaw and
chuck jaw are removed from
the chuck body then repositioned
as required.*

3. *PRECISION-GROUND TEETH
on the driving and chuck jaws are mated ▶
to accurately adjust chuck capacity.*

Peifer asserts that it once took two or more hours to change over a chuck. Now it takes just a few minutes to change not only the chucking diameter, but hard jaws to soft as well. And the chuck holds a high accuracy — 0.001" (0.03 mm).

Changeover. In order to change jaws, a button is pushed on the control panel of the GE Mark Century® 1050 L&S Optimizer II control. It effects an override that disengages the wedge hook from the clamping piston. An oil cylinder is utilized in the chuck for jaw actuation, engagement, and disengagement.

A safety release button on each jaw is then depressed. The round driving jaw and the chuck jaw can then be slipped out of the chuck body for transposition, reversal, or for a complete change, as illustrated in *Figure* 2.

The driving and chuck jaws feature racks. This permits jaw sizes to be accurately changed by simply meshing the precision-ground teeth, as shown in *Figure* 3.

Utilizing shop air to blow out any metal particles that may have become lodged in the teeth before reassembly, Peifer points out that inaccuracies in

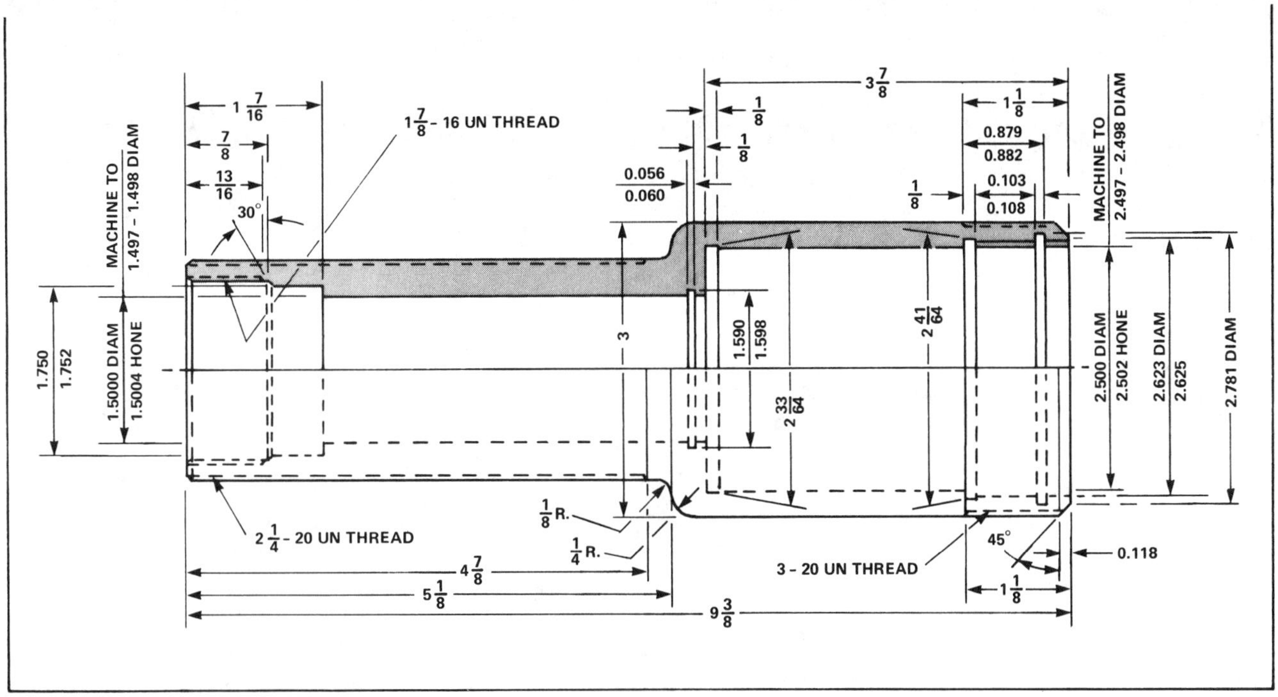

4. DESIGN OF SHOCK ABSORBER BODY machined in three different chuckings. Concentricities are held within 0.001″

chucking result from errors in putting improperly cleaned jaws together.

Wedge Hook. Not only is the high accuracy achieved because of the precision-ground teeth, but also because the KTNC chuck utilizes the wedge-hook principle rather than a lever-type design.

In the lever-type chuck, the lever has only a small contact point on a spherical land. All of the chuck's force goes through that point, so it can break down quickly and become inaccurate.

The wedge hook has a long, straight contact area. As a result of a patented design, the Forkardt chuck has a high gripping force; there's direct power transmission with little friction loss.

Application. As an example of the accuracy of the chucking, Peifer holds up a cylindrical shock absorber body. A drawing of the part is shown in *Figure* 4. The original barstock was 9½″ (241 mm) long and 3⅛″ (79.4 mm) in diameter. The 1213 cold drawn steel, a free-machining material, is finished on the turning machine to measure 9⅜″ (238.1 mm) in length with two OD's, one 2¼″ (57 mm) and the other 3″ (76 mm). Including grooves for retaining rings, there are nine different ID's. All concentricities are held within 0.001″ (0.03 mm).

Here's how the part is cut.

► The bar slug is chucked, then one cut ⅜″ (9.5 mm) deep and 5⅛″ (130.2 mm) long is made at 900 sfpm (274 m/min).

► A hole 1⅜″ (34.9 mm) in diameter and 5⅝″ (142.9 mm) deep is drilled in the bar with a carbide drill.

► The part is turned end for end and chucked on the 2¼″ OD.

► A 2″ (51-mm) hole is drilled into the workpiece.

► The end of the bar is faced and chamfered, and the 3″ OD is finish turned.

► 3-20 UN Class 3 threads are cut 1⅛″ (28.6 mm) in from the chamfer. Threading is performed with a single point tool in four passes at 730 rpm. The last pass takes 0.002″ (0.05 mm) for a good finish.

► The ID is generated by boring the 2″ hole to 2.497-2.498″ (63.42-63.45 mm) diameter (it's later honed to 2.500-2.502″, 63.5-63.55 mm).

► 4″ (102 mm) into the ID of the bar, a groove 0.056 to 0.060″ (1.42 to 1.52 mm) wide and 1.590 to 1.598″ (40.39 to 40.59 mm) in diameter is required. The ID is machined and the groove is cut for a retaining ring.

► Three other grooves are required in the 3⅞″ (98.4-mm) long, 2.500 to 2.502″ diameter ID. They are finish cut.

► The part — for a second time — is turned end for end and chucked on the 3″ OD. Since that OD has been finish machined, the hard chuck jaws are replaced with soft jaws made out of 1020 steel. Peifer says the switch is made because the soft jaws wrap around the workpiece and clamp a wider area than the hard jaws. This reduces vibrations and prevents the part from being distorted, a particular problem since the ID has been finished.

► The 5⅛″ (130.2-mm) long OD that was rough cut during the first step is finished. Two radii, the first ⅛″ (3.2 mm) and the second ¼″ (6.35 mm) are generated so the 2¼″ OD curves up to the 3″ OD.

► 2¼-20 UN Class 3 threads are cut on the finished OD.

► The ID is rough and finish bored to 1.497-1.498″ (38.02-38.05 mm) diameter; 0.002″ (0.05 mm) is left for hone stock.

► Two counterbores, the first ⅞″ (22.23 mm) long and the second ½″ (13 mm) long with a 30° chamfer separating them, are finished. The outermost counterbore, 1.750 to 1.752″ (44.45 to 44.50 mm) in diameter, must be concentric with the ID finished in the previous step to within 0.001″ (0.03 mm) TIR.

► 1⅞-16 UN Class 3 threads are cut into the ⅞″ long counterbore, thus finishing the part.

Time required for machining and setups: Just 25 to 30 minutes.

Cost. But the accuracy and versatility provided by the Forkardt KTNC is not without a price. Peifer admits that a conventional power chuck would cost several thousand dollars less.

But with good business sense — as important in the job shop field as machining expertise — he points out that the quickness of jaw reversal or replacement and the ability of the chuck to handle an assortment of workpieces makes up for the premium in a short period of time.

Confidence. Bill Peifer is convinced that the chuck has helped his company perform its short-run operations. As he puts it, "If somebody walked in with 10 pieces, we could do them in two or three hours. Before getting the chuck, it would take two or three hours to set up the machine alone."

Peifer has so much confidence he has specified a 15″ (actually 400-mm) Forkardt chuck on a new Lodge & Shipley 15/40 he has ordered. ■

Reprinted Courtesy of Positrol Inc.

Typical Applications of Hydraulic-Controlled Holding Devices

By Jonathan T. Weber
Positrol Inc.

Principle of Operation

Hydraulic holding systems are designed to accurately locate and securely grip parts in position while performing various operations in manufacturing. The basic concept consists of expanding or contracting a steel sleeve within the elastic limit of the metal by means of hydraulic pressure. In so doing, the sleeve expands or contracts uniformly about the axial centerline of the workpiece.

Hydraulic chucks and arbors generally have accuracy and repeatability within .0002 TIR as standard. Chucking diameters as small as .062" and as large as 50" or more are possible limited only by part and tolerance variations. The amount of expansion is relative to sleeve diameter and hydraulic tools are designed to accommodate most any part tolerance.

In the schematic chuck diagram, the piston is manually advanced by clockwise rotation of the actuator screw. The hydraulic fluid is displaced in the piston chamber and moves through the ports exerting pressure under the steel sleeve. This equalized internal pressure is contained within the hydraulic seals compressing the sleeve radially about the workpiece to be chucked.

The sleeve is further compressed and conforms to the shape of the workpiece providing accurate centering and maximum holding power. Counter-clockwise rotation of the actuator screw releases the hydraulic pressure and relaxes the compressed sleeve for loading and unloading the piece part.

Methods of Actuation

There are numerous ways in which hydraulic arbors and chucks may be applied. As a result, there are various methods of actuation. The most common is manually by an actuator screw which can be located at the most convenient place to suit the application. The actuator may also be cam operated or from an external source such as an air cylinder for automatic actuation.

Absolute Centering

The action of the expanding or compressing sleeve is radial and concentric with the true axial centerline, thus assuring positive centering of the workpiece. Within expansion limits, the sleeve conforms to the shape of the part regardless of inaccuracies due to part tolerance, taper, eccentricity or "bell month" conditions. Odd shaped parts also are accomodated including most importantly splined or threaded parts which can be located and held on the minor, major or pitch diameters.

Typical Application

The following diagrams and pictures are examples of hydraulic arbor and chuck applications where accurate centering, repeatability and positive clamping are necessary.

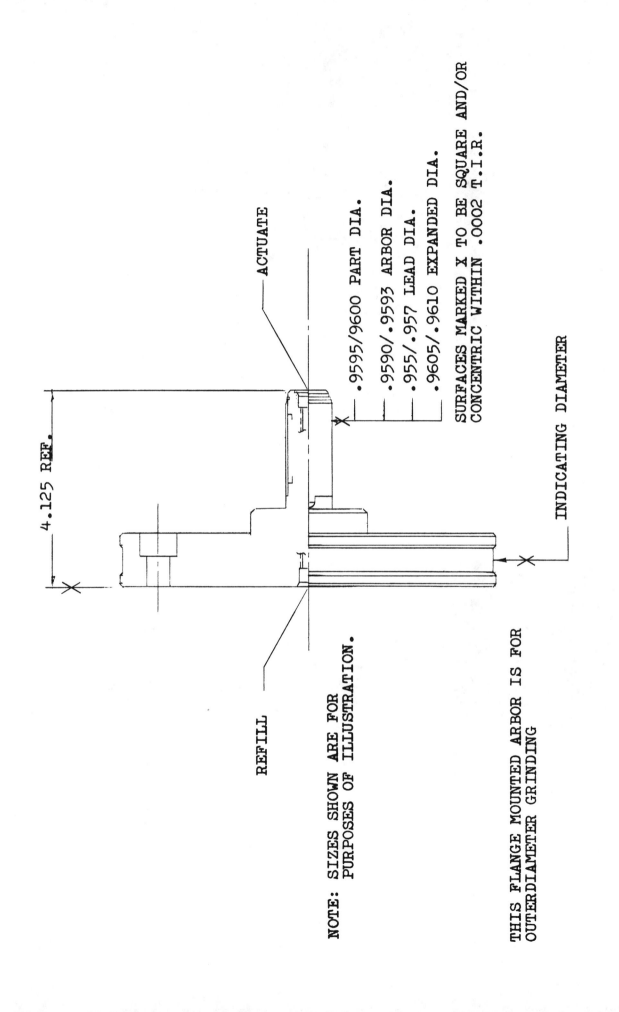

ACTUATE

.9595/.9600 PART DIA.

.9590/.9593 ARBOR DIA.

.955/.957 LEAD DIA.

.9605/.9610 EXPANDED DIA.

SURFACES MARKED X TO BE SQUARE AND/OR
CONCENTRIC WITHIN .0002 T.I.R.

INDICATING DIAMETER

4.125 REF.

REFILL

NOTE: SIZES SHOWN ARE FOR
PURPOSES OF ILLUSTRATION.

THIS FLANGE MOUNTED ARBOR IS FOR
OUTERDIAMETER GRINDING

FLANGE MOUNTED ARBORS

For turning, grinding or inspection.
Split collets may be used to cover a
wide range of part sizes.

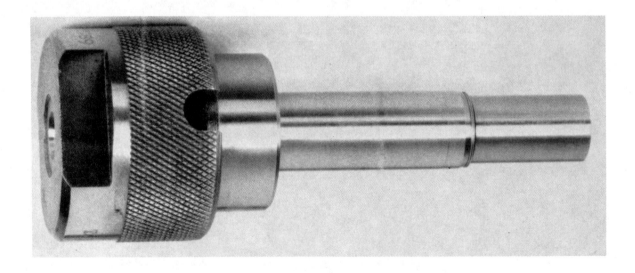

STEPPED ARBOR

This is a hydraulic workholding arbor which has
two expanding areas of different diameters.

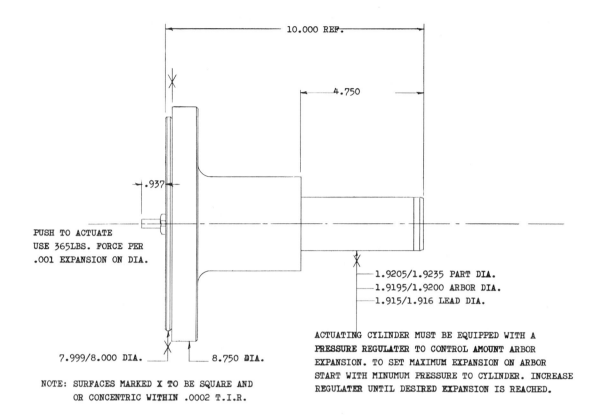

10.000 REF.

4.750

.937

PUSH TO ACTUATE
USE 365LBS. FORCE PER
.001 EXPANSION ON DIA.

1.9205/1.9235 PART DIA.
1.9195/1.9200 ARBOR DIA.
1.915/1.916 LEAD DIA.

7.999/8.000 DIA.   8.750 DIA.

NOTE: SURFACES MARKED X TO BE SQUARE AND
OR CONCENTRIC WITHIN .0002 T.I.R.

ACTUATING CYLINDER MUST BE EQUIPPED WITH A
PRESSURE REGULATER TO CONTROL AMOUNT ARBOR
EXPANSION. TO SET MAXIMUM EXPANSION ON ARBOR
START WITH MINUMUM PRESSURE TO CYLINDER. INCREASE
REGULATER UNTIL DESIRED EXPANSION IS REACHED.

SPLINED ARBOR

This arbor locates on the pitch diameter for turning, grinding or inspection. ALL POSITROL arbors are round and concentric within .0002 TIR.

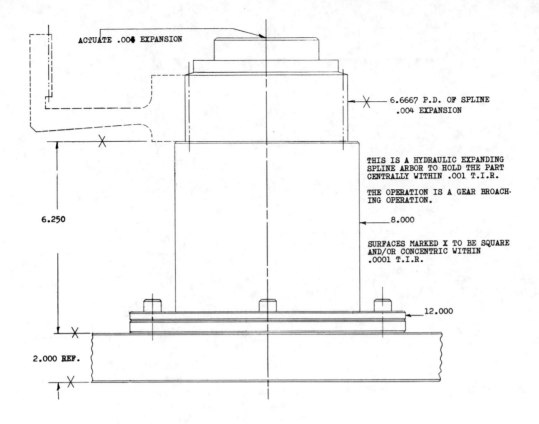

ACTUATE .004 EXPANSION

6.6667 P.D. OF SPLINE
.004 EXPANSION

THIS IS A HYDRAULIC EXPANDING
SPLINE ARBOR TO HOLD THE PART
CENTRALLY WITHIN .001 T.I.R.

THE OPERATION IS A GEAR BROACH-
ING OPERATION.

6.250

8.000

SURFACES MARKED X TO BE SQUARE
AND/OR CONCENTRIC WITHIN
.0001 T.I.R.

12.000

2.000 REF.

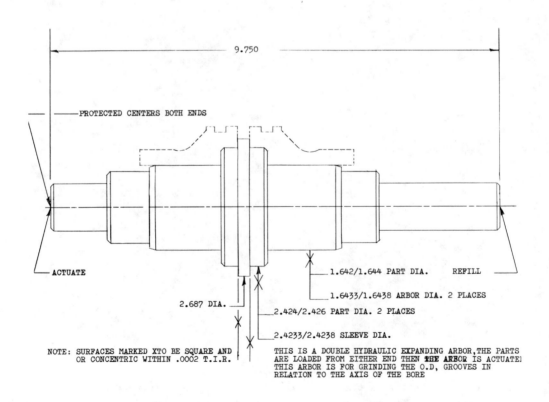

9.750

PROTECTED CENTERS BOTH ENDS

ACTUATE

2.687 DIA.

1.642/1.644 PART DIA.   REFILL

1.6433/1.6438 ARBOR DIA. 2 PLACES

2.424/2.426 PART DIA. 2 PLACES

2.4233/2.4238 SLEEVE DIA.

NOTE: SURFACES MARKED X TO BE SQUARE AND
OR CONCENTRIC WITHIN .0002 T.I.R.

THIS IS A DOUBLE HYDRAULIC EXPANDING ARBOR,THE PARTS
ARE LOADED FROM EITHER END THEN THE ARBOR IS ACTUATE
THIS ARBOR IS FOR GRINDING THE O.D, GROOVES IN
RELATION TO THE AXIS OF THE BORE

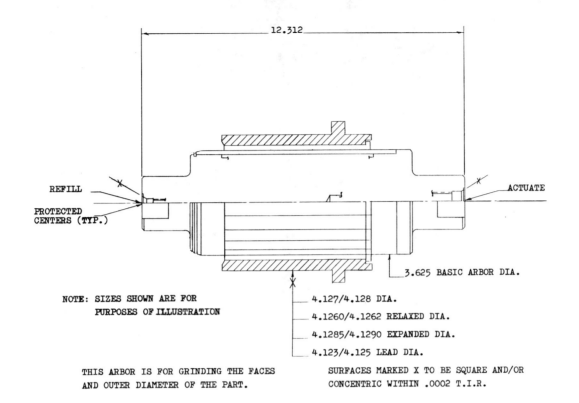

12.312

REFILL

PROTECTED
CENTERS (TYP.)

ACTUATE

3.625 BASIC ARBOR DIA.

NOTE: SIZES SHOWN ARE FOR
PURPOSES OF ILLUSTRATION

4.127/4.128 DIA.

4.1260/4.1262 RELAXED DIA.

4.1285/4.1290 EXPANDED DIA.

4.123/4.125 LEAD DIA.

THIS ARBOR IS FOR GRINDING THE FACES
AND OUTER DIAMETER OF THE PART.

SURFACES MARKED X TO BE SQUARE AND/OR
CONCENTRIC WITHIN .0002 T.I.R.

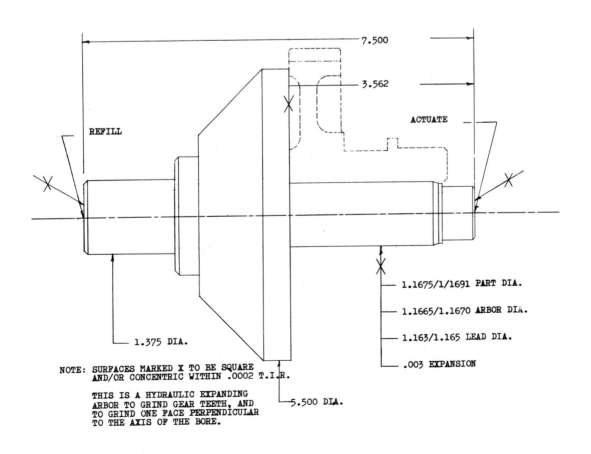

7.500

3.562

REFILL

ACTUATE

1.1675/1/1691 PART DIA.

1.1665/1.1670 ARBOR DIA.

1.163/1.165 LEAD DIA.

.003 EXPANSION

1.375 DIA.

NOTE: SURFACES MARKED X TO BE SQUARE
AND/OR CONCENTRIC WITHIN .0002 T.I.R.

THIS IS A HYDRAULIC EXPANDING
ARBOR TO GRIND GEAR TEETH, AND
TO GRIND ONE FACE PERPENDICULAR
TO THE AXIS OF THE BORE.

5.500 DIA.

SPLIT COLLET

This is a typical split sleeve used to accommodate various
size parts with one arbor or chuck.

ARBOR WITH SPLIT COLLET

This is an example of an arbor which has a series of
split collets to accommodate various bore sizes.

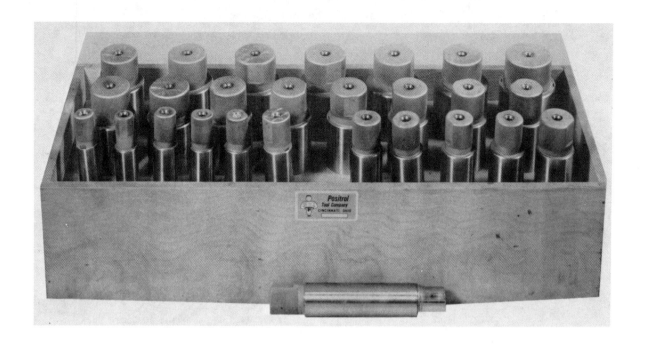

HYDRAULIC ARBOR SET

This is a set of arbors we use for manufacturing in our shop. For further details, contact our representative.

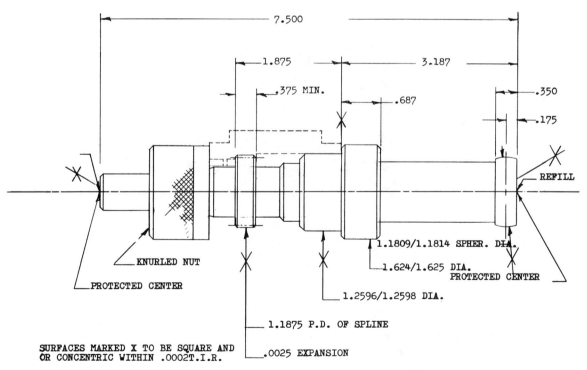

7.500

1.875

3.187

.375 MIN.

.687

.350

.175

REFILL

KNURLED NUT

PROTECTED CENTER

1.1809/1.1814 SPHER. DIA.

1.624/1.625 DIA.
PROTECTED CENTER

1.2596/1.2598 DIA.

1.1875 P.D. OF SPLINE

SURFACES MARKED X TO BE SQUARE AND
OR CONCENTRIC WITHIN .0002T.I.R.

.0025 EXPANSION

THIS ARBOR IS DESIGNED TO HOLD
ON AN INTERNAL SPLINE PITCH
DIAMETER FOR GRINDING ON A
REISCHAUER GEAR GRINDER

111

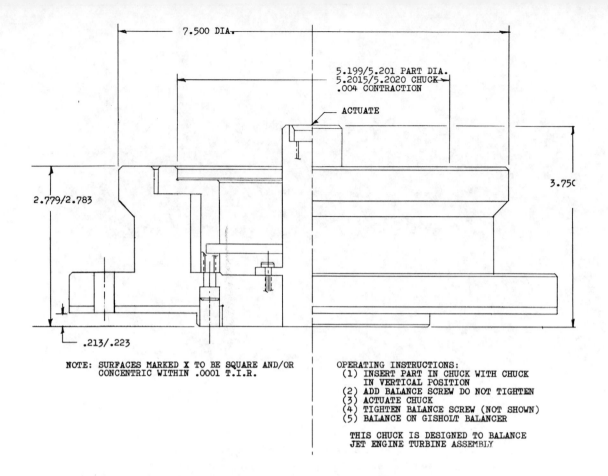

7.500 DIA.

5.199/5.201 PART DIA.
5.2015/5.2020 CHUCK
.004 CONTRACTION

ACTUATE

3.750

2.779/2.783

.213/.223

NOTE: SURFACES MARKED X TO BE SQUARE AND/OR
CONCENTRIC WITHIN .0001 T.I.R.

OPERATING INSTRUCTIONS:
(1) INSERT PART IN CHUCK WITH CHUCK
    IN VERTICAL POSITION
(2) ADD BALANCE SCREW DO NOT TIGHTEN
(3) ACTUATE CHUCK
(4) TIGHTEN BALANCE SCREW (NOT SHOWN)
(5) BALANCE ON GISHOLT BALANCER

THIS CHUCK IS DESIGNED TO BALANCE
JET ENGINE TURBINE ASSEMBLY

HYDRAULIC CHUCK

One of many types of
chucks used for turn-
ing or grinding as a
separate unit for one
part or with collets
for two or more parts.

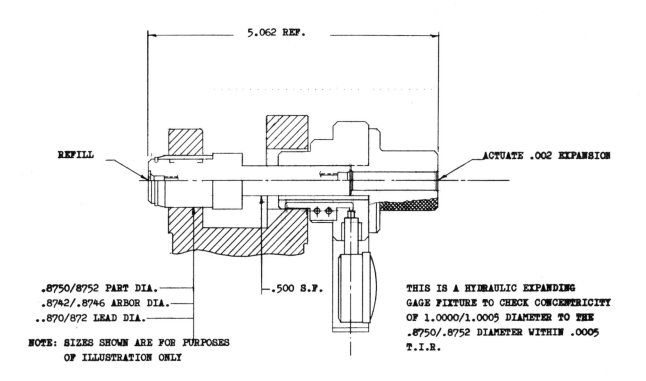

5.062 REF.

REFILL

ACTUATE .002 EXPANSION

.8750/8752 PART DIA.
.8742/.8746 ARBOR DIA.
..870/872 LEAD DIA.

—.500 S.F.

NOTE: SIZES SHOWN ARE FOR PURPOSES
OF ILLUSTRATION ONLY

THIS IS A HYDRAULIC EXPANDING
GAGE FIXTURE TO CHECK CONCENTRICITY
OF 1.0000/1.0005 DIAMETER TO THE
.8750/.8752 DIAMETER WITHIN .0005
T.I.R.

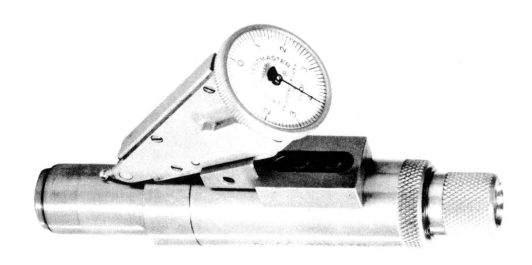

SQUARENESS AND CONCENTRICITY SWEEP GAGE

This gage expands into a bore and checks
squareness to a face and concentricity
of a counterbore.

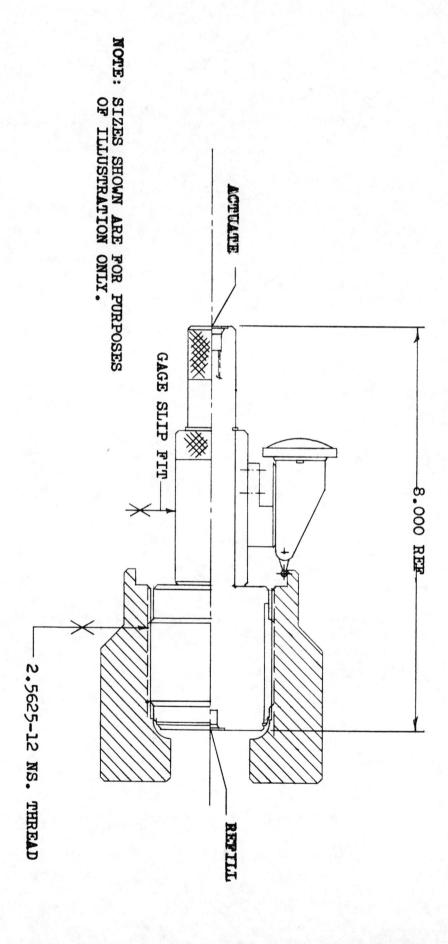

ACTUATE

GAGE SLIP FIT

8.000 REF

REFILL

2.5625-12 NS. THREAD

NOTE: SIZES SHOWN ARE FOR PURPOSES
OF ILLUSTRATION ONLY.

THIS SWEEP GAGE IS FOR CHECKING THE
OUTER DIAMETER OF THE COUNTERBORE TO
THE PITCH DIAMETER OF THE THREAD IN THE PART

SURFACES MARKED X TO BE SQUARE AND/OR
CONCENTRIC WITHIN .0002 T. I. R.

114

SWEEP GAGE

This gage checks the concentricity of two "o" ring grooves with the bore. The hydraulic arbor is expanded into the bore and the handle on the right releases the probes from the retracted position for measurement.

SWEEP GAGE WITH PART

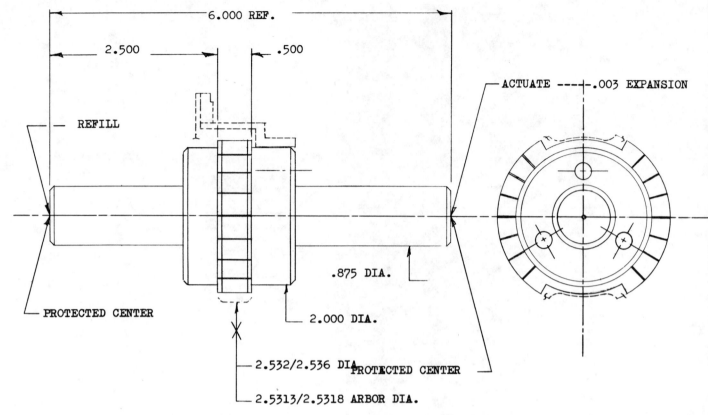

6.000 REF.

2.500

.500

ACTUATE .003 EXPANSION

REFILL

.875 DIA.

PROTECTED CENTER

2.000 DIA.

2.532/2.536 DIA. PROTECTED CENTER

2.5313/2.5318 ARBOR DIA.

NOTE: SURFACES MARKED X TO BE SQUARE
AND OR CONCENTRIC WITHIN .0002 T.I.R.

THIS IS A HYDRAULIC EXPANDING ARBOR TO
INSPECT ALL OUTER DIAMETER GROUND SURFACES

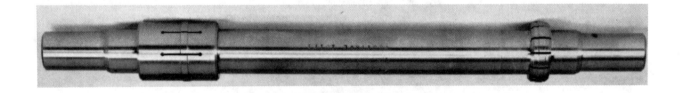

SPECIAL INSPECTION ARBOR

Inspection arbor with two expanding
areas, one with a split collet.

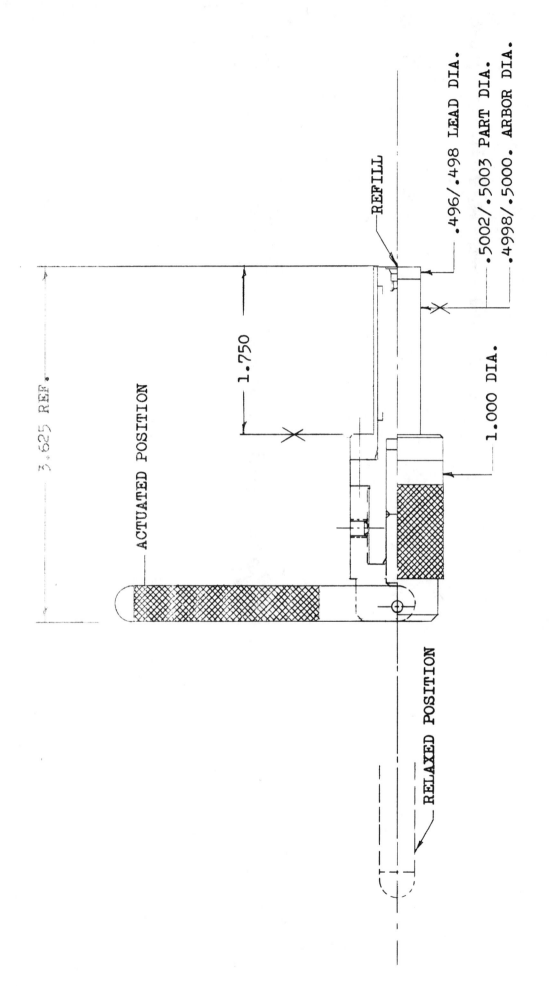

REFILL

.496/.498 LEAD DIA.

.5002/.5003 PART DIA.

.4998/.5000. ARBOR DIA.

1.750

3.625 REF.

ACTUATED POSITION

1.000 DIA.

RELAXED POSITION

THIS IS A CAM OPERATED HYDRAULIC LOCATING PIN. THE PIN IS DESIGNED FOR THE ACCURATE LOCATION OF TWO .5003/.5003 DIA. HOLES.

NOTE: SIZES SHOWN ARE FOR PURPOSES OF ILLUSTRATION ONLY. SURFACES MARKED X TO BE SQUARE AND/OR CONCENTRIC WITHIN .0002 T.I.R.

117

Reprinted from Tooling & Production, May 1980

Positive clamping
for multistation fixtures

On this four-station rotating fixture (three work, one load and unload), eight parts are drilled, milled, reamed and tapped. One Sperry Vickers power unit supplies hydraulic power to all eight clamping cylinders controlled by one directional valve at each workstation. Drawing shows clamp design. Inset shows the workpiece, which is a hydraulic pump body.

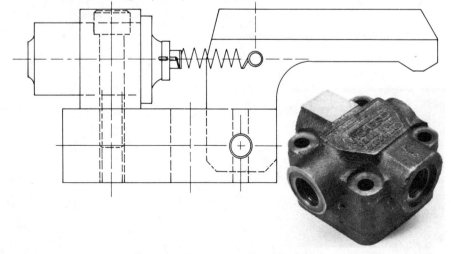

A unique method of assuring that hydraulically clamped parts remain clamped and rigid during multifixtured milling and drilling operations has been devised by Sperry Vickers tool engineers, according to the firm. Positive clamping force is especially important when an operator is unloading and loading one fixture while machining continues on other parts clamped in another fixture in the same hydraulic circuit. Each part being machined is held securely by ten operating cylinders, eight on the circumference and two on top for hold-down.

With the old clamping method, the ten fixture cylinders dropped to unclamp pressures during unloading. This reduced the clamping force below that required to resist the tool force. The result was that a part moved while being machined and thus became a reject or required reworking.

To combat this problem, Sperry Vickers, Div Sperry Rand Corp, Omaha, NE, incorporated one of its pressure-compensated, variable-volume hydraulic piston pumps in the machine's hydraulic system, with the compensator set to provide the maximum pressure necessary for clamping. At each machining station, a two-position valve and a pilot-operated check valve are installed. The check valve is the key component, because it allows free oil flow into the fixture cylinder during clamping, but will not allow reverse flow from the cylinder unless the two-position valve on the fixture is shifted.

As soon as the cylinder moves to its extreme clamping position, system pressure rises to maximum. At this point the variable pump downstrokes to supply only enough flow to maintain clamping pressure. The engineers say the only caution necessary is to assure that the tool force acting on the workpiece (and, in turn, against the holding cylinders) does not generate enough pressure to burst the cylinders or hydraulic lines between the cylinders and check valve.

Other components that contribute to the clamping system are a Sperry Vickers C-10 power package and Dueblin rotating couplings for rotating fixtures. On nest-type stationary fixtures, where the toolhead moves toward a workpiece, subplate-mounted directional flow valves are used in conjunction with the pilot-operated check module. The arrangement is assembled on the power unit and allows an operator to clamp and unclamp with spaced push buttons.

Pressure-reducing valves and sequence valves can also be added if required.

The engineers say this system works equally well on spring-return clamping cylinders. Double-acting cylinders are recommended because of their more positive action.

Hydraulic circuit shows connections to Wilton No 230 spring-return cylinders, Vlier Model ASC-3B swing clamps and Model ASC-10 swing clamps. The power pack at bottom of drawing is furnished by Vickers.

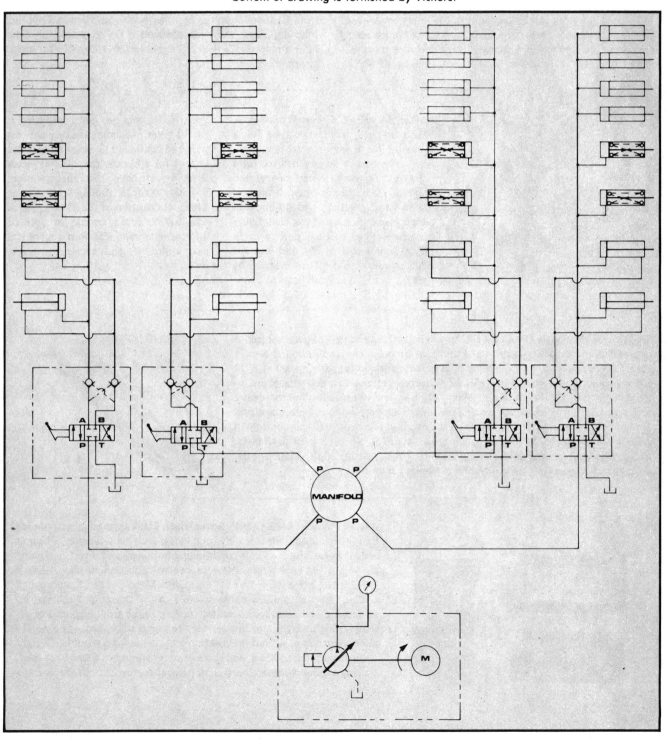

Reprinted from Modern Machine Shop, March 1981

Getting The Grip On Workholding

Part One of a Two-Part Series

Gripping On The ID

An integral part of good machining is good workholding. One of the nation's leading experts discusses the finer points of gripping workpieces on the inside diameter.

KEN GETTELMAN, Editor, interviews
DAVE BEAVER, General Manager
Erickson Tool Company
Solon, Ohio

Mr. Beaver, what are the major problems of ID (inside diameter) workholding?

There are three: (1) A small ID in relation to the OD, (2) holding a workpiece square to the ID centerline, and (3) holding a workpiece so that all faces can be machined.

What causes these problems?

Consider the problem of a large-diameter workpiece with a small ID mounted on a solid mandrel. The torque generated when cutting on a relatively large OD creates a slippage problem. Likewise, a short-length ID creates a problem of holding the workpiece square to the centerline. The use of end drivers or other assists to gain either holding capability or to insure part accuracy may create a problem of machining end surfaces. In addition, serrated or tooth-type end drivers could leave marks on the ends of workpieces.

What are the answers?

Simply stated, the answers are found in a wide range of mandrel designs and applications that have been developed to meet virtually every need. However, these items are not always known to those who face a new problem.

In any discussion of mounting mandrels, where would be a good place to start?

With the very simple plug-type solid mandrel, which can be very effective when used with certain types of workpieces.

What are its strengths and weaknesses?

Its strength lies in its very simplicity. It is nothing more than a plug of suitable length with a threaded end upon which is slipped a workpiece which is then bolted into place with some kind of an end clamp.

Its weaknesses are twofold. The first is the inability to accommodate workpieces with any variation of in-

side diameters. For example, if the mandrel is one inch in diameter and the workpiece is a casting with an oversize 1 1/8-inch ID, the workpiece cannot be end clamped. It, obviously, would locate off center due to its own weight, and concentricity between the ID and OD would be lost. Secondly, any kind of an end clamp either covers part of the workpiece end, which may be a machining area, or end drivers may leave unacceptable indentations on the workpiece.

How are most workholding mandrels mounted on today's machine tools?

Most machines are equipped with the standard mounting shown in Figure 1. The spindle nose has a 7 1/2-degree taper and holds the mandrel plate to the nose with bolts. This is the American Standard Spindle Nose (ASSN).

As you stated, the simple plug-

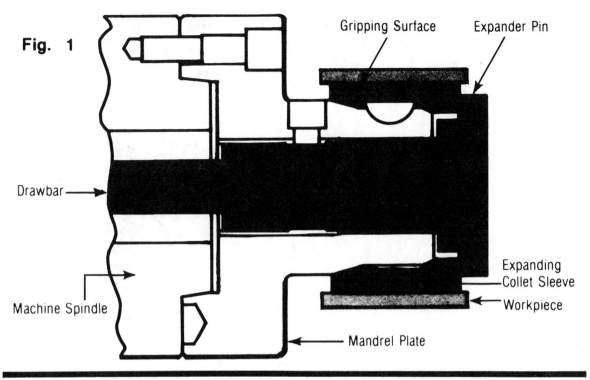

Fig. 1

Gripping Surface

Expander Pin

Drawbar

Machine Spindle

Expanding Collet Sleeve

Workpiece

Mandrel Plate

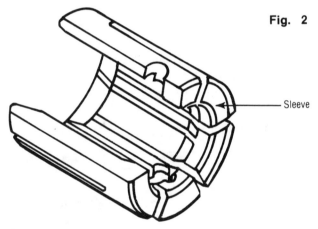

Fig. 2

Sleeve

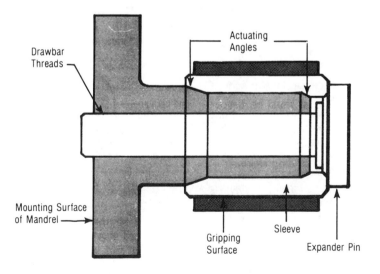

Drawbar Threads

Actuating Angles

Mounting Surface of Mandrel

Gripping Surface

Sleeve

Expander Pin

type solid mandrel cannot cope with any variation of workpiece ID. What is the answer to that problem?

The most prevalent approach is the use of an expanding-sleeve mandrel as shown in Figure 2.

How does it work?

It usually is used on a machine equipped with either an air or hydraulically actuated drawbar. The mandrel is most often threaded into the drawbar. Frequently, the drawbar will be equipped with a setscrew to prevent the mandrel from rotating. Pulling the drawbar (and faceplate) pulls a slotted sleeve up an incline, which in turn expands the sleeve into the ID of the workpiece.

What is the variable range of such sleeves?

They have a normal working range of about 0.030 inch, which will accommodate most variations in bores that have been previously machined. Other styles can accommodate very rough, cast, or forged bores. In addition, the sleeve action squares the workpiece with the natural centerline of the hole.

What is the best slope actuating angle to use?

Experience has shown that a slope of at least 15 degrees is needed to

121

obtain a natural release action when the drawbar pressure is removed.

As the drawbar is tightened, is there any axial motion of the sleeve?

Yes, it tends to pull the workpiece in the direction of the drawbar pull. However, this is desirable since workpieces are often drawn back against an end locator or a positive drive. In some instances, where a very strong gripping force is needed, the sleeve or end driver may be serrated. But it should be understood that this latter approach can only be utilized when the residual marks in the workpiece made by the serrations are acceptable. Serrated sleeves are often employed when machining must be done on the outer diameter and face of workpieces that have relatively small ID's.

You mentioned positioning the workpiece against a locator. What factors are to be considered in designing an end locator?

The first consideration is whether the locating surface of the workpiece has been previously machined square with the bore to be gripped. If it has been machined square, a flat square-type locator is used. If not, then a compensating locator must be used. Both types are shown in Figure 3.

What if the ID is a casting surface with a variation wider than that handled by the expanding sleeve?

Then a segmented-sleeve is used, and this situation occurs more frequently than one would assume. The segmented sleeve can handle ID variations over a range of 1/8 inch or more. On very large workpieces, a range up to 3/4 inch is possible.

How does it differ from the expanding sleeve?

The segmented sleeve (Figure 4) is composed of three individual segments held together with spring bands. It operates on the same principle as the slotted expanding sleeve except that the segmented-sleeve range of expansion is considerably wider. The segmented sleeve may have serrated segments or it may be utilized with end locators. This type of ID holding is ideal where the outer surfaces of a casting are to be machined and those machined surfaces are then used to locate and

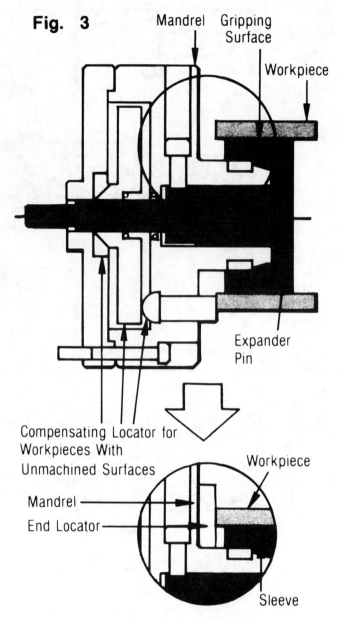

Fig. 3

Mandrel Gripping Surface

Workpiece

Expander Pin

Compensating Locator for Workpieces With Unmachined Surfaces

Mandrel

End Locator

Workpiece

Sleeve

Flat Square-Type End Locator for Workpieces With Machined Surfaces

hold the workpiece for finish machining the bore.

With both the expanding and segmented sleeve there is the possibility of a small ID and a large OD, which would create torque problems in the event of heavy machining cuts. Obviously, a sleeve with an end stop or a serrated surface would help solve that problem, but what if the workpiece configuration prevents that type of an approach?

The alternative is some type of a positive driver on the end stop, which will project into some recessed area of the workpiece as far out as possible from the ID. This technique is frequently used to allow heavy machining rates without the workpiece slipping on the mandrel.

There has been the implication that the expanding mandrel is used mostly for turning work. Is this true?

Although most applications do in-

Fig. 4

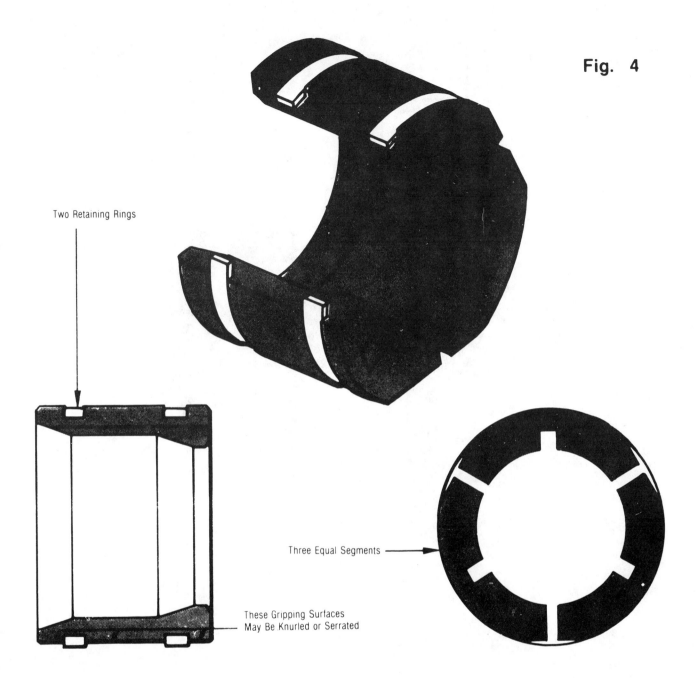

Two Retaining Rings

Three Equal Segments

These Gripping Surfaces
May Be Knurled or Serrated

volve lathes, there are plenty of examples whereby round workpieces are held on machining-center and milling-machine tables. The drawbar principle can be applied to nonrotating work as well. Figure 5 shows the typical arrangement for use on a lathe and its variation for use in a nonrotating application. These are often powered by shop air and they can develop several tons of drawbar pressure. There are two things to keep in mind about the drawbar and

expanding sleeve or segments. First, a preload pressure is always maintained. Even when the drawbar is released, it never moves forward so far that the slope of the sleeve loses contact with the slope of the mandrel body. Secondly, while expansion is achieved with air or hydraulic pressure, the return is achieved with springs built between the mounting and mandrel plates.

It seems that a separate mandrel must be designed for each individual

part number.

Not necessarily. Many have been designed for a full family of parts. By changing only the sleeve or end locator, it is possible to accommodate many workpieces within a family with one expanding-mandrel set.

To this point, our discussion has been on workpieces with straight internal diameters. How about workpieces with internal splines?

This presents no problem. Expanding mandrels have been built

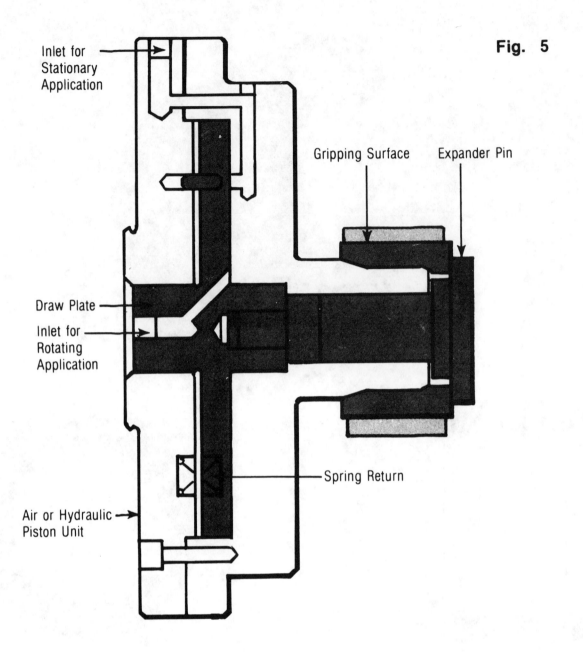

Fig. 5

Inlet for
Stationary
Application

Gripping Surface

Expander Pin

Draw Plate

Inlet for
Rotating
Application

Spring Return

Air or Hydraulic
Piston Unit

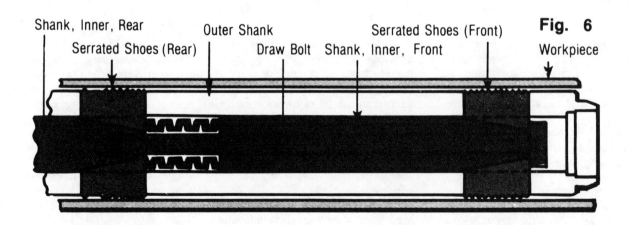

Shank, Inner, Rear

Serrated Shoes (Rear)

Outer Shank

Draw Bolt

Serrated Shoes (Front)

Shank, Inner, Front

Fig. 6

Workpiece

Fig. 7

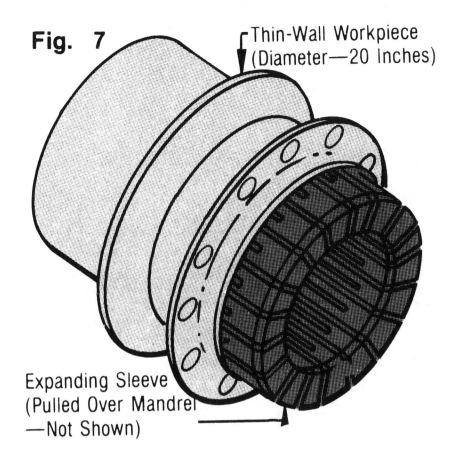

Thin-Wall Workpiece
(Diameter—20 Inches)

Expanding Sleeve
(Pulled Over Mandrel
—Not Shown)

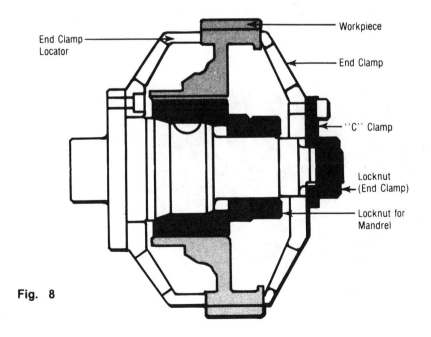

End Clamp
Locator

Workpiece

End Clamp

"C" Clamp

Locknut
(End Clamp)

Locknut for
Mandrel

Fig. 8

with sleeves or segments that will expand and grip the major, minor, or pitch diameter of the spline. It all depends upon the design of the workpiece, the optimum way to hold the workpiece, and similar factors. But there is a wide flexibility in the approach to holding workpieces with internal splines or gear teeth.

How about long shapes where the OD is not much larger than the ID?

Figure 6 shows a workpiece that is 34 inches long. The method of holding it on the ID for turning was a special expanding mandrel that utilized serrated shoes to actually grip the workpiece.

Do delicate or fragile workpieces lend themselves to this type of workholding?

Yes, the expanding sleeve is ideal. A thin-wall aerospace component nearly 20 inches in diameter and with two flanges is shown in Figure 7. It was held by an expanding mandrel to drill multiple holes in the flanges. By utilizing this holding method, the gripping pressure was evenly distributed across the entire interior surface of the workpiece. It was most important that no distortion of the workpiece took place, and the expanding mandrel was the solution to this particular holding problem. While the drawing shows only a limited number of holes, there were actually 360 in each flange.

What about special situations, such as a relatively small ID and a larger OD with full rigidity wanted on the OD?

That, too, is a situation occurring more frequently than realized. Often, the answer is the expanding mandrel to grip the ID while an auxiliary clamp supports the outer diameter. Figure 8 shows a gear blank that is typical of this type of holding problem. The solution was found by taking the expanding-mandrel concept and adding a set of outboard clamps. Notice that the inner threaded locknut forces the expanding sleeve up the mandrel. To simplify removal of the OD-support for workpiece mounting, a plate with a swingout "C" clamp was employed. Workpiece loading is a three-step operation. The operator mounts the work-

piece on the mandrel, he tightens
and locks the threaded locknut, and
then end clamps the workpiece. Of
course, workpiece removal is just the
opposite. Usually, two mandrels are
used for this type of operation. While
one workpiece is being machined,
the second mandrel is being unloaded and reloaded. This clamping
method provided rigid holding and
positive positioning for critical
machining of gear components.

*What about those special designs
with a tapered ID? Do they yield to
this type of mandrel holding?*

In many instances they do. Figure
9 shows an expanding tapered
sleeve arrangement that was developed for a workpiece with an ID
taper. An expanding sleeve was utilized and it held the workpiece very
well. However, this would not work
well for heavy machining operations.

*What amount of taper can you
satisfactorily hold by this method?*

This is a variable, depending on
the amount of work to be accomplished. Usually, any taper less than
ten degrees (on a side) can be accommodated for light machining
operations.

*Where there are long production
runs or a thrust toward special automation, is it possible to mount more
than one workpiece on a mandrel?*

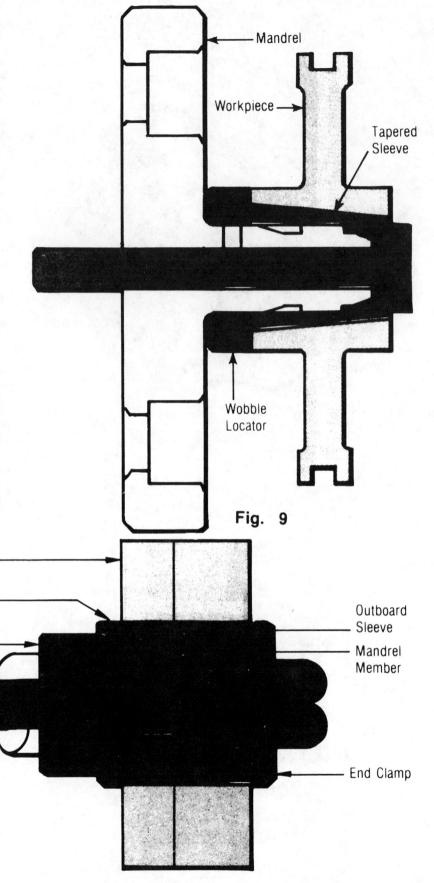

Fig. 9

Fig. 10

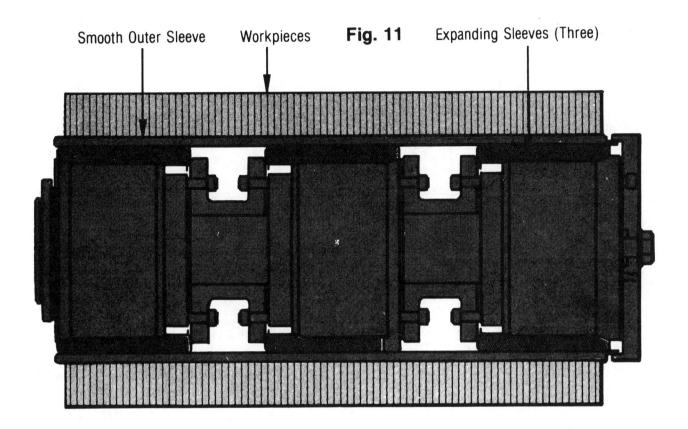

Smooth Outer Sleeve   **Workpieces**   **Fig. 11**   **Expanding Sleeves (Three)**

Yes, this is a good possibility and, it is done in many instances.

Figure 10 shows a mandrel with two expanding sleeves to mount two gear blanks that are machined simultaneously. As the end clamp is drawn back, it squeezes both the outboard sleeve and the mandrel member against the inboard sleeve so that both are tightened and both gear blanks are drawn together. This is a completely automatic operation on a gear hobber with automatic loading and unloading. Notice, in this instance, that the gear blank ID's are slightly larger than the end clamp of the draw bar; thus, the blanks slip over the contracted sleeves. The concept enhances creative automation.

Are there other significant variations of the multiple-sleeve approach?

A very good variation is the mandrel that was developed for holding electric motor laminations comprised

of dozens of discs. Since the discs are relatively thin and each must be held, a special expanding-sleeve approach was utilized as shown in Figure 11. However, the expanding sleeves themselves do not grasp the lamination discs. The sleeves expand and stretch a smooth outer sleeve, upon which the discs are actually mounted. This provides a smooth even grip along the entire length which may be as much as 24 inches. This approach has worked extremely well for the electric motor industry.

Is the concept limited to the motor industry?

No, the smooth outer sleeve for gripping has numerous other applications. It is ideal for grinding operations where it is desired to hold the workpiece on the ID (see Figure 12). In this situation, instead of using a sleeve that mechanically expands with cams, a hydraulic force is used

to stretch and expand the sleeve. The operator mounts the workpiece and tightens a screw, which forces oil into the space between the mandrel and the sleeve. As a result, the sleeve expands to grip the workpiece.

The range of expansion would be quite limited, would it not?

Yes, that is true. Practically, it is limited to about 0.001 inch per inch of ID, but it has three very distinct advantages for ground parts: (1) There is no opening or slots; thus, grinding grit, swarf, and fluids are eliminated from the working areas of the mandrel. (2) There are no gripping marks; thus, if the ID is ground, it can be gripped by this type of mandrel with no degradation to the workpiece ID surface while still providing a powerful grip. (3) It gives the best overall accuracy of any type of expansion gripping since there are no moving parts. The sleeve is merely

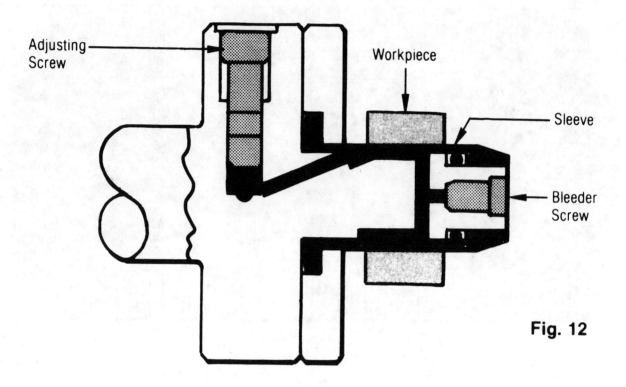

Adjusting Screw

Workpiece

Sleeve

Bleeder Screw

Fig. 12

stretched.

All of our discussion has been based on mounting the workpiece, then expanding the sleeve to an ID. Is there ever a case where the sleeve is larger to start with and the workpiece is forced over it to squeeze the sleeve for holding purposes?

This, too, has its area of application, although it is quite limited. The approach shown in Figure 13, is most often used to mount workpieces for some type of inspection. This approach does not provide enough holding force for any kind of significant machining, but it is a viable application in limited areas and a solu-

tion to certain inspection needs.

Are there any other internal holding methods that should be discussed?

The holding method shown in Figure 14 is rather unique. It very closely resembles the simple collet chuck so often used for OD gripping. It has a collet-type action but produces an expansion rather than a contraction.

Please describe it.

It is a four-segment, collet-type shank. In fact, it is the same as the standard 5C collet. It fits directly into the spindle of a machine such as a small precision lathe. Each has a cam

surface. As the drawbar pulls the segment, the cam surfaces riding on the slope of the holder collapse inward. However, they push the opposing segment outward so there is a net expansion. It is an excellent means of quickly mounting small workpieces by gripping them on their ID.

You have pointed out through this entire discussion that there is some type of ID workpiece holding to meet virtually any machining or inspection requirement.

Yes, and the right holding technique can be a powerful factor in economical and productive machining. **MMS**

Fig. 13

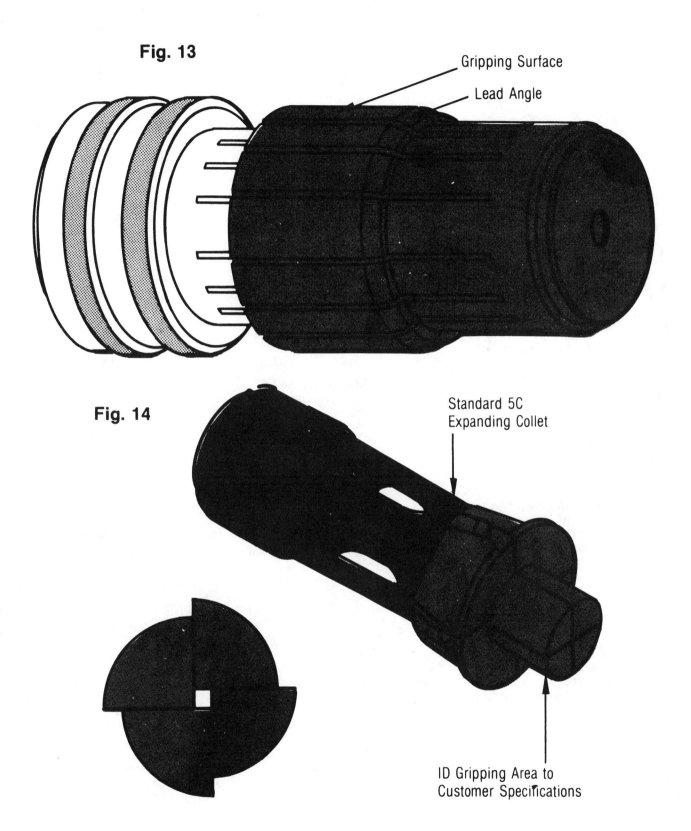

Gripping Surface

Lead Angle

Fig. 14

Standard 5C
Expanding Collet

ID Gripping Area to
Customer Specifications

Reprinted from Modern Machine Shop, April 1981

Getting The Grip On Workholding

Part Two of a Two-Part Series
Gripping On The OD

The many and creative means of holding a workpiece on the OD (outside diameter) are discussed with a leading tool expert.

KEN GETTELMAN, Editor, interviews
DAVE BEAVER, General Manager
Erickson Tool Company
Solon, Ohio

Mr. Beaver, how many different ways are there of holding workpieces on the OD?

Basically, there are two. The first makes use of the jaw chuck while the other employs some type of a collet chucking approach.

I presume there are many variations of each.

Yes, the number is quite large although eight or ten cover the major types of holding requirements.

Since we left the discussion on ID workholding with the collet method, let's pick up on that theme. What are some of the main advantages of the collet?

It has the ability to provide both an accurate and powerful gripping force, and it grips around the entire outside diameter, which is absolutely essential for some applications. It can also be employed for holding a workpiece off the centerline.

Can you discuss a good example?

Figure 1 shows a workpiece with several diameters to be ground. The portion that is gripped is offset 14½ degrees from the axis of the portion that is to be ground. The collet principle was employed for this workpiece-holding requirement.

How is the chuck opened and closed?

A threaded worm wheel is used. The operator turns the worm gear with a key. This in turn rotates the threaded wheel, which either pulls the collet back to collapse it and grip the workpiece or advances the collet to release it. As the collet is pulled back, it pulls the workpiece against a plate for positive location. This is a special approach, but it was the best answer for this particular holding problem. It supplied the means to obtain precision ground surfaces in proper relationship to one another.

It seems that most collet approaches utilize a pulling-type col-

lapse. Is the reverse type ever used?

There are times when a push closing is preferred. This is especially true when the workpiece reference is the face of the workpiece projecting from the collet.

Could you give us an example?

The collet chuck shown in Figure 2 is a good example. The workpiece is located against a flag stop for positive referencing. The stop is not shown as it is part of the machine tool itself. Once the workpiece is positioned, the stop retracts. In this instance, there were some turning and threading operations to be performed on the end of the workpiece. To insure positive gripping, there are two collets in the chuck. Notice that the outer end of the first collet is rounded so that the workpiece will not be marred as it is inserted into the chuck. The push tube is a simple and effective approach where it applies.

If the push tube could be used to

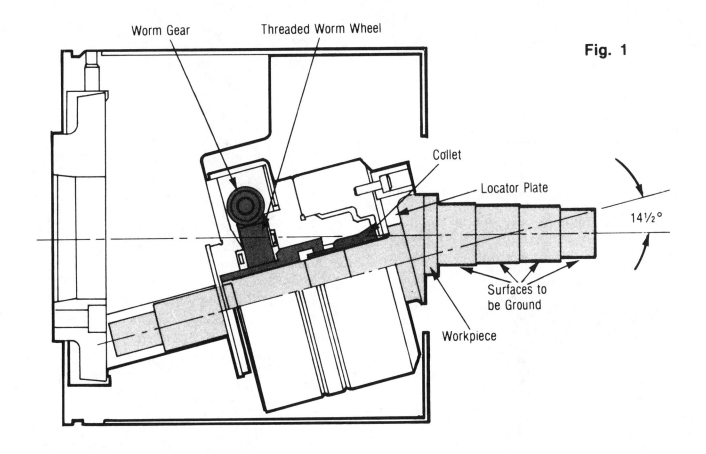

Fig. 1

Worm Gear

Threaded Worm Wheel

Collet

Locator Plate

14½°

Surfaces to be Ground

Workpiece

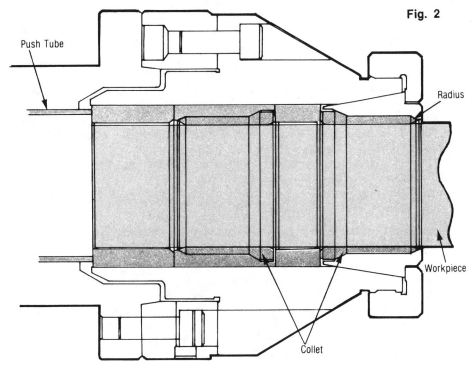

Fig. 2

Push Tube

Radius

Workpiece

Collet

collapse a collet chuck, then the opposite could be applied, could it not?

That is true. The drawbar-type collet chuck for external gripping is a good workholding technique for some types of parts.

In general, what would they be?

Those that are not too long and must have operations performed on the face. The drawbar is easy to actuate on a lathe with either an air or hydraulic cylinder. Unlike the three-jaw chuck, the collet chuck grips on the entire periphery of the workpiece rather than at three points. This can be most important for some workpiece designs.

Could you explain the workings of the external gripping collet chuck?

Figure 3 shows a good example. A drawbar, actuated by a rear-mounted cylinder, is attached to a drawbar head. This in turn pulls a single-angle master collet, which collapses in upon the workpiece. However, the collet itself does not grip the work-

131

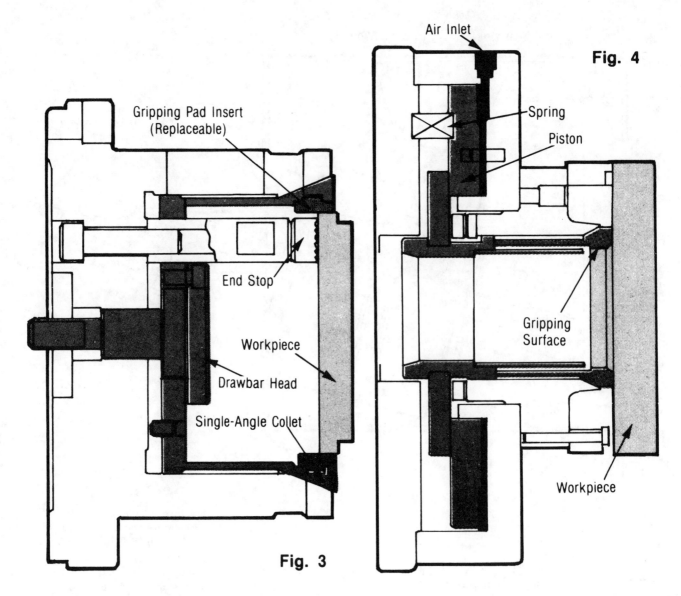

Fig. 4

Air Inlet

Spring

Piston

Gripping Pad Insert
(Replaceable)

End Stop

Workpiece

Drawbar Head

Single-Angle Collet

Gripping
Surface

Workpiece

Fig. 3

piece. There is a replaceable gripping insert that does the actual workpiece holding. One of the additional features of this chuck design is an adjustable end stop. The workpiece is pulled square and firmly against the end stop for positive locating and to make certain the workpiece is square in the collet chuck. The serrated end stop also helps give a positive drive to the workpiece. These end stops are often made of high speed or tool steel for strength and wear resistance.

Thus far we have discussed collet chucks that function on some type of turning equipment where a rear mounted cylinder for drawbar push or pull is utilized. How about the use of the collet-type chuck on machines without this feature?

There are two types of collet chucks for this purpose. The first is a chuck for a nonrotating application where there is work to be done on the end of the workpiece. As shown in Figure 4, air is introduced through a connector and actuates a spring-loaded plate, which is connected to the collet assembly. The introduction of air causes the collet to collapse and grip the workpiece.

What about a turning machine without a rear-mounted cylinder?

The same fundamental approach is used except that it becomes a little more complex. Obviously, there is a nonrotating outer housing for the air connection and a bearing system to allow the chuck itself to rotate. A typical design is shown in Figure 5. In

this instance, the pressure plate pushes against a lever with a 3:1 mechanical advantage and the lever pushes the collet closer. With this design, standard shop air pressure can provide a tremendous gripping force.

What keeps the housing from rotating?

A clevis rests against some rigid portion of the machine tool, but the air hose itself is often sufficient to keep the housing in place.

Is this the limit of the collet-chuck approach?

No, there are numerous specials for a variety of applications.

What represents one of the more interesting approaches?

The adjustable offset collet chuck

is a prime example of a special approach that takes care of an individual workpiece that must be held off center. It can be adjusted for an entire family of workpieces.

Is it infinitely adjustable?

There are two types. One is infinitely adjustable within a range. The actual amount of offset is determined by a calibrated adjustment screw. The chuck shown in Figure 6 has an offset range of 5/8 inch and it is infinitely variable across that range. Others have been constructed with a selection of three or four discrete offsets that may be changed by the machine operator.

Do these eccentric chucks come in both the push and pull models?

Yes, they can be configured either way.

The most common type of holding is the traditional three-jaw chuck. Is the manually operated type becoming obsolete?

Although the trend is toward a power-actuated chuck, the traditional scroll-type chuck, which has been around for more than 100 years, is still popular. The Archimedes-type spiral offers simplicity, reasonably long life, and an effective wedging action. It does not offer super ac-

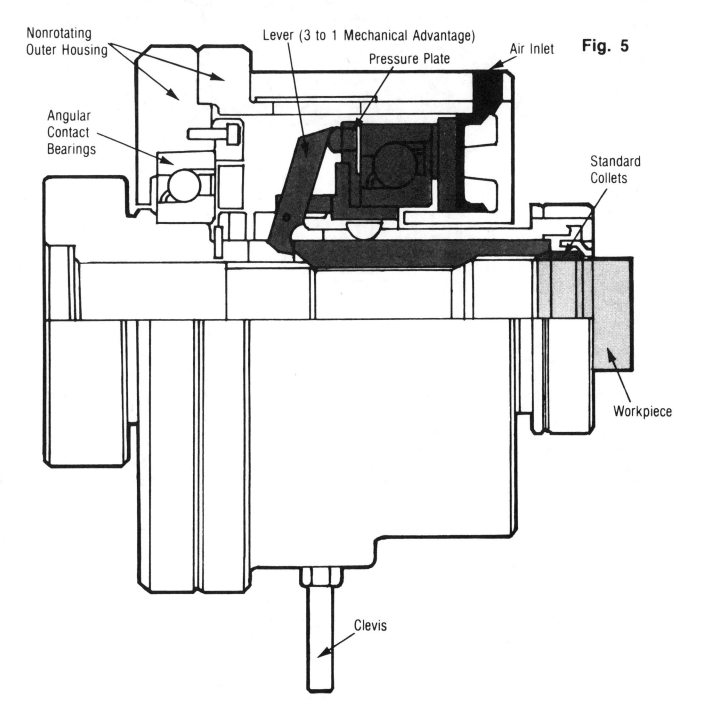

Nonrotating Outer Housing

Lever (3 to 1 Mechanical Advantage)

Pressure Plate

Air Inlet

Fig. 5

Angular Contact Bearings

Standard Collets

Workpiece

Clevis

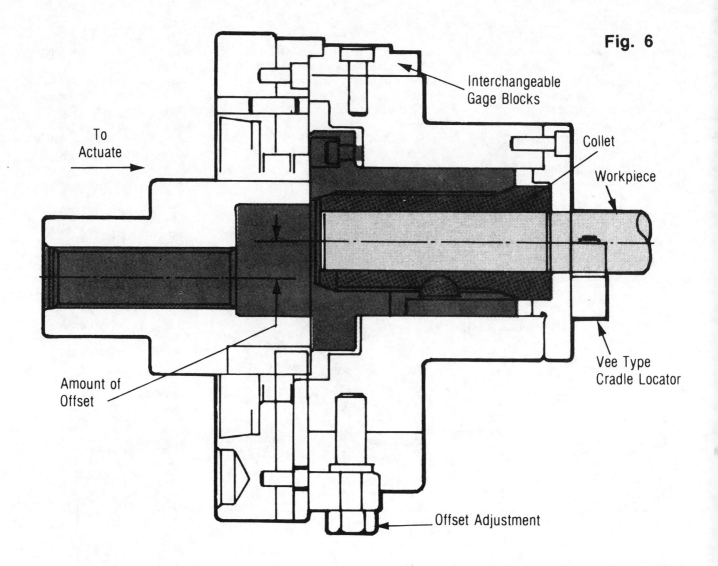

Fig. 6

To Actuate →

Interchangeable Gage Blocks

Collet

Workpiece

Vee Type Cradle Locator

Amount of Offset

Offset Adjustment

curacy. The wedge-type actuation is the most accurate but it has not yet gained in popularity as has the simple three-jaw scroll chuck.

Since the power-actuated chucks are gaining in popularity, let's discuss the types available today.

There are two basic types: the lever type and the wedge type. With the lever type, shown in Figure 7, the powered actuator (either air or hydraulic) pulls the individual jaw levers. As the levers pivot, the eccentric on the thrust end forces the master jaws toward the centerline. Obviously, a top jaw would then grip the workpiece. A reverse action releases the jaws. The action is positive and firmly controlled. With the mechanical advantage of the lever and the power operation, gripping forces can be very substantial.

Does the wedge concept also apply to the powered three-jaw chuck design?

Yes it does and, as with the mechanical wedge type, it does tend to offer a finer degree of precision. This is because the wedge, shown in Figure 8, can be made to a higher degree of overall precision more easily than the lever can.

How about holding power?

The wedge and the lever can be built with approximately the same mechanical advantage, so they can be built with the same gripping power.

As higher turning speeds are coming into general use, does the centrifugal force of turning tend to offset the clamping force of the jaws?

Indeed it does. For this reason, the countercentrifugal chuck is gaining in

acceptance. The counterweights offset the outward thrust of centrifugal force in high speed turning. Remember that when the rotational speed is doubled, the centrifugal force goes up by a factor of four. A tripling of the speed increases the centrifugal force nine times. Many countercentrifugal chucks employ counterweights directly above or behind the jaws. The design shown in Figure 9 has the counterweights beside the jaws, which provides more room for heavier counterweights without increasing the length of the chuck or the outside diameter.

Another type of countercentrifugal chuck employs a wedge-type design with a double-safe jaw-locking mechanism as shown in Figure 10. This provides an extra margin of safety if there is a power failure or if the draw-

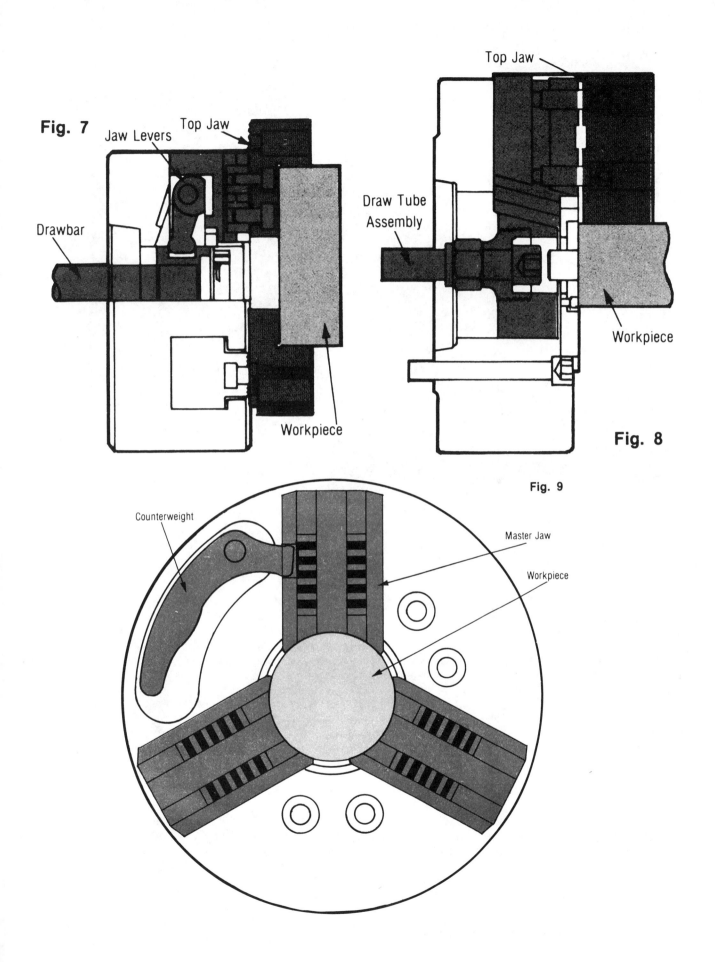

Fig. 7

Jaw Levers

Top Jaw

Drawbar

Workpiece

Top Jaw

Draw Tube
Assembly

Workpiece

Fig. 8

Fig. 9

Counterweight

Master Jaw

Workpiece

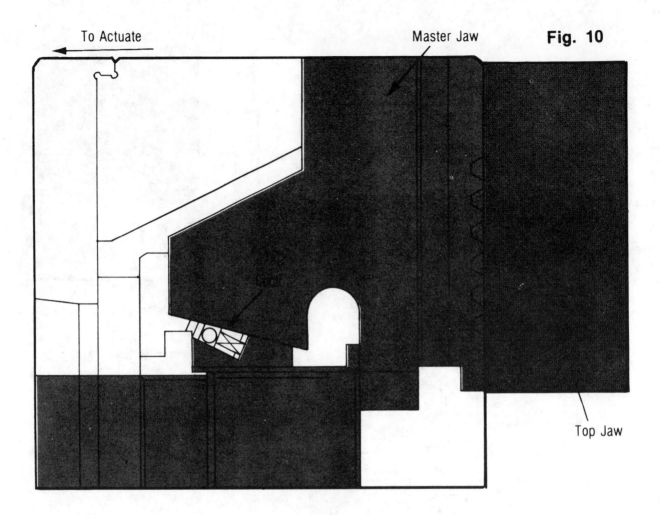

To Actuate

Master Jaw

Fig. 10

Top Jaw

bar threads should become stripped.

Obviously, the three-jaw chuck offers flexibility, a wide range of holding power, and the ability to mount either bored jaws or soft jaws to meet specific job requirements.

Yes, the manual or powered three-jaw chuck does have a lot going for it. But it is a mechanical device with moving parts and in a hostile environment, it does have its weaknesses. Therefore, other types of chucks such as the diaphragm, spring jaw and geared types are employed for special roles.

Could we discuss the more common workpiece holding problems and their solutions?

Consider the problems of grinding. The generated swarf could play havoc with the mechanical three-jaw chuck. Therefore, most grinding operations make use of the diaphragm chuck, as shown in Figure 11.

What is the principle of operation?

The chuck jaws are mounted on a diaphragm, which completely covers the chuck face and provides a seal over the actuating push bar. The diaphragm is a metal plate that works on the oil-can principle. It has a natural shape and when pushed, it wants to return to its rest position. If the diaphragm is properly made and strong enough, the desire to return to the natural rest position can generate a strong force, which is about 90 percent of the force required to open it. When the diaphragm is flexed by the power-actuated push bar, the flexing will open the chuck's jaws. When the force is relaxed, the diaphragm returns to normal and the jaws grip the workpiece.

This flexing does not produce much jaw motion does it?

No, and this is the reason why it is usually limited to grinding applications where the workpiece is already machined to a close tolerance. But it is a sealed chuck and it works well within its limited sphere. However, manufacturing a strong diaphragm that will give several million flexings over its lifetime and maintain a firm grip is a difficult process. That is why there are relatively few diaphragm-chuck manufacturers.

As the diaphragm pivots, doesn't it tend to create a point-holding rather than a full jaw-face gripping?

This is true, and it is also true that the diaphragm action does not produce a natural workpiece squaring action as the jaws close. For this reason, there should be a squaring

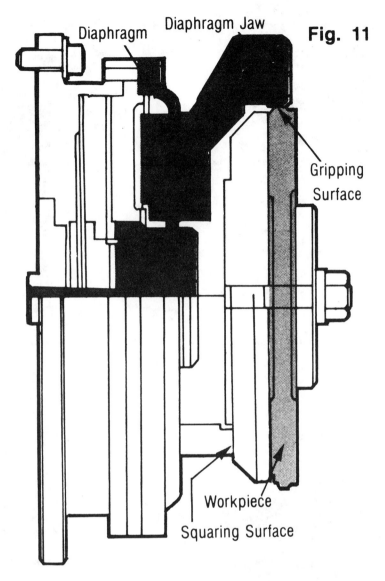

Diaphragm **Diaphragm Jaw** **Fig. 11**

Gripping Surface

Workpiece

Squaring Surface

surface to properly orient the workpiece. Also, the jaws should be faced so that they grip the part with a full face action. Even though there are drawbacks, the very simplicity and the durability of the diaphragm chuck in hostile environments make it ideal for holding certain types of workpieces, especially for grinding.

How are the jaws mounted to the diaphragm face?

They may be bolted, mounted with dowels, or even welded to the face. In some instances, as many as eight jaws are mounted on a chuck and the diaphragm works as well as if only three were mounted.

Can the diaphragm chuck be built to special configurations?

Figure 12 shows a double-dia-

phragm chuck that was built to hold a workpiece in two places. The workpiece was long and the two-level gripping did offer a natural squaring of the workpiece with the centerline of rotation. It also provided a double gripping force.

You mentioned a spring-jaw chuck. How does it function?

This is somewhat of a second cousin to the diaphragm chuck. It operates in much the same fashion. An air or hydraulic cylinder provides a push on a bar, which forces open the spring jaws. When the pressure is released, the jaws return and grip the workpiece. Here too, the gripping power is about 90 percent of that required to release the jaws. The main difference between the diaphragm

chuck and the spring-jaw chuck is the individuality of each jaw. For this reason, the spring-jaw chuck can better accommodate some variations in workpieces and it does a better job of naturally squaring the workpiece as it is gripped. The spring-jaw chuck is often used to hold gears. The jaws are equipped with pins to locate on the pitch diameter of the gear as shown in Figure 13. Another variation is the use of spring jaws to hold cages, which are equipped with locating pins. With this variation, gears can be held on two different levels—such as cluster gears.

Are there other major types of external holding?

One very important concept for the gear industry is the Garrison, or

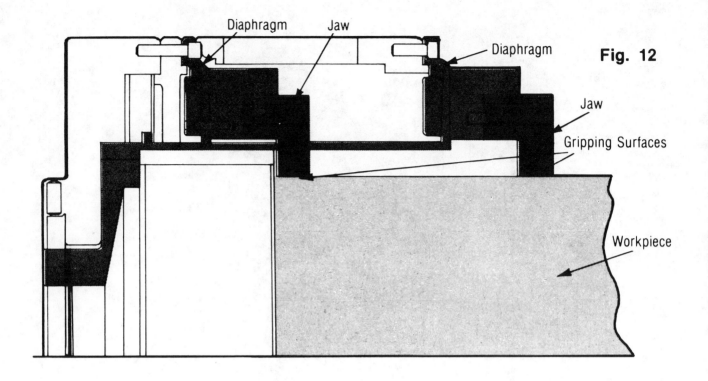

Fig. 12

Diaphragm | Jaw | Diaphragm | Jaw | Gripping Surfaces | Workpiece

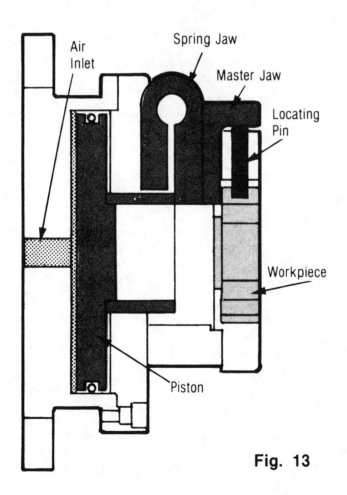

Air Inlet | Spring Jaw | Master Jaw | Locating Pin | Workpiece | Piston

Fig. 13

trunnion-type chuck shown in Figure 14. In this design, the holding is achieved by a number of jaw members that are rotated into and out of contact with the workpiece by an internal plate with a ring gear. The jaws, of course, must match the tooth form of the gear to be held. The advantage of this design is that 12 or more jaws will find the best pitch circle of the gear so that any ID machining is concentric with the circular pitch of the gear. There are many variations of the concept so that the holding principle can be applied to bevel gears, spur gears, or even internal gears when machining will take place on the OD. Any one set of jaws will have a working range of about 0.015 inch on either side of the nominal pitch diameter. Thus, the concept is quite flexible. It has been applied to just about any gear or spline that has been manufactured.

And the last important category of external holding chucks you have mentioned is the indexing chuck.

Yes, this type of chuck is shown in Figure 15. It is used to hold valve bodies, fittings, or other types of workpieces with intersecting axes where work must be performed on two or more of the faces or bores.

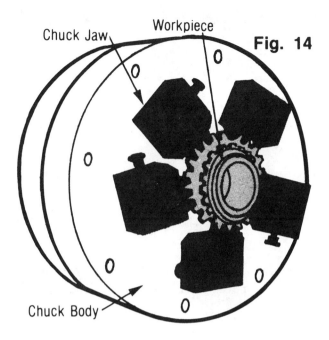

Chuck Jaw   Workpiece   **Fig. 14**

Chuck Body

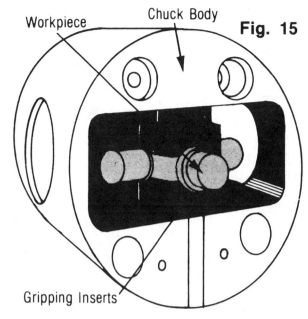

Workpiece   Chuck Body   **Fig. 15**

Gripping Inserts

Although there are standard indexing chucks, many of them are specials that have been designed for individual workpieces. Their advantage lies in the fact that several machining operations on a workpiece can be handled with one chucking.

Can they be applied to NC or automatic machines?

Definitely. The actuation of the chuck indexing cycle can be tied in with an automatic machine cycle or an NC command. Thus, the operation can be automated.

How many degrees can the chucks index?

Standard models can index in 45-, 60-, 90-, or 120-degree increments. However, specials can be constructed for any degree.

Is the indexing actuated by hydraulic or air pressure?

Either. The actuating cylinders may be powered by either method, depending upon the type of machine on which the chucks are installed and the availability of shop utilities.

So there is a great deal of flexibility in the use of indexing chucks as well as the others we have discussed.

Yes, there are many creative approaches to workpiece holding that expedite both machining efficiency and the quality of the workpiece being machined. Creative approaches to workpiece holding can be a powerful resource to improve machine tool efficiency and to increase overall shop floor productivity. **MMS**

ABOUT THE AUTHOR

Dave Beaver was born in 1932 in Cleveland and has spent his entire educational and working career in the area. Thirty years ago he joined the Erickson Tool Company as an apprentice and moved steadily through the ranks to his present position. He has studied engineering at Cleveland State University, Fenn College, and Western Reserve University. He is a member of the ASTME, NC Society and has lectured before professional groups.

Case Histories—Power Workholding Systems

In Machining Operation -
HYDRAULIC CLAMPING CONTRIBUTES
TO CONSIDERABLE DOLLAR SAVINGS

As a manufacturer whose precision hydraulic equipment relies on close tolerance accuracies, Owatonna Tool Company has adopted the power work-holding equipment of one of its divisions to help insure those accuracies and to save considerable money in the process.

As a result, the company is saving nearly $4,000 annually on one machining operation it measured in recent cost comparison studies.

Hytec, the division whose products are being employed by the parent company, produces a matched line of hydraulic power sources, rams, clamps and accessories designed to secure workpieces during various metalworking operations. OTC is a leading manufacturer of hydraulic equipment and maintenance tools.

The operation studied by the company is one in which three parts -- 2-inch diameter piston barrels for several of OTC's high-pressure hydraulic pumps -- are machined. Annual production of these components totals 4,170 pieces.

The old production method required four operations. Manually tightened bolts clamped each part which was machined in two steps by a turret drill. A portable drill press countersunk the holes, and finally the part was belt sanded and inspected.

The new method requires only one operation. A specially designed fixture using a Hytec hydraulic clamping system, and mounted on an Erickson indexing head, holds four workpieces for machining by a Monarch N/C machine.

Time requirements under the old manual method totaled 10.438 minutes for each part and included:

1) Handling time to load and unload each piece: 1.404 minutes.
2) Machine time for each piece: 8.276 minutes.
3) Inspection of each part: .758 minutes.

The new method requires a total of 6.390 minutes to accomplish the same task--drilling 11 holes in each workpiece. The breakdown of work includes:

1) Handling time: .335 minutes per piece.
2) Machine time (including simultaneous countersinking, deburring and inspection): 6.055 minutes per part.

Compared to the old method, OTC is now saving 4.038 minutes in the machining of each part. With labor and burden costs figured at $.233 per minute, the savings on each part is $.94. And with 4,170 of these parts machined each year, the annual

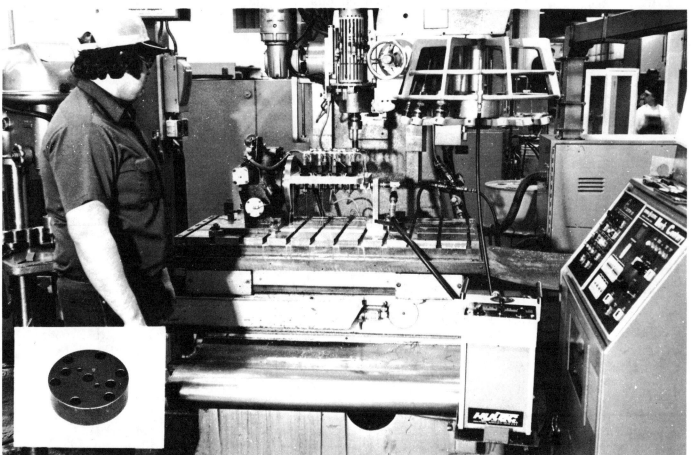

Machining Operation Benefits from Power Work-Holding System of Hytec Air/Hydraulic "Booster-Pac" (lower right), Threaded Rams, Clamps and Special Fixture with Indexing Head (upper photos). Inset Shows Part.

saving amounts to $3,919.80. With the cost of the fixture and hydraulics estimated at approximately $3,000, the company will turn a profit on this single operation in less than a year.

The key to the efficiency of the new method, of course, is the hydraulic clamping system which makes it possible to execute the work on a numerically controlled machine.

Eight threaded-body rams -- screwed into the fixture designed for this job -- clamp four workpieces in place simultaneously. The rams are actuated by a Hytec air/hydraulic Booster-Pac, a pre-engineered, self-contained source of two-stage power. The indexing head rotates the fixture 180 degrees so parts can be worked on both sides sequentially.

For Precise Finishing -
HYTEC THREADED-BODY RAMS ARE KEY
COMPONENTS IN HYDRAULIC COLLET

Finishing of the inside diameters of bronze bushings has met the close-tolerance requirements of Angelus Sanitary Can Machine Co. since the firm's tool designer devised an ingenious hydraulic collet to secure the workpieces.

The Los Angeles manufacturer uses myriad bushings in the manufacture of specialized can-closing equipment. Two inches in diameter and 2 to 3 inches in length, they are purchased with an undersized ID for in-house finishing to precise concentricity. Reaming removes a minimum of .005 of an inch, with typical cuts ranging from 1/64 to 1/32 of an inch.

To properly support and position the bushings during finishing, a custom-designed fixture (collet) was required to fit existing engine or turret lathes. Harry Van Dam, the tool designer, chose hydraulic over mechanical means because a mechanical collet would still have to be custom designed without any cost advantage and the clamping force would necessarily have been unevenly distributed over the circumference of the bushing.

The resulting fixture employs two threaded-body rams (products of Hytec Division, Owatonna Tool Co.) to pressurize hydraulic fluid and create the clamping action.

The stainless steel tool -- 7 11/16 inches long and 5 5/8 inches in diameter -- has a large ringnut to facilitate clamping and unclamping of a workpiece. As the operator turns the ringnut clockwise one-quarter turn, its movement is transmitted through a thrust washer to the crowns of the rams, which in turn pressurize oil within the tool body. (The rams are threaded into axial bores 180 degrees apart.)

The collet itself -- 2.50 of an inch OD and 2.005 of an inch ID -- is encased in a central bore of the tool body. Its open end terminates in a flange which is bolted and gasket-sealed to the tool body. The outer surface of the collet has two 7/8-inch-long bands from which most of the original 1/4-inch-thick cylinder material has been removed, leaving only a .031-inch wall thickness. The resulting peripheral spaces between the collet's OD and the tool body's ID are connected with each other, and with the main oil reservoir (1/2 pint capacity) within the tool body, by axial flow passages machined into the collet's body.

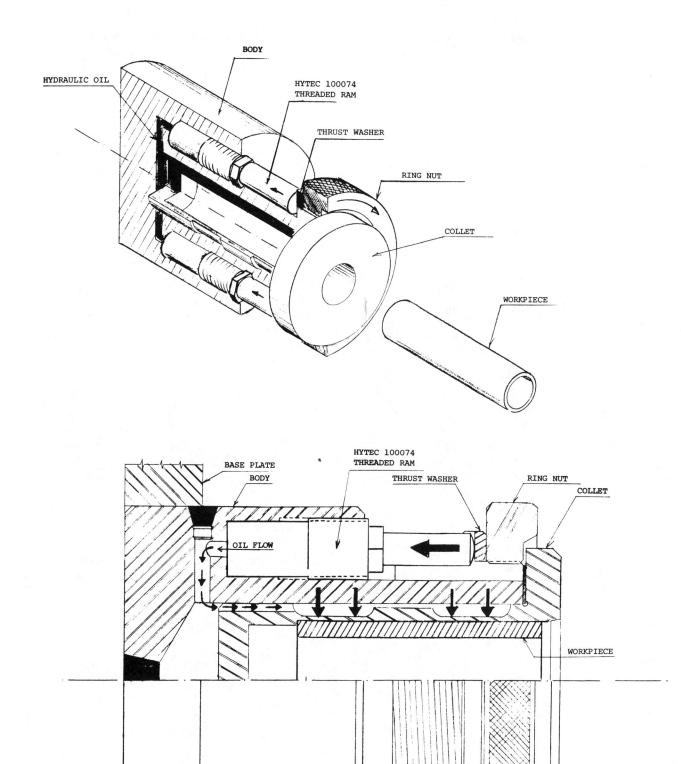

BODY

HYDRAULIC OIL

HYTEC 100074
THREADED RAM

THRUST WASHER

RING NUT

COLLET

WORKPIECE

BASE PLATE
BODY

HYTEC 100074
THREADED RAM

THRUST WASHER

RING NUT

COLLET

OIL FLOW

WORKPIECE

Drawings Depict Design of Hydraulic Collet and how Hytec Rams Actuate Precise Clamping Action

143

As the operator generates hydraulic pressure through the ringnut, thrust washer and the Hytec rams, the oil compresses the thin-walled sections of the collet. This reduces the collet's inside diameter by slightly more than .005 of an inch and clamps the bushing with an evenly distributed peripheral force over nearly 14 square inches of the bushing's surface.

When the finishing pass is completed, the operator backs off the ringnut (counter-clockwise) one-quarter turn. The inward-compressed collet sections spring back, releasing the bushing and acting through the oil to keep the ram crowns firmly in contact with the receding thrust washer.

Setup time for this operation, according to Van Dam, is approximately 10 minutes. Production rate is about 60 units per hour, with a typical run consisting of 300 finished bushings.

A HYDRAULIC COLLET (upper left) employs two Hytec threaded-body rams (vertical elements partially visible at each side) to pressurize the fluid and create clamping action in the ingenious tool. The adjacent photo shows the collet mounted on a standard chuck of a lathe. A bronze bushing is inserted in the collet (lower left), after which a boring bit finishes the workpiece to precise concentricity and close-tolerance inside diameter. The tool was custom-designed by Angelus Sanitary Can Machine Co. using hydraulic rams from Hytec Division, Owatonna Tool Co.

Simple to use and economical (about $275) to build, this successful hydraulic collet will be replicated by Angelus and used for bushings of different sizes.

Rate up 300 Percent –
"TOOLING GROUP" DEVISES FIXTURE, ADDS
HYDRAULICS TO INCREASE PRODUCTION

Saving costs through improved production efficiencies is the objective of a "tooling group" at Clayton Manufacturing Company, El Monte, Calif. If an idea recently generated and implemented by the six members of this group is any indication, their objectives are being met.

That idea -- which combines a unique fixture and hydraulic clamping with an NC machine -- is saving the company 300 percent in production time on one process, compared with the previous method, according to Don Muncy, production engineer.

Clayton is a 45-year-old company with about 1,000 employees whose products include steam generators and cleaners, dynamometers, auto diagnostic equipment and emission analyzers. The part on which these economies are being effected is a cast-iron injector body for steam cleaning equipment in which five ports are drilled and thread-cut.

With the new process, 16 parts are machined in 77 minutes (including loading and unloading) in a Milwaukee-Matic 200, which employs nine tools making 240 entries in the workpieces. Previously the parts were machined one at a time in a special fixture on a turret drill press, a method which required considerably more machining time, materials handling and setup time.

Following the acquisition of a Milwaukee-Matic 200, the tooling group devised a fixture for the machine's indexing table and equipped it with threaded-body rams powered by an air-hydraulic pump. The hydraulic components are products of Hytec Division, Owatonna Tool Company.

The fixture is an excellent example of functional design and intelligent use of hydraulic work-holding components. Its compact, custom design was made possible by the unique Hytec rams. Threaded into tapped holes of the fixture and self-sealed with built-in O-rings, the rams receive hydraulic fluid directly through passages drilled in the fixture which serves, in effect, as an oil-feed manifold.

Basically a hollow steel octagon, the fixture measures approximately 18x18x18 inches and weighs about 300 pounds. Bolted to its four opposite sides are tool units, each of which contains four "balconies" in a vertical stack to accommodate the workpieces. The floor of each "balcony" has three locating stubs which engage counterbored recesses machined into three bosses of the workpiece to facilitate placement of the parts.

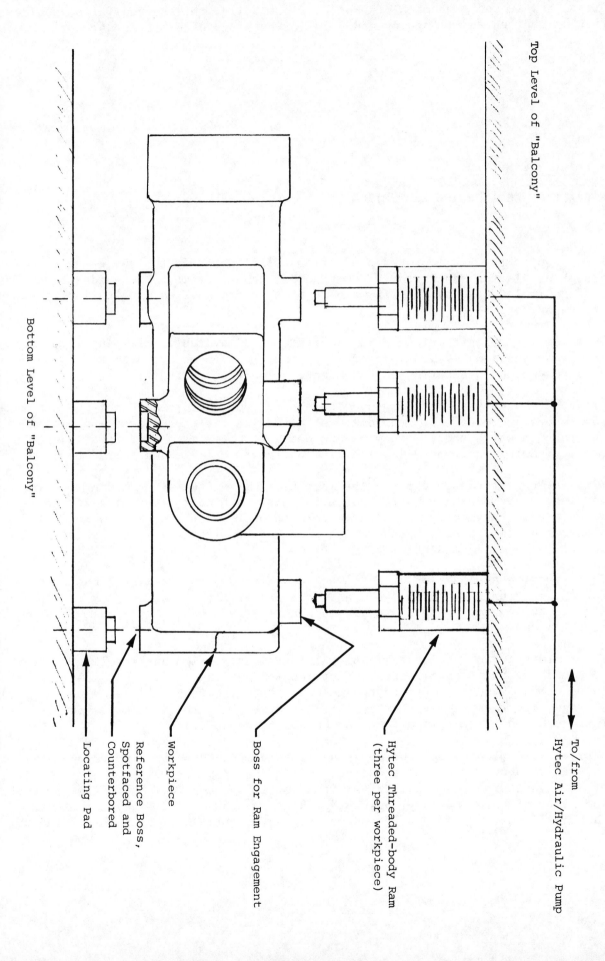

Sketch Shows Positioning of Threaded-Body Rams in Fixture "Balcony" Relative to Workpiece

Top Level of "Balcony"

Bottom Level of "Balcony"

To/from
Hytec Air/Hydraulic Pump

Locating Pad

Reference Boss,
Spotfaced and
Counterbored

Workpiece

Boss for Ram Engagement

Hytec Threaded-body Ram
(three per workpiece)

Three Hytec Threaded-Body Rams (inset) Clamp Each Workpiece in Clayton's Unique Fixture

HYDRAULIC WELDING FIXTURE ASSURES
PRODUCT QUALITY, SPEEDS PRODUCTION

A hydraulic workholding technique commonly applied in various machining operations is saving time and assuring product quality in a welding process at Dataproduct, Inc., Van Nuys, California.

The automatic clamping system -- which secures an assembly that forms part of the main frame for high-speed line printers -- consists of a custom-made welding fixture and standard hydraulic components mounted on a mobile stand.

Production time has been cut in half with this new method, according to Tool Engineer Lennart Erikkson, but more importantly the quality of the assemblies is constant.

The flat, rectangular fixture is equipped with positioning stops and mounting brackets in which nine Hytec threaded-body rams are strategically located. Because welds have to be made around the perimeter of the member-to-member contact, the fixture and turn-table on which it is mounted have cutouts to permit access to the underside of the weldment. The welder manually turns the table 180 degrees as required.

After the operator places pieces of the assembly in the fixture, he activates a Hytec "Booster-Pac" to supply hydraulic pressure to the tubing manifold and the nine rams. Time delays were designed into the system so the rams position the pieces sequentially before clamping them in place for welding. Parts of the assembly consist of seven square and round sections of 1010 stainless tubing which combine to form a frame approximately 12 by 24 inches.

(The Booster-Pac is a self-contained compact source of power which operates on shop air supply of from 60 to 125 psi. Low pressure, high volume oil quickly advances and positions the rams, after which high-pressure, low-volume oil clamps and holds the workpieces at up to 4,125 psi.)

To protect the welder and meet OSHA regulations, the hydraulic system was equipped with non-flammable fluid to prevent possible fire hazards in case a line is cut in the welding operation.

Previously the operator positioned components manually and secured them individually in separate clamps. A time-consuming process, it also was subject to human error in the relative placement of the parts. The high-tolerance requirements of the main frames of precision printers necessitate correspondingly precise fabrication of sub-assemblies which serve as positional references as well as support.

With the system designed by Erikkson, the assemblies are turned out in less than four minutes -- about half the time of the previous method -- and precise alignment is assured. The time is saved in setup.

The ingenious, self-contained workstation carries all the elements required for the operation, except shop air supply. The fixture, turntable, Booster-Pac and associated hydraulics are all mounted at table-height on a steel stand equipped with four casters. Trunnions permit easy rotation of the fixture, while the casters permit movement of the entire unit as desired.

Erikkson, who says that resulting quality assurance is well worth the estimated $3,500 required to design and build the setup, is presently designing a more complex hydraulic welding fixture that will further facilitate production of components used in the company's peripheral computer equipment. Dataproduct, which headquarters in Woodland Hills, CA, and employs 2,500, also manufactures computer core memories.

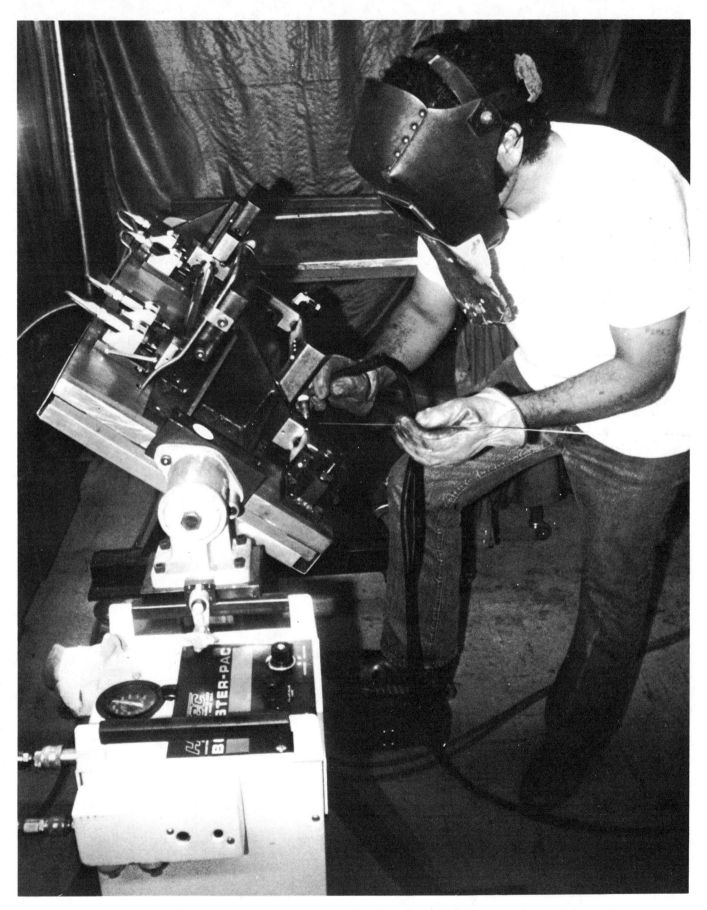

Turntable Enables Welder to Rotate Hydraulic Welding Fixture 180 Degrees as Required

Welding Fixture on Mobile Stand Employs Hytec Hydraulic "Booster-Pac" (left) and Threaded-Body Rams (inset)

For Quality Control –
RETRACTING CLAMPS SECURE PART
IN SOPHISTICATED TEST MACHINE

Hydraulic retracting clamps of the type that normally secure workpieces in metal-working applications proved to be equally effective in holding automatic transmission valve bodies for testing in a sophisticated device.

Advanced Technology & Testing of Livonia, Mich., which designs and builds test equipment for the automotive and heavy industries, used those clamps in four machines it devised for a car manufacturer to test valve bodies that go into the transmissions of two popular models.

Seven retracting clamps, manufactured by OTC Hytec Division of Owatonna Tool Co., are fixed in each machine to hold a valve body at the points where it would be bolted to the car's transmission. Each clamp applies pressure at 1,000 psi to approximate the force with which the bolts are torqued. Limit switches assure proper clamping.

150

Computer-Controlled Machine Tests Operation of Automatic Transmission Valve Body, Clamped on Subplate Below Panel

Seven OTC Hytec Retracting Clamps Secure Automatic Transmission Valve Body in Test that Simulates Actual Operation.

"We're really simulating installation and operation of the valve body," according to Tom Stovel, senior project engineer for Advanced Technology & Testing.

Retracting clamps were required, Stovel said, because they draw back so the operator can load a part on the fixture and remove it after the hydraulic test. They are automatically cycled and evenly apply pressure around the perimeter of the valve body to seal it as it would be in a transmission. "They work extremely well," according to Stovel, who said about 40 parts can be tested per hour at each station.

Custom-made for the car manufacturer's quality control program, the test stands are computer controlled and have their own hydraulic power sources. Oil is introduced into a valve body through a manifold, after which various pressures are applied by a servo control in the hydraulic system to assure proper operation.

Each test stand, which measures about 6 by 8 feet, has a CRT that displays system status and test results to the operator and reject printer for recording test results on failed parts. The 54 by 28-inch subplate on which the OTC Hytec clamps and test assembly rest in a one-inch steel plate, surface ground.

In the testing process a valve body is positioned upside down with the shifting mechanism underneath, Stovel said, so the appearance of the setup is in keeping with the clean design of the machine and the simplicity of the OTC Hytec clamping system. (Because of its internal design, a valve body will work in any position.)

Advanced Technology & Testing is a division of Allen Automated Systems whose 260 employees operate in five buildings to produce test equipment for worldwide markets.

In Milling Process –
3 HYDRAULIC WORK SUPPORTS HELP
MANUFACTURER SAVE $20,000 a YEAR

Three hydraulic work supports with a total price tag of $477 are saving a Toledo, Ohio, manufacturer over $20,000 a year in the production of automotive clutch housings.

Spicer Transmission Division of Dana Corporation, a large producer of medium and heavy-duty truck transmissions, saves 48 cents on each part because the work supports enable a vertical milling process to be accomplished in one pass instead of two. Last year 43,000 of the parts were milled on the supports.

The work supports -- from OTC Hytec Division of Owatonna Tool Company -- are small spring-loaded units whose support plungers accept the level of a workpiece and are then automatically locked into place hydraulically to prevent sagging or bending under machining forces.

They proved to be a perfect solution for Spicer, which previously could not hold flatness on the convex center portion of the round housings in one milling pass, according to Bill Pouter, process engineer.

"As the cutter reached the center section it would depress the part slightly, requiring the operator to back up the cutter, lower it and make another pass," Pouter explained.

With the work supports, which are secured at strategic points on the special fixture, sagging does not occur even under the 2,000-pound load created by the 18-inch-diameter cutter. (Each support has a 5,000-pound load capacity at 3,000 psi.) Although the amount of stock on the underside of the castings varies, the spring-loaded supports "find" the proper height and lock into position to provide positive support.

"Product quality has also improved since the supports were added," Pouter said, "and setup time has been reduced by about a half hour." The OTC Hytec supports have not failed in the year since they were installed.

The supports are powered by an OTC Hytec air/hydraulic pump and operate independently of the air clamps which secure the part to the fixture. They are locked in place after the operator has clamped the perimeter of the workpiece.

Since the clutch housings vary in size depending on the series, the work supports are mounted in T slots so they can be moved in or out at 120-degree angles to the points where proper support is required.

At least 1/8" of material is removed from each cast iron or aluminum housing during the minute it is under the mill. If the mill runs continuously, approximately 600 parts can be processed in 24 hours.

Hytec Hydraulic Work Supports (one visible in center) Prevent Sagging of Part in Milling, Save Company $20,000 a Year

Reprinted from Manufacturing Engineering, March 1978

Toolholder Subs as Workholder

...and trims machine time while improving workpiece accuracy

*1. VALVE BODIES
like these are bolted to a special
mounting plate (background)
which is then attached to
a shell mill adapter.*

WORKHOLDING AND TOOLHOLDING are two different things. Sometimes, however, the equipment used for these functions winds up in the opposite camp. At G.W. Dahl Co., for example, the solution to a workholding problem turned out to be a toolholding system.

Located in Bristol, RI, Dahl builds flow and direction control valves, electrical and electronic process control devices, and electronic monitoring and indicating mechanisms. The unusual workholding method is used in valve body machining operations. Kwik-Switch® toolholders supplied by Universal Engineering Div., Houdaille Industries, Inc., Frankenmuth, MI, are used to hold the bodies during machining on a Pratt & Whitney Starturn 8-15 NC lathe.

Concentricity Required. The rough castings shown in *Figure* 1 are two of four different valve styles that have a mounting flange on one end and either a nipple or bore at the other end. The

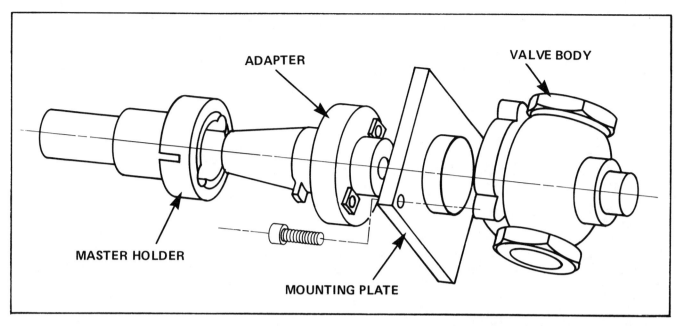

ADAPTER

VALVE BODY

MASTER HOLDER

MOUNTING PLATE

2. STRAIGHT-SHANK MASTER HOLDER
is chucked in lathe to serve as workholding device.

nipple, or bore, must be machined concentric with the flange-end opening. In the first operation, the flange end is machined completely, with the part chucked at the other end on either the nipple or the bore. The workholding problem is encountered in the next operation, where the finish-machined flange end must be chucked to allow machining the nipple or bore.

On the valve body at the left in *Figure* 1, the nipple is faced, drilled, and threaded. In the earlier setup, the lathe was fitted with a special face plate to which the finish-machined casting flange was bolted. Although this arrangement did a satisfactory job of making production, there were two drawbacks. According to production engineer Stan Wilbur, it took 70 sec-

onds, or more, to unbolt one casting and mount another, and it was difficult to get consistent concentricity between the flange-end bore and the finish-machined nipple.

"We have used Kwik-Switch for a number of years as a toolholding system," Wilbur says. "We use it regularly on our Starturns, and it occurred to me that we ought to be able to adapt the system for workholding as well."

The Setup. Wilbur's solution, shown in *Figure* 2, involves use of a straight-shank Kwik-Switch master holder and a standard system shell mill adapter. The adapter is equipped with a special mounting plate to which the casting flange is bolted.

The straight-shank end of the Kwik-Switch holder is chucked in the lathe.

With this setup, the switch from one adapter-mounted casting to another takes less than 10 seconds. Compared with the former method, this provides a machine time saving of 60 seconds, or more, per part. It is all machine time, because the casting is mounted to the adapter off the machine while another casting is being machined. At $24.33 per hour for Starturn machine time, the Kwik-Switch workholding system paid for itself in short order.

In addition, Wilbur points out that the new system is much simpler to use, and the operators like it much better. Concentricity between the flange-end bore and the nipple is now achieved consistently with a minimum of effort. ■

Power Chucking Manual

Most power chucks are purchased for use on NC Lathes, and will be used to grip a variety of workpieces. This manual will describe the various types of chucks and accessories which are available, and the advantages and disadvantages of each.

CHOOSING A POWER CHUCK

The most popular type of chuck is the 3-jaw design, so we will concentrate on this chuck and the accessories to be used with it. Further on in the manual you will find a description of other chucks which can be used to meet specific workholding problems.

Your power chuck can be purchased with or without a thru hole, with four types of master jaws, with different jaw movements, and with either self-locking or non-locking design.

The thru hole type is desirable if you want to grip bars or slugs, or if you want to put part of the workpiece back into the hole in the chuck. Usually a chuck with no thru-hole is used only for chucking work.

The type of master jaws selected will determine what type of top jaws can be used. The four common types (see Figure 1) are:

1. American Standard Tongue & Groove
2. American Standard Serrated
3. Square Serrated
4. Fine Serrated

The tongue and groove type top jaw is mounted on the master jaw in a fixed position. It can be reversed, but it cannot be adjusted in or out. Soft top jaws, once they are machined to hold a specific diameter, can only be used to hold that diameter or a larger diameter. This type of master jaw is not recommended when you need the ability to grip a wide variety of workpieces machined in small lots.

The American Standard serrated master jaw has serrations which look like an Acme thread, and tee nuts in the tee slot of the master jaw. The top jaws have serrated keys, which bolt to the top jaws. The serrated keys engage the serrations in the master jaws and the top jaw fastening screws lock the top jaws and master jaws together. The serrations are 1/4″ apart, so that the position of the top jaw on the master jaw can be adjusted increments of 1/4″. A further adjustment can be made by reversing the key; this will permit a 1/8″ adjustment.

A disadvantage inherent in American Standard serrated jaws is limited adjustment, particularly in thru hole chucks. The top jaws have 2 bolt holes, and adjustment is limited to 3 serrations.

The square serrated jaws are similar to American Standard serrated, except that the serrations are square; they have the same limitations.

Fine serrated jaws provide the most flexible arrangement. The serrations have 1/16 pitch, although some foreign chucks have 1.5 mm. pitch. The standard top jaws have 3 bolt holes.

The combination of fine serrations, plus the ability to use any two of the 3-bolt holes provides maximum flexibility to locate the top jaw where you want it.

Sometimes a shop may have many sets of special top jaws in stock which can only be used on one type of chuck. Top jaw adaptors (Figure 2) are one solution to this problem. These are intermediate plates which go between the master jaw and the top jaw. The most common type of top jaw adaptors will adapt American Standard tongue & groove top jaws to fine serrated master jaws.

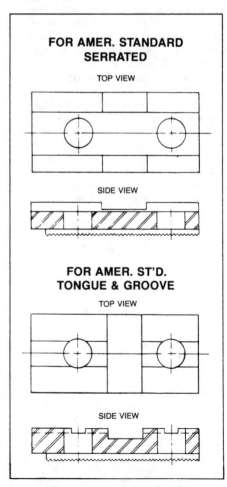

FOR AMER. STANDARD SERRATED

TOP VIEW

SIDE VIEW

FOR AMER. ST'D. TONGUE & GROOVE

TOP VIEW

SIDE VIEW

FIG. 1 TYPES OF MASTER JAWS.

FIG. 2 TOP JAW ADAPTOR.

JAW MOVEMENT

Another factor to consider in selecting a power chuck is jaw movement. Most wedge type and lever type chucks have either 1/4″ or 3/8″ jaw movement, depending on the chuck diameter, and the leverage is usually 4:1 (jaw movement is 1/4″ of the stroke). The Gamet chuck has a cam lever, which reduces the jaw movement, and increases the gripping power. When a large jaw movement is required, a medium-opening or wide-opening chuck can be furnished. For example, the Gamet chuck can be furnished in these 3 types:

12″ Normal	.177″ jaw movement per jaw
12″ Medium Opening	.394″ jaw movement per jaw
12″ Wide Opening	.984″ jaw movement per jaw

Remember that the total movement of the jaws is twice the jaw movement per jaw.

SELF-LOCKING CHUCKS

The last factor to consider in choosing a chuck is whether or not the chuck is self-locking. The power to operate a chuck comes from an external source, either compressed air or hydraulic pressure. If the chuck is not self-locking and the pressure drops, the chuck will not hold the workpiece. For this reason, fail-safe rotating air cylinders are now frequently used to operate power chucks. However, if the chuck is self-locking, a drop in pressure will not affect the chuck's ability to hold the workpiece. Most wedge type chucks and Gamet cam lever chucks are self-locking.

CHOOSING A CHUCK ACTUATOR

Two types of actuators are available: Rotating cylinders with no thru hole and chuck actuators with a thru hole. Both types are available either air-operated or hydraulic.

If no thru hole is required, a rotating cylinder will do the job and is considerably less expensive than a thru-hole chuck actuator. If the machine tool hydraulic system can be used to operate a rotating cylinder, either air or hydraulic pressure can be used. If the chuck chosen is not a self-locking type, a cylinder with a built-in fail-safe valve should be used.

Cylinders equipped with a fail-safe valve usually will not work satisfactorily at low air line pressures. This is also true of some types of hydraulic chuck actuators (Figure 3).

Choosing a rotating cylinder to operate a specific chuck usually depends on the draw bar pull recommended by the manufacturer of that chuck. If a chuck actuator with thru hole is selected, then the choice depends on maximum draw bar pull, maximum spindle speed of the lathe, and the I.D. of the thru hole in the actuator.

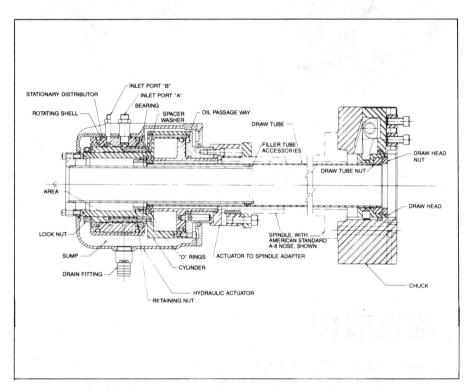

FIG. 3 LAYOUT GAMET CHUCK & HYDRAULIC ACTUATOR.

DETERMINING THE BAR CAPACITY OF A SPECIFIC LATHE

A lathe equipped with a thru hole chuck and a thru hole chuck actuator will have a draw tube connected to the chuck and to the actuator, running thru the I.D. of the lathe spindle. Draw tubes are usually made from steel tubing with 1/8″ walls, so

the O.D. of the tube will be about 1/4″ larger than the I.D. With the I.D. of the chuck, of the chuck actuator, and of the spindle known, the O.D. of the draw tube must be slightly smaller than the smallest of these inside diameters.

The I.D. of a draw tube is not the bar capacity of the lathe. Draw tubes are not perfectly straight or round, and neither are the bars which pass through the tubes. If the desired bar capacity is 2″, the minimum I.D. of the tube should be 2-1/16″.

CHOOSING TOP JAWS FOR YOUR CHUCK

There are many varieties of top jaws in use, but these three are the most popular: Soft blank top jaws, hard stepped top jaws, and collet pad top jaws.

Soft blank top jaws (Figure 4) are made of steel or aluminum, are rectangular in shape, and may have either pointed or square ends. They are usually machined to accept a particular size workpiece.

FIG. 4 TYPICAL SOFT BLANK TOP JAW.

Hard stepped top jaws (Figure 5) have hardened gripping serrations, and have either one or two steps so that a variety of diameters can be gripped. This type of jaw is usually not used on American Standard tongue & groove master jaws, because it is not adjustable, and therefore cannot hold a range of diameters. Hard jaws are most commonly used to grip forgings, castings, or bar slugs on a diameter which will later be finished machined.

FIG. 5 TYPICAL HARD STEPPED TOP JAW.

Collet pad top jaws (Figure 6) are made to accept standard collet pads, which in turn will grip a bar or a slug. The bars to be gripped can be round, square, or hex. The pads can be furnished either smooth or serrated.

Collet pads are readily available, inexpensive, and easy to store. They also pro-

FIG. 6 TYPICAL COLLET PAD TOP JAW.

vide a very practical way to grip square bars with a 3-jaw chuck. Standard pads are produced by 1/16″ increments; special and blank pads are also available. The pads are easy to install and remove, and should provide accurate gripping and concentricity within .001″ T.I.R.

OTHER USEFUL ACCESSORIES

1. **Chuck and Collet Combination.** As shown in Figure 7, this combination allows mounting a collet on a power chuck to grip bars or slugs. The changeover from chuck to collet, and vice versa, can be made in 15-20 minutes. Since a collet does not have as much gripping power as a chuck, it is recommended that a hydraulic chuck actuator be used to provide the additional draw bar pull required when using a collet.

2. **Gripping Work Between Centers.** If the workpiece has to be positioned between centers, some method for driving it must be provided. A lathe dog with a center mounted on the face of the chuck can be used, but a better method is to use a driving center held in chuck jaws. A driving center has a spring-loaded center point, surrounded by 4, 5, or 6 chisel pointed drive pins. The workpiece is positioned between centers; the tailstock

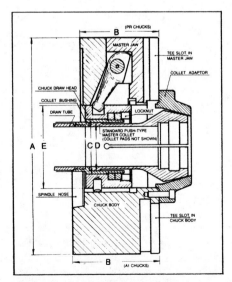

FIG. 7 CHUCK AND COLLET COMBINATION.

is used to push the spring-loaded center back until the end face of the workpiece contacts the drive pins. Continued pressure forces the chisel points to bite into the workpiece, providing a means for driving it. The principle advantage of the driving center is that the entire O.D. of the workpiece can be machined in one set-up, reducing set-up time and insuring that all diameters turned will be concentric with each other.

If the workpiece is a piece of heavy wall tubing, a special bull nose center can be made to fit the driving center. This arrangement is particularly good if the finished O.D. of the tubular workpiece must be concentric with the I.D.

BAR FEEDS

Devices used to feed bars through a chuck vary from very simple feeds to elaborate multiple bar feeders. Since most shops will be feeding bars only part of the time, just one of these devices will be described.

The primary type of bar handler is a simple bar feed, (Figure 8) designed to push a bar through the spindle and chuck. It is basically an open-end air cylinder mounted in back of, and in line with, the lathe's spindle. When loading, the bar feed tube is swung out to clear the headstock. The operator then uses the bar to push back the piston inside the feed tube until the bar is completely inside the tube. The tube is pushed back to center and locked in place. Low pressure air is introduced into the tube, and the bar is pushed forward through the spindle and chuck. A flow control valve, mounted on the bar feed, is used to control the speed of feeding, while a pressure regulator, also mounted on the bar feed, is used to reduce the force exerted against the bar so that only enough pressure is used to move the bar. The bar is centered by a cup mounted on the front end of the piston. The bar feed tube has a bleed hole in it near the front end of the tube. When the piston moves past this hole, air pressure bleeds through the hole, and the piston will no longer move forward.

When the first bar is completely fed out of the bar feed tube, a remnant of the bar remains inside the lathe spindle. Either a second bar, or a pusher bar, can be used to push this remnant through the spindle.

We recommend that a pusher bar be used, since this bar can be cut to the correct length to insure that all of the bar will be used, except for a short length of scrap.

FIG. 8 BAR FEED.

BAR FEED ACCESSORIES

To feed a range of different diameter bars (e.g. from 1/2″ to 2-1/2″ round), a set of filler tubes is required. A filler tube (Figure 9) is used to guide the bar through the draw tube to the chuck and to reduce the whipping of the bar when it revolves. Each tube has two flanges; one at the back end to bolt the tube to the actuator, and one near the front end to support and locate the tube. The inside diameters of the filler tubes are usually made in 1/4″ increments, so that no bar being fed will be more than 1/4″ smaller than the I.D. of the filler tube.

A set of guide bushings (Figure 10) is recommended when using one bar to push the remnant of the preceding bar through the filler tube. The O.D. of the guide bushing will be a slip fit in the front end of the bar feed tube, with an O-Ring inserted in the O.D. to hold the guide

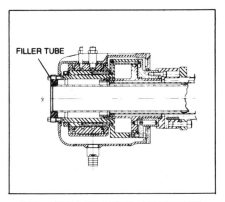

FIG. 9 FILLER TUBE, MOUNTED.

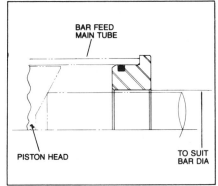

FIG. 10 TYPICAL GUIDE BUSHING.

bushing in place. The guide bushing used for a specific bar diameter will have a hole the same size as the hole in the filler tube. The bar will then be supported by the pis-

ton at the back end of the bar and by the guide bushing at the front end of the bar. The bar will feed through the guide bushing and filler tube to the chuck.

WORK STOPS AND WORK LOCATORS

Work stops are devices located in the I.D. of the chuck or inside the draw tube to position a bar slug or long workpiece before gripping it with the chuck jaws. The work stop may be fixed (Figure 11A) or adjustable (Figure 12B). If the bar slug is long and slender, the work stop should provide a means to center the slug as well as to locate it.

Work locators are generally mounted on the face of the chuck and are used to end locate a workpiece by some means other than the chuck jaws. Either the Tee Slots in the face of the chuck or drilled and tapped holes can be used to fasten a work locator to the chuck body. Combination work stops and work ejectors, either air (Figure 11C), or hydraulic (Figure 11D), are also available.

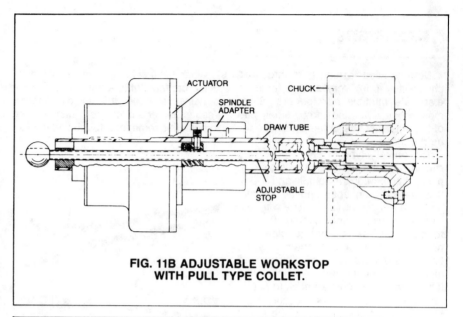

FIG. 11B ADJUSTABLE WORKSTOP WITH PULL TYPE COLLET.

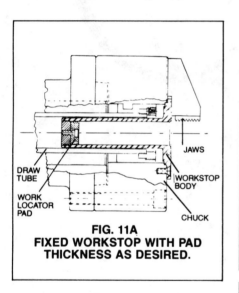

FIG. 11A
FIXED WORKSTOP WITH PAD THICKNESS AS DESIRED.

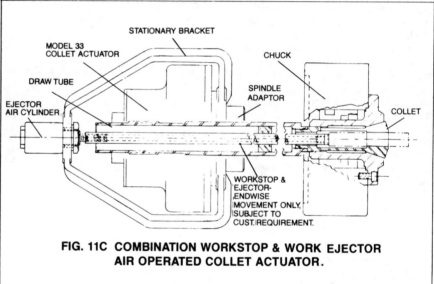

FIG. 11C COMBINATION WORKSTOP & WORK EJECTOR AIR OPERATED COLLET ACTUATOR.

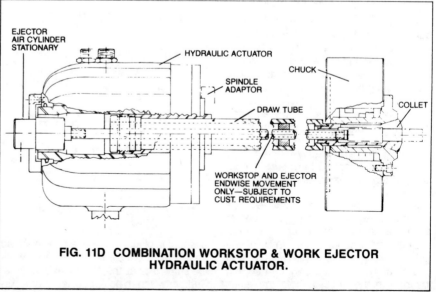

FIG. 11D COMBINATION WORKSTOP & WORK EJECTOR HYDRAULIC ACTUATOR.

OTHER TYPES OF POWER CHUCKS

While 3-Jaw chucks are the most popular, there are several other types which are commonly used, including the 2-3 Jaw Chuck, Compensating Chuck, 4-Jaw Chuck, and Index Chuck.

The 2-3 chuck (Figure 12) is a power chuck with four master jaws. It can be used either as a 2-Jaw or as a 3-Jaw chuck. Its principal advantage is that it can be used for either type of chucking, with a minimum amount of set-up time. It's principal disadvantage is that this type of chuck cannot be balanced, and therefore will create vibration if revolved at the high speeds commonly used for 3-Jaw chucks. All of the accessories listed in this manual for 3-Jaw chucks can also be used with 2-3 Jaw Chucks.

COMPENSATING CHUCKS

A compensating chuck (Figure 13) has a center point, either fixed or spring-loaded, and 2, or 3, jaws which float. The jaws grip the workpiece, but do not center it. The center in the chuck and the tailstock center position the workpiece between centers. The spring-loaded center point is most useful if the end location of the workpiece is important. If the workpiece is a tube, a bull nose center point can be used.

4-JAW CHUCKS

If the workpiece is square, octangular, or rectangular, a 4-jaw power chuck (Figure 14) may be used to center and grip it. However, only one of the two pairs of jaws will actually grip the workpiece, due to variations in the dimensions of the work. If it is important that all four jaws self-center and grip, a Gamet 2+2, 4-jaw chuck can be used. This chuck is so designed that each pair of opposed jaws operates independently of the other pair. It requires a double piston actuator, connected to the chuck by a draw bar and a draw tube.

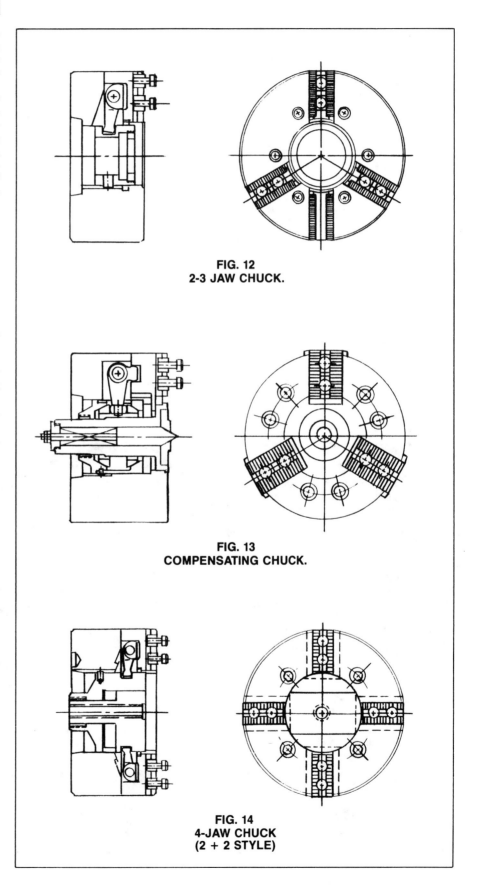

FIG. 12
2-3 JAW CHUCK.

FIG. 13
COMPENSATING CHUCK.

FIG. 14
4-JAW CHUCK
(2 + 2 STYLE)

INDEX CHUCKS

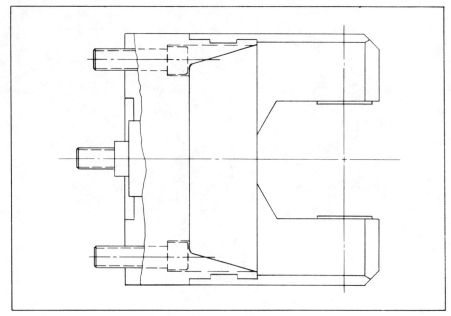

FIG. 15 INDEX CHUCK.

An index chuck (Figure 15) is used when two or more faces of a workpiece located on a common center line must be machined in one set-up. A common example of this type of workpiece is a valve body. The two gripping jaws are mounted on spindles, so that the workpiece may revolve, or index, from one position to the next. Four 90° indexes are usually provided, but other angles may be specified. Index chucks can be either manual or power operated, but the actual indexing is almost always manual. Standard index chucks are made in sizes from 5″ to 15″ diameter; special index chucks are made in sizes up to 60″ diameter. Index chucks always require special gripping jaws which are fitted to a particular workpiece.

SAFE WORKHOLDING

Chucks must hold the workpiece securely without slipping or coming out of the chuck jaws. It is important that each chuck be checked periodically to make sure that its gripping power is maintained. The most practical method is to use a chuck pressure gage (Figure 16) to check the chuck when first installed, and then periodically as long as it is in use. The chuck jaws grip the chuck pressure gage and register the resulting force in pounds or kilograms. If any reduction in gripping force is seen, the maintenance department should be notified, and the chuck should be inspected.

FIG. 16 CHUCK PRESSURE GAGE.

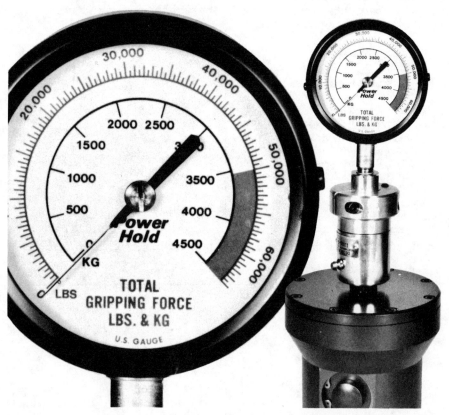

FIG. 17 STATIC TESTING FOR STATIONARY CHUCKS.

162

CHAPTER THREE

INDEXING AND POSITIONING DEVICES

Reprinted from Modern Machine Shop, September 1978

The ABC's of X-Y Positioning

The X-Y positioning table can be used to solve materials handling, transfer and positioning problems. But the type of bearings used, the specifications desired, the drive options and whether to buy or build must be considered for optimum performance.

By RICHARD DOUGANS, Chief Engineer
Design Components, Inc.
Medfield, Massachusetts

X-Y positioning tables have been integral parts of machine tools for 75 years. Still, many engineers and designers are unfamiliar with the new positioning tables that bear little resemblance to the comparatively massive tables found on large lathes, milling machines, and other production equipment.

These highly accurate tables use anti-friction bearings—instead of metal-to-metal sliding contact—to provide a low-cost, easily implemented means of transfer and positioning. They permit linear and, in a few special cases, rotary motion with very low friction. The motion is accurate and repeatable, and it can be controlled by open-loop or closed-loop electronics.

Types Available

With new applications for X-Y positioning tables being discovered almost daily, it is prudent to be aware of the types of tables that are available. The main difference that sets table styles apart is the bearing technology incorporated.

The ball-bushing, perhaps the most common linear bearing, is the basis for many X-Y positioning tables. A positioning table configured in this manner consists of two sets of ball-bushings set at right angles to each other. Each set rides on two polished and ground shafts. Because the ball-bushings can travel along a shaft of any practical length—up to many feet if required—this approach to X-Y positioning is unusually flexible in regard to physical size. However, the time required to adjust the assembly for optimum results, when a series of machines are required, may be unacceptable.

In contrast to the ball-bushing type of linear bearing, the nonrecirculating bearing type does not use a system of recirculating balls. Instead, the balls are held in a retainer and roll against precision shafts or V-ways mounted to the stationary section of the table. The moving surface of the table forms the outer "race" of the bearing. Several different variations are currently available, using roller elements or ball elements. The advantages of the nonrecirculating type of linear bearings are many. From a design point of view, they offer somewhat less friction than those based on recirculating anti-friction elements. But in a more general sense, the constrained-bearing positioning table eliminates the headache of designing, building, and testing a positioning subsystem.

Another approach to X-Y positioning is based on the use of inexpensive chassis-slide type mechanisms. These are essentially light-duty linear bearings, usually constructed from sheet metal. They are often used in electronic chassis or in some hand-driven microfilm readers, where loose tolerances can be maintained manually by the operator. The strong point of such mechanisms is the low price and satisfactory performance where high precision and repeatability are not required. They should NOT be considered for high duty cycle applications, or those involving even moderate loads or shock.

The increasing trend among manufacturers is to purchase a complete subassembly for incorporation into a larger assembly. Under these conditions, it might make little sense to get involved with build-it-yourself projects, especially if several machines are required. In essence, the manufacturer is buying a guaranteed performance specification when he buys a preassembled positioning table.

Of the various types available, those using balls offer the advantage of self-cleaning—pushing dirt out of the way, rather than having to run over or crush it. This makes them less susceptible to deterioration from silica and alumina dust, and other contaminants typically found in many electronics and industrial manufacturing facilities. Accuracy and repeatability are typically very high in spherical element positioning tables. Tolerances of 0.0001 inch per inch of travel are typical accuracies that can be expected.

Positioning tables based on roller elements offer high load ratings as their strong suit. Much like a roller bearing, the rollers in the linear bearing offer line, rather than point contact, thus increasing their load rating. It should be noted, however, that roller-based linear bearings tend to require very precise assembly and the elements must be protected from contamination.

What About Specifications?

With the different types of X-Y tables thus placed in perspective, it is useful to list the key specifications that should be determined before purchasing a particular type. Let's take a look at these one by one.

■ Accuracy: This specification refers to the table's ability to travel in a straight line relative to some known reference plane. Accuracy is usually specified as deviation from perfect straight line travel, in inches per inch of travel.

The constrained nonrecirculating type of linear bearing offers the highest accuracy, measuring down to 0.0001 inch/inch travel. The recirculating ball-bushing type also offers high accuracy, but final system performance will depend largely on careful selection, installation and testing of the individual components. The chassis-slide type of bearing, basically a nonprecision type of mechanism, offers accuracy in the range of 0.01 inch/inch travel.

■ Repeatability: This specification refers to the positioning table's ability to return to the same place under repeated cycling. Constrained type bearings can typically offer a repeatability of less than 0.0001 inch. Once again, the repeatability of recirculating ball-bushing bearings depends largely on installation.

■ Lifetime: This specification refers to the cumulative number of linear inches of travel guaranteed by the manufacturer of the positioning table. Lifetime is correlated to load rating. If the positioning table is applied properly, its lifetime normally exceeds that of the machine in which it is placed.

■ Load rating: This specification refers to maximum permissible load of the positioning table, expressed in pounds. Load rating is usually not a major constraint on small format, nonmachine-tool table designs.

■ Friction: In general, friction is a direct function of linear bearing preload. Nonrecirculating bearings will offer the lowest friction, especially if it is designed with balls. Roller-based, nonrecirculating bearings offer somewhat higher friction because of their line contact. Ball bushings have low friction characteristics, but they are also very sensitive to preload adjustment.

■ Mass: This characteristic should be considered in servo designs, especially in high speed applications. Modern designs based on lightweight alloys minimize inertia; however, for maximum dimensional stability Meehanite castings should be considered.

A micrometer lead screw provides extreme accuracy for small single-axis inspection table. Unique marriage of micrometer and table helps insure accuracy and repeatability.

Drive Options

After selecting a positioning table, the manufacturer will have to consider the drive options. Positioning tables can be indexed using either open-loop or closed-loop control systems. Let's take a look at these two different philosophies.

Open-loop control uses stepping motors to position the table. Stepping motors respond to pulse commands from a controller. The term "open-loop" refers to the absence of a feedback sensor to tell the controller the exact position of the table. Instead, the stepping motor moves in small angular increments, each increment being a response to a pulse command from the controller. Commonly, the stepping motor turns 1.8 degrees with every pulse; thus, 200 pulses will rotate the motor a total of 360 degrees. By coupling the motor to the positioning table with a lead screw, the linear position of the table becomes a direct function of the angular position of the stepping motor.

For many applications, open-loop control is easily implemented and relatively inexpensive. Controllers tend to be simple electronic interfaces. Open-loop control has the added advantage of adapting well to microcomputer control. Computers work with a series of on/off commands, and these commands can be converted into position commands for the stepping motor with very little additional electronics.

Although open-loop control is inherently simple, linear speeds beyond about five inches per second are not feasible. The stepping motor's lower torque at high pulse rates can cause the motor to "miss" pulse commands. If the stepping motor misses a command, there is no way for it or the controller to know that this has occurred. Furthermore, the absolute positioning accuracy of an open-loop system is inherently less than that possible with a closed-loop system because of the absence of a position feedback sensor.

A closed-loop control system is based on the use of a servomotor and a position sensor. Servomotors offer precise, high-speed movement of lead screws on an X-Y positioning table. The positioning sensor can also cancel out tolerance errors and backlash. High performance systems and large positioning systems generally mandate the use of closed-loop.

This performance advantage comes at a price. System control electronics are inherently more complicated and expensive. Final choice between the two types of controls will depend largely on the performance requirements needed.

Make or Buy?

One of the biggest decisions facing the user of a positioning system is whether to buy a complete positioning machine from an OEM manufacturer or build a custom machine using in-house capability. In other words, how does a user resolve the make or buy decision?

The first question to answer is a simple one: Is in-house capability available? Designing a positioning system is not difficult, but it does require competent engineering and follow-up. Often, manufacturers have in-house engineering departments that are experienced in building production machines. In such cases, their expertise can be applied to the building of a positioning system. Manufacturers of positioning tables and control systems can help by supplying engineering departments with design assistance.

The question of proprietary information also enters the picture. The application of the positioning system may involve sensitive or classified manufacturing processes. In such cases, the manufacturer may not wish to involve outside suppliers. Furthermore, new or proprietary processes may not appear profitable enough to an equipment supplier. In such cases,, the manufacturer will have no choice except to build a positioning system in-house.

Costs can often be kept low by designing in-house as well. Slack engineering time that would otherwise be wasted can be applied to the design of a positioning system. The manufacturer need not pay the profits of an outside supplier.

On the other hand, the time factor may suggest buying from an outside source rather than spending several months designing and debugging an

Low mass aluminum X-Y viewing stage moves at speeds of 500 steps per second (120 miles per hour) in search of white blood cells in Corning Larc system. The table travels in a picket fence pattern of 300 microns width in 5-micron increments. Mechanical components of table—cross roller ways, harmonic drive stepping motors, ¼-inch ball screw—respond to computer command to provide 0.0001 inch positional accuracy.

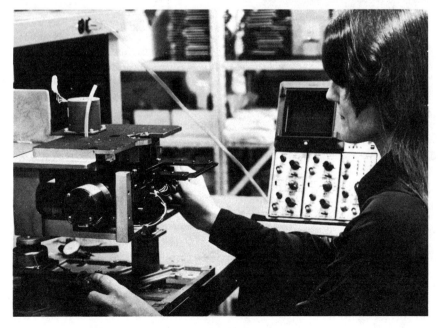

Almost faster than the eye can follow, Autosplice's numerically controlled machine inserts 9,000 to 12,000 terminals an hour into printed circuit boards. NC electronics command the insertion head and a positioning table to maintain extreme accuracy and repeatability.

Manual joystick-controlled X-Y positioning accuracy to within ± 0.0005 inch. The control, a completely solid-state, two-axis digital translator, responds to joystick maneuvering in a slew mode for coarse positioning, then a jog mode for precise location. The joystick-controlled X-Y table is used in electronic and optical assembly, instrumentation and other areas where ultra-close positioning must be accomplished manually.

in-house system. If many machines are required, it may tax the in-house capabilities of a user. In such cases, it may make more sense to buy from an outside supplier.

Whatever decision is reached, the end user should be well-informed about the various approaches to X-Y positioning. Alternatives are available; the goal is to refrain from buying more sophistication than needed, while still obtaining a positioning system that meets all requirements and specifications.

Applications

Perhaps the largest user of small positioning tables is the electronics industry. In the manufacturing of integrated circuits, positioning tables are used for scribing, dicing, and slicing of substrate material. In these applications, the positioning table moves a laser beam or a mechanical cutting tool across the wafer, or vice versa, in controlled patterns.

Quite often, after integrated circuits are cut and packaged, other positioning tables are used for microassembly, for hybrid circuit manufacturing, and for quality control checking. In these kinds of manufacturing operations, the positioning table moves the circuits in small increments, allowing examination under a microscope. The accurate and repeatable movement of the X-Y positioning table, coupled with its low friction, allows assembly and quality control operations that would have been impossible just a decade ago.

In other areas of electronics manufacturing, positioning tables are found in automated insertion operations. For example, several companies are now offering automatic terminal insertion machinery which will insert hundreds of terminals in printed circuit boards in a fraction of the time required with conventional methods. Application of positioning tables in automated components will grow at an increasing rate in the next few years. This growth will be accelerated by the availability of inexpensive, computer-controlled automation equipment.

In addition to the electronics manufacturing marketplace, X-Y positioning tables are making in-roads in material handling and assembly. An entirely new automation market is developing around numerically controlled handling equipment coupled to positioning tables.

For example, a typical application is inserting hundreds of small pharmaceutical vials in the small compartments of shipping containers. Eliminated are slow, tedious manual operations and frequent breakage due to human error.

Likewise, positioning tables are ideally suited for drilling and tapping operations in machine tool applications. They are not substitutes for the heavy duty tables found on large machine tools; instead, they provide a way to position small parts for light machining.

It's obvious that the above examples just scratch the surface on applications for X-Y positioning tables. Numerous other applications in metrology, optics, and even medical applications, among others, could be described.   MMS

Presented at SME's Ohio Valley Tool and Manufacturing Engineering Conference, October 1981

A New Approach to Positioning and Holding Workpieces

By Wendell F. Smith
Technology Research Associates

Castings and forgings often cause problems in the manufacturing process because their irregular shapes and surfaces may not be easily accommodated by conventional fixtures and clamps. Also, a workpiece of this sort is often distorted under clamping pressures and, following machining operations, returns to its original unstressed state when the clamps are released. The net effect is well known!

Niedecker GmbH of Frankfurt, Germany, has developed and employed a new tooling concept which largely eliminates this warping problem by providing "floating ball" supports at all points of contact with the workpiece. This paper describes the basic concept of the SAFE™ (Self Adapting Fixture Element) System and shows why its use is likely to provide significant cost savings.

INTRODUCTION

In most industries, engineering designs continue to impose difficult requirements upon manufacturing operations. In particular, castings and forgings are often very difficult to position and clamp without producing unwanted stresses. Also, it may not always be possible for the machine operator to visually detect the effect of such stresses. Figure 1 clearly demonstrates how conventional clamping can distort the workpiece. Here, the machined surface springs out of flat when the clamps are released. Costly re-work or rejection is the usual result in this situation.

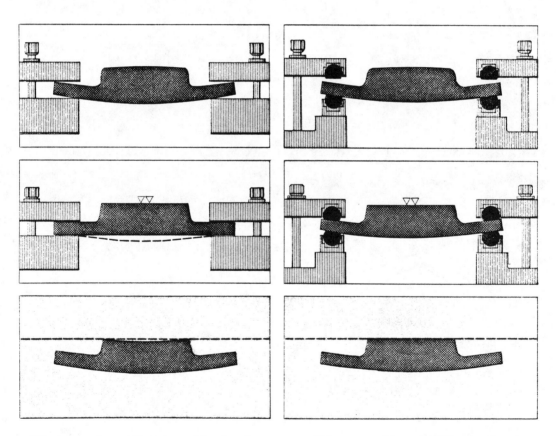

Figure 1. Common Method   Figure 2. SAFE Method

Figure 2 shows the SAFE™ (Self Adapting Fixture Element) System in use. The workpiece is supported by the flat contact areas of hardened steel balls. These balls are free to swivel within their sockets and automatically adapt to the warped under-surface of the workpiece. The upper arms of the clamps are also provided with matching balls and sockets. There is absolutely no distortion of the workpiece under heavy clamping pressure and the machined surface will remain true. Figure 3 shows a cut-away model of a production strap clamp in the SAFE product line.

Figure 3. Cut-away Model of A
Production Strap Clamp

SYSTEM DEVELOPMENT AND APPLICATION

The "floating ball" concept was developed in the company's manufacturing plants to eliminate the problem of workpiece distortion and also to produce significant savings in the overall cost of tooling and machining. The heart of the SAFE System, of course, is the movable ball which is incorporated in each one of a family of different fixture elements at every point in contact with the workpiece. Since these balls adapt to any rough or uneven surfaces, it is unnecessary to perform the usual preliminary machining at tooling points.

In normal practice, the workpiece is positioned in the first plane by three points. In the SAFE System, this plane is established by the three "flatted" balls incorporated in the three fixture elements selected. By adding precision locating pins (Figure 4), the capabilities of the system were extended to locating and holding the workpiece in the second and third planes as well. Longitudinal and lateral "float" is built into the elements of the SAFE System to compensate for the usual small variations in the hole pattern of the workpiece. Depending on the application, the workpiece may be held by strap clamps or hold-down screws.

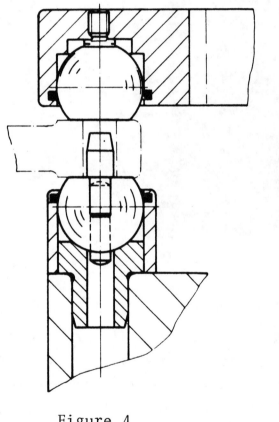

Figure 4

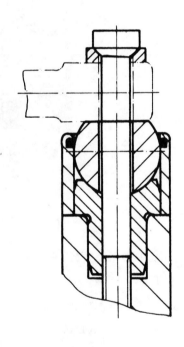

Figure 5

As shown in Figure 5, a hold-down screw can also serve as a locating pin because of its precisely ground body. As in the previous example, the spherically ground head and matching washer adapt to the workpiece automatically. Note that the ball beneath the workpiece is also free to move "around" the screw.

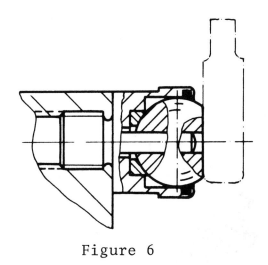

Figure 6

End stops (Figure 6) incorporating the floating ball concept may also be used to position a workpiece in situations where the standard fixture elements may be difficult to apply. Also, should a workpiece require more than three support points, because of its own lack of stiffness or because of heavy cutting forces, vertically adjustable fixture elements may be placed under the workpiece as required and locked solidly in place.

A conventional fixture is usually designed and fabricated to permit machining of a particular workpiece and considerable cost may be involved. Contributing to this cost are the efforts of design engineers and toolmakers along with overhead expenses associated with storage and handling. Such a fixture may require many weeks to design, fabricate and check out before it can be employed with confidence in machining operations.

Using the SAFE System usually provides substantial savings in cost and time and often results in enhanced machining quality. In the case of 25 different workpieces being produced by the company, substitution of the SAFE System for conventional fixtures led to savings of 77 percent. The specially designed fixture elements are truly available on an "off-the-shelf" basis. They may be mounted directly on the machine table using T-nuts or on various intermediate base plates providing precisely located hole patterns which exactly match the hole patterns of the fixture elements themselves. Master grid plates, in various sizes, incorporate a 50mm grid pattern and each hole contains a bushing for locating purposes and threaded inserts for holding down the fixture elements (Figure 7).

Figure 7. Typical SAFE Set-up With Grid Plate, Riser Blocks
and Standard Fixture Elements

If the standard 50mm grid pattern is not compatible with
the workpiece, simple base plates or angle plates may easily
be drilled through a template and threaded to permit attachment of the fixture elements. Differential elevations can be
handled with standard riser blocks (Figure 7). After the fixture elements have served their purpose and machining has been
completed, the elements may be removed from the base plate for
use on another job. Setting up for the new workpiece is as
simple and inexpensive as previously.

Of special value in NC machining, a master reference cube
may be located on the grid plate or base plate in very precise

relationship to the workpiece and is used as the zero point for programming machine operations.

The availability of the SAFE fixture elements, along with grid plates and other accessories, can produce real time savings in the design and assembly of a particular fixture. To assist the tool designer even more, overlays of each fixture element in three views are available in both full and half scale for direct application to the drawing.

CONCLUSIONS

The SAFE System provides a complete and practical solution to work-holding problems involving castings and forgings for prototype, short run and medium production operations. The major advantages are:

1. Distortion-free clamping
2. Machined reference surfaces unnecessary
3. Very fast set-up of fixture elements
4. Fixture elements usable on several fixture set-ups
5. Minimum cost for fixture design and assembly
6. System very compatible with CAD/CAM
7. Improved machining quality
8. Professional support available from factory

Reprinted from Power Transmission Design, November 1980. Copyright Penton/IPC, 1980

Ten nifty devices for converting rotary motion into linear motion

PETER KATZ, P.E.
Consulting Engineer
Nashville, Tenn.

Transforming rotary motion into linear actuation is, for some, more art than science. Here are ten different devices – with their advantages, limitations, and typical applications – that will help you evaluate and select the right device for your job.

EVALUATING and selecting alternatives for converting rotary motion into linear motion can be confusing without guidelines. Key criteria you will want to apply are:
- The output force you need,
- The length of travel required, and

Fine Screw Motion

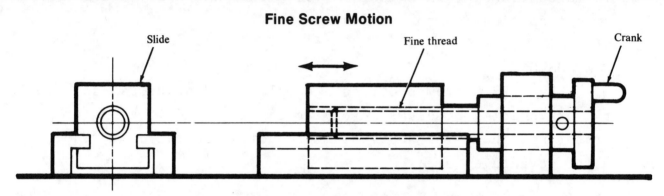

Slide · Fine thread · Crank

The crank-driven fine thread produces a uniform linear motion, allowing exceptional control — measurable in small graduations of the crank wheel. It is the most common type of fine adjustment, and is used for machine tables, instrument slides, and electronic devices.

Select the screw on the basis of required force, stroke length, and degree of adjustment. A standard fine thread may be adequate. If used primarily for fine adjustment, the thread pitch should correspond to the dial markings (*e.g.,* 1 mm). If screw is motor- rather than crank-driven, carriage motion must be controlled with limit switches, revolution counters, or similar means. If the carriage or slide must be firmly fixed after adjustment, use an external clamp on the slide.

Advantages	Limitations:
• Simplicity	• Limited travel
• Accuracy	• Travel requires many turns
• Ease of calibration	• Requires a precision thread
• Low cost	• May require clamping after adjustment

Coarse & Fine Screw Motion

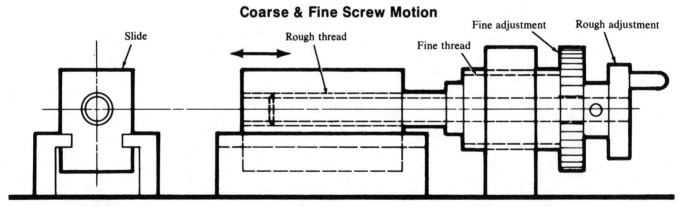

Slide · Rough thread · Fine thread · Fine adjustment · Rough adjustment

This variation of the first device combines the fine adjustment capability of the fine thread with the faster motion of a coarse thread. Notes on engineering and applications prallel those for the previous example.

Advantages:	Limitations:
• Faster travel than with fine screw alone	• Limited travel
• Good accuracy	• Requires clamping after adjustment
• Simple to produce	• Precision threading required

- The cycle-to-cycle repeatability.

To help you apply these guidelines to your particular problems, we use ten mechanisms as examples — each showing a different approach. Dimensional estimates of maximum force, travel capabilities, and repeatability are given for each mechanism in the table. In the following detailed discussion of each mechanism you will find their advantages, limitations, and typical applications.

Key characteristics			
Device	Maximum practical output force	Maximum stroke length	Travel repeatability from cycle-to-cycle
Screw			
Fine motion	100 lb	1 in.	0.001 in.
Coarse & fine motion	100 lb	6 in.	0.001 in
Differential adjustment	100 lb	1/2 in.	0.001 in.
Hydraulic	100 to 150 lb	1/4 in.	1/32 in.
motion transfer		(Diaphragm)	
Friction roller	3 to 4 oz	*	1/2% of stroke
Rack and pinion	Up to 100 lb*	*	About 1% of
with worm gear			crank turn **
Cam and gear	10 lb	2 to 3 in.	2% of stroke
Toggle	Up to tons*	Up to 3 in.*	1/32 in.
Crank and slide	Up to tons*	18 in.	0.1% of total travel
Wedge	50 lb	1/2 in.	0.001 in.
Bent spring	10 lb	1/4 in.	0.01 in.
Drum and cable	Up to tons*	*	5%, with
or endless belt			moderate travel (drum dia. changes)

* Depends on size or length   ** Depends on play in gears

Differential Screw Adjustment

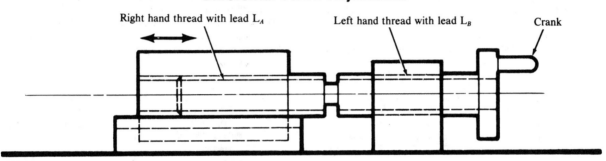

Right hand thread with lead L_A Left hand thread with lead L_B Crank

This second variation on screw motion and adjustment uses two slightly different threads to give a very fine linear motion. One thread is right handed with a lead of L_A; the other is left handed with a lead of L_B, where $L_B = L_A \pm 10\%$. One turn of the adjustment wheel will advance or retract the sliding block by $L_A - L_B$, or 10% of the lead. This design has wide application in appliances and light machinery; it is also excellent for instrument slides.

Advantages:	Limitations
• Low cost	• Limited travel
• Extremely sensitive	• Requires many turns
• Standard threads can be used	• Limited to small devices

Hydraulic Motion Transfer

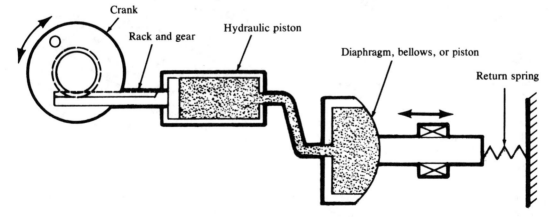

Crank   Rack and gear   Hydraulic piston   Diaphragm, bellows, or piston   Return spring

This device is used in machine tools and farm machinery when it is necessary to remotely actuate the small movement of a heavy machine part (for example, when there is an obstruction between the powered shaft and the driven piece). The basic principle is that the product of the piston area and stroke in the first piston must equal the same product in the second piston. Thus, a fairly long stroke of the gear-driven small piston will produce a small linear movement (but large force) in the large diaphragm. Maximum precision (repeatability) is obtained with a diaphragm because, unlike a piston, it does not leak. A bellows would be as tight as a diaphragm, but would enable a somewhat longer stroke.

Advantages:	Limitations:
• Permits linear motion around an obstruction	• Expensive
	• Space consuming
	• Difficult to calibrate

converting rotary motion into linear motion

Friction Roller

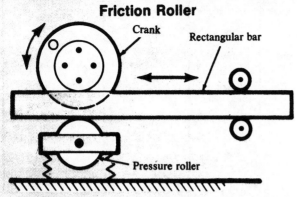

This roller device can be used to obtain long linear motion from a crank without using gears. It is useful for moving long parts, such as scales, on machine tools and business machines — or wherever long, quick travel with good control is required.

For maximum accuracy, the drive roller must be very accurately machined, hardened, and precisely mounted; also, the spring force must be high. Inaccuracies in machining the drive roller will lead to cumulative errors in rack travel.

Advantages:	Limitations:
• Long stroke • Self-adjusting for wear • Durable • Noiseless; good appearance	• Low force capability • Requires very accurate machining and hardening of drive roller

Rack & Pinion with Worm Gear

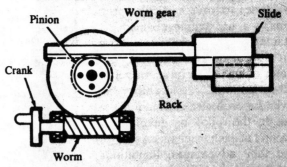

This device combines appreciable force with fine control. Since the gear rack can be long, it is often used for long strokes with steady force and velocity, and for powered motion in both directions. Typical applications include opening and closing valves, gates, and other long-stroke mechanisms. It is also used for moving machine parts along their beds or ways.

The worm drive can be motorized, in which case a fine-pitch worm and motion sensing should be used. Either fine or coarse adjustment can be obtained, depending on the force required.

Advantages:	Limitations:
• Powerful and rugged • Long stroke • Powered in both directions • Good control	• Fairly expensive • Space consuming

Cam and Gear

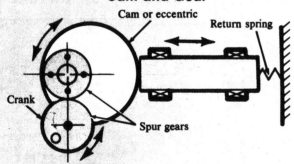

This mechanism is widely used in appliances and business machines, and in clamping and locking mechanisms.

Using a cam allows you to optimize the variations of force or velocity over the cycle time. In this application, an eccentric wheel is only a cheap cam. The return spring can be eliminated by pinning the slide to the cam (the pin would slide within a groove at a constant distance from the cam's periphery) but this arrangement would have more play than spring-loading.

Advantages:	Limitations:
• Low cost • Moderate force • Movement can be optimized by cam shape	• Slide movement limited to cam throw • Pushing power depends on cam shape • Difficult to calibrate slide motion in terms of crank rotation • If used for adjustment, requires locking

Toggle

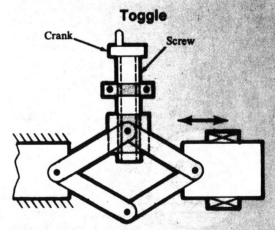

The toggle is used primarily where high pressure is needed, as in molding presses and in clamping. It is ideal in these applications because the maximum force occurs when the slide is at its maximum extension. It is also self-locking. For quicker cycling, a cylinder may be substituted for the screw mechanism.

Advantages:	Limitations:
• High pressure • Self-locking	• Pressure varies linearly with toggle extension • High pressure available only at a small portion of the total travel • Elaborate

Crank & Slide

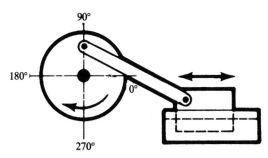

This centuries-old device is used for converting rotary motion into reciprocating linear motion. Representative applications include engines and locomotives, office machinery, vending machines, appliances, pumps, etc. It is probably the most widely used device for converting rotary to linear motion (and vice versa).

A simple crank-arm may be substituted for the crank-wheel shown. If the connecting rod is at a large angle relative to the sliding block — as when the block would be close to the wheel — the vertical force component may cause binding. Binding can be alleviated by lubrication, lengthening the connecting rod (reducing the angle), and by substituing linear bearings for plane bearings.

Advantages:	Limitations:
• Strong • Easy to produce • Crank easily driven with gears • Versatile	• Motion is sinusoidal; hard to calibrate end-motion into graduations • Power varies from a maximum at 0° and 180° to a minimum at 90° and 270° • Reciprocating slide may bind.

Bent Spring

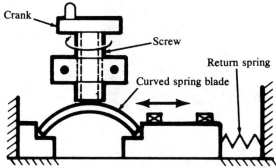

This is a crude but inexpensive device to obtain a small linear movement requiring no lock. It is used on high-volume products, such as appliances, where small adjustments are needed. The spring need only overcome sliding friction. Since the slide is spring-loaded in both directions, the opportunity for damage is minimized. Spring sizing is done experimentally.

Advantages:	Limitations:
• Very low cost • Self-locking • Minimizes accidental damage	• Difficult to calibrate • Small travel

Wedge

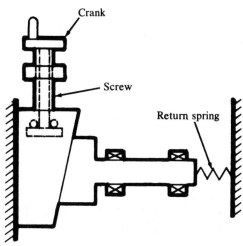

The wedge is a low-cost device for obtaining fine adjustment in short-distance (especially parallel) linear motion. It is often used to take up play in slides or bearings, or as a low-cost clamping device. The wedge angle is always small — typically ranging from about 5° down to as little as 0.04°. Movement from crank turns is readily displayed on a concentric dial or digital readout.

Advantages:	Limitations:
• Low cost • High pressure • Good accuracy with no play • Requires little space	• Very small travel • Requires accurate machining

Drum and Cable or Endless Belt

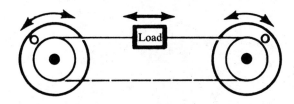

This device, a cable or belt wound on a drum, has both light and heavy-duty applications. In business machines, it is used for ribbon or film transport. Heavy-duty applications include winches, elevators, cranes, conveyor belts, etc. Except for moderate travel, calibrating cable/belt travel with crank turns is difficult because the effective drum diameter is not constant.

Advantages:	Limitations:
• Simple and versatile • Small space requirement • Long travel	• Belt/cable can transmit force only when in tension • Crank calibration is difficult

Reprinted from Manufacturing Engineering, March 1978

Universal Rotary Indexer

*. . . it makes predetermined angular
moves at the push of a button*

Form grinding steel rollers that are 12″ (304.8 mm) in diameter, 87″ (2209.8 mm) long, and weigh 4000 lb (1800 kg) presents a real challenge in terms of holding and positioning. The rollers in question are used to form the convoluted medium that is glued between two sheets of liner to form boxboard. The hardened surface of these rollers is form ground in large surface grinders to a fluted configuration resembling gear teeth.

The size and weight of the corrugating rollers — some go as high as 6000 lb (2700 kg) — complicate the holding problem. Likewise, the flutes are form ground one at a time, so the roller must be precision indexed to grind each flute. For Corrugating Roll Corp. (CRC), New Hudson, MI, these problems prompted a study and design program that led to the development of an automatic angular positioner with the potential for many applications beyond the one involved here.

As a supplier to the boxboard industry, CRC takes old worn-out corrugating rollers and rebuilds them to a like-new condition. Flute grinding is done on large surface grinding machines. In the old method, the roller was held between centers and coupled to an index head by means of a "dog." The difficulties encountered in this approach were numerous.

Indexing Requirements. In tooling up

PRECISION AUTOMATIC INDEXER handles all index counts electronically without need for index plates. A chuck, face plate, or other workholding device can be mounted on the spindle.

a new flute grinding machine, the company required a heavy duty precision index head capable of covering a range of 60 to 250 divisions — a capacity beyond that of available pin/plate indexing devices. Also, the mass of the large rollers dictated that the indexing action be smooth and without shock.

CRC's existing tooling consisted of older indexing heads utilizing a pin/plate system. Excessive shock in this setup resulted in frequent index pin and index plate failures. In addition, the wide variety of different corrugating rollers handled by the company required a costly inventory of index plates, since each plate was only usable for a specific number of divisions. A further problem was that each index change required that the operator change the plate, which took time and provided opportunity for error.

Based on the preceding situation, it was established that any new indexing device should: (1) avoid the use of mechanical index plates; (2) index with smooth acceleration and deceleration; (3) operate with minimum wear to maintain indexing accuracy; and (4) be failsafe in terms of providing an appropriate error signal in the event of an incorrect index.

New Design. The payoff of CRC's two-year development program — the URAAP™-Universal Rotary Automatic Angular Positioner — is shown in the photo as set up for a typical roller grinding application. Essentially, the system consists of a stepping motor, a heavy duty precision gearbox, and an electronic controller. All indexing requirements are handled electronically rather than by changing index plates. A feedback system and cross-checking electronics verify the correctness of all moves.

For manual rotary positioning, the versatile URAAP automatically rotates the work a given number of degrees, minutes, and seconds according to commands entered on the keyboard. For indexing, the device accepts keyboard entries controlling moves of 2 to 500 divisions. Moves can be made in either direction, although for maximum precision it is recommended that the destination be approached from the same direction.

Absolute positioning is within ±0.0015°, and repetitive positioning is within ±0.001°. The time required to index one step is approximately three seconds. Maximum output torque of the unit is 5000 lb/in. (565 N•m), maximum rate of operation is 30 rpm, and the radial and thrust load capacity of the spindle is 4000 lb (17 792 N).

The configuration of the URAAP package is based on the requirements of the grinders used at CRC. Specifically, it is necessary to maintain a centerline height of 4″ (101.6 mm) and a maximum spindle diameter of 8″ (203.2 mm). This results in a unit that is close-coupled to the machine table for maximum rigidity. The use of a horizontal spindle with the gear box located off the table at one end removes the precision indexing mechanism from the wet and abrasive environment near the grinding wheel.

Tramming Simplified. In a mechanical positioning system, the coupling between the part and the indexing head incorporates a "tramming," or phase adjustment. This permits the part to be adjusted rotationally relative to the index plate. This is difficult to do mechanically where large, heavy parts are involved. Small adjustments are hard to make due to the inherent windup in the system.

Phase adjustments are now done electronically with the URAAP. Once the index is set up, and the operator determines that he wants to move the part relative to the indexer, it's a simple matter of dialing in the change on a thumbwheel switch. At the next index, or on a separate command basis, the rotational offset is taken into account. Also, since the offset has considerable capacity, it's possible to accomplish a multiple of any index by making the appropriate intermediate angular move.

The basic simplicity and versatility of the URAAP suggest that it can be used in many applications requiring rotary motion control, such as in gear and spline manufacturing. Used on a machine with an NC system, it could provide an additional axis of motion. It also serves as a memory package to program repetitive moves of any angle. The flute grinding application described here uses a horizontal spindle, but the basic principles can be used to provide similar performance in alternate configurations such as a rotary table. ∎

CHAPTER FOUR
PARTS FEEDING
AND HANDLING

Feeders Solve A Lot Of Problems

Among the unexplainable in manufacturing is the expensive automatic machine that performs erratically—or not at all—because it has been fed a dirt-cheap part of poor quality. A simple qualifying check of parts in feeders can keep those machines producing.

Rules for applying parts feeders bring to mind the old Indian's recipe for grizzly bear stew. "The first step," he would advise, "is to capture a grizzly bear." Those most familiar with parts feeding have equally intriguing advice for the uninitiated.

Experts stress that successful applications depend upon a supply of orientable parts that consistently meet design and quality specifications. Other requirements include the selection of the right type of feeder among the dozen or so types available, the creation of soundly-engineered tooling and mechanisms to integrate the feeder into some higher-order machine system, and a motivated operator cum mechanic to keep the equipment running.

Some engineers in industry are puzzled by two of the requirements. What has a motivated operator to do with automatic feeding? And, surely, parts to be fed are made to a proper engineering drawing and then inspected somewhere in the system—why the emphasis on consistent part quality?

The answer to these questions is to be found in numbers. Feeders are associated essentially with large-quantity, high-production rate processes like machining, forming, assembly, testing, and packaging items for the consumer market. The products are of a relatively stable design over very long periods or some predictable life cycle of a year or so. Production runs range from hundreds of thousands to millions. Hourly production rates are measured in the thousands, machine cycles in seconds.

In such an environment, minutes of downtime translate into considerable dollars worth of lost production. And any loss stands out in sharp relief when traced to a malfunction caused by something like a small burr on some inexpensive component. Economics dictate the use of an operator or attendant of some kind to seek out and correct the trouble when a machine goes down or a malfunction is detected. But even the best efforts of a highly motivated, trained operator can be defeated by a batch of poor quality parts. In short, automatic parts feeding and

associated processes are more than just another engineering problem. Both people and parts play a critical role in successful applications.

In explaining just how critical that role is, Warren Burgess, president, Burgess & Associates Inc., refers to a study of an assembly machine made years ago in his company. And, he notes, "the lessons are as timely today as then."

Parts to be handled in the assembly machine were defined by drawings giving dimensions and tolerances. The basic engineering of the machine was based on those drawings. However, it became apparent that all parts furnished by the buyer for use in the final development stage did not meet specifications. And since final development becomes a can of worms if you can't distinguish between trouble caused by mechanism designs and that caused by bad parts, a time-consuming 100% inspection of parts was necessary to get the large quantity needed for final engineering.

When the machine was completed, acceptance tests using the fully inspected parts achieved a

185

production rate 7% above specifications. When the same acceptance tests were run using "as received" parts, the production dropped to 50% to 60% of the specified rate.

Similar results should be expected in any relatively simple machine where there has been no focused effort to obtain consistently high-quality parts. Considerably worse results should be expected in complex systems involving large numbers of feeders.

Burgess relates that stoppages were caused by "... screws without slots, without heads, without threads, with partial shanks, with deformed heads, even screws of a different size ... brackets deformed and out of tolerance ... scraps of metal, pieces of anode from a plating process, chips of granite from a tumbling process, and a variety of other parts apparently mixed in from tumbling, pickling, and plating processes."

After what Burgess calls "production development" (high-level talent studies a new machine's operation using production parts and makes modifications to enhance performance) this particular machine was able to produce 120% more than specified. This phase of development included modifications allowing for rapid clearing of jams, automatic ejection of various tramp material and out-of-tolerance parts, rework of alignment mechanisms, increasing the machine rate, use of signal lights to indicate malfunctions—and working with parts suppliers to improve the quality of their shipments.

An interesting conclusion of the study was that a qualified production development team could probably contribute to the increase in machine productivity over the life of the machine but its major contributions will probably occur during the manufacture of the first 400,000 assemblies. When Burgess says "qualified", he is not referring to the typical shop maintenance types of skills. He means the type of talent that understands, for example, where a track might be adjusted to ensure that a given type jam always occurs and then modified to provide for quick access.

Unknown quality

There's an important, but generally overlooked, message embedded in equipment manufacturers' emphasis that efficient performance depends upon consistently good parts. It appears that, in general, using companies simply are unaware of the true quality conditions of most of the smaller, hjgh-volume components used in their products.

In short, whatever is now done in the belief that it leads to a consistent quality level—tolerances and notes on part drawings, purchase specifications, sample inspections, whatever—simply is not getting the job done. Every feeder and assembly machine manufacturer can recount scores of examples in which users were

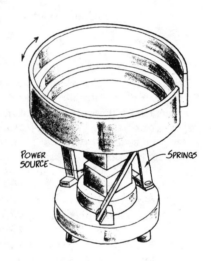

Vibratory bowl feeders are mounted on springs set at an angle. The bowl is vibrated by a physical force, typically from an electromagnet or a pneumatic cylinder. In effect, bowl surface jerks back from under parts at each vibration and they fall back on the surface at a new point; with repetitive vibrations, parts will move. Circular feeders are available in cylindrical form with inside fabricated tracks; spiral form with inside cast tracks; columnar form with outside tracks; or some combination of the three forms. Linear vibratory feeders are also available.

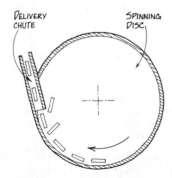

Centrifugal hopper throws parts to periphery of disc where they contact discharge point designed to accept items of proper shape or in desired orientation. Suitable for fragile parts which can't be tumbled.

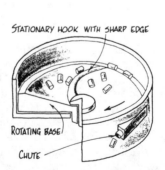

Stationary hook hopper separates mass of parts and orients and guides the parts upwards and outwards to edge of rotating base for feeding into a discharge chute. Suitable for delicate, simple forms.

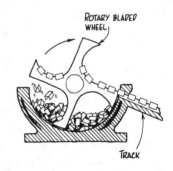

Rotary centerboard hopper has bladed wheel which picks up parts as it rotates through a mass of parts. Suitable for orienting channel-shaped parts, like stampings. The blade rotates only fast enough to supply parts.

not simply surprised but incredulous when told that significant numbers of their components did not meet their own specifications. "But our workers have been turning out assemblies for 10 years using those parts and the percentage of rejected parts hasn't been anywhere near what you say!"

That type of remark indicates that a manual assembly operation has never been critically analyzed. If so, it would have been obvious, for example, either that specified part tolerances were insignificant to the assembly, or that operators were selectively fitting components, or some unusual prying and hammering were used in assembling, or, perhaps, some combination of the three. In short, human "feeders" are extremely flexible. Automatic feeders are inflexible, being built to handle some stated fixed configuration and characteristics.

In various cases, close study will show that the fixed specifications are in fact spurious and that a feeder can be modified to accept what was thought to be "out of tolerance" parts. Where specifications are critical, component manufacturing must be revised to ensure that only parts consistently meeting specifications are used.

Automatic inspection is one answer. Equipment can inspect parts for characteristics and features such as color, weight, diameter, depth, location, length, taper, out-of-round, concentricity, center distance, straightness, thickness, alignment, hardness, squareness, flatness . . . to name only the obvious. Considering the general run of parts usually handled, more often the solution is a qualifying operation rather than a rigorous inspection—parts are either suitable or they are rejected. These qualifying systems can be simple or complex, depending upon the characteristics being examined.

Feeder misuse

Some users experiencing trouble with high speed assembly equipment have come to realize that their performance problems

can be traced to the fact they're using the part feeders on the assembly machine in a dual mode. One use is to feed and orient parts. But part jamups can be interpreted as the feeder "discovering" bad parts—an on-line qualifying process that separates the bad from good parts. Where frequent interruptions of a process are caused by poor feeder performance, an elegant solution is to separate the two modes by using two similarly tooled feeders. One can qualify parts off-line where it can be easier to attend its operation. Parts successfully fed by the first feeder become the input to the on-line feeder.

Kahle Engineering Co.'s vice president of sales and engineering, Carl A. Napor, notes that the twin-feeder approach is particularly useful for problems involving inexpensive, small, complex forms—like stampings. Such parts are almost impossible or extremely costly to qualify by a special machine. The twin-feeder technique won't solve all operating problems but results can be dramatic. Napor cites a case in which the use of a duplicate feeder

resulted in a machine running for upwards of an hour between shutdowns compared to an earlier history of only 3 to 5 minutes between shutdowns. The resultant productivity improvement gave the user a distinct advantage in a highly competitive market.

Vibromatic Co. Inc's sales manager, Neal Graham, agrees that the twin-feeder technique can, indeed, be an effective solution to keeping certain expensive, complex machines up and running. He says, "When you see a money-wise company come back and buy a whole bank of 15 feeders that duplicate 15 on a machine, you know that there are applications where it pays off."

Generalizing from applications seen in industry, Graham notes that twin feeders are most useful when the basic feeding and orienting tooling does just about everything that needs to be done, or reasonably can be done, to also qualify parts. But when additional or different characteristics than those required for feeding need to be checked, the off line feeder should be designed as a special parts qualifier. An example would

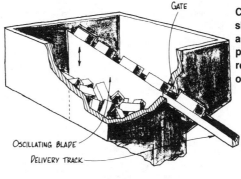

Centerboard hopper has blade with shaped top that picks up some parts as it is oscillated through a mass of parts. Improperly oriented parts are rejected by blade shape and design of gate leading to delivery track.

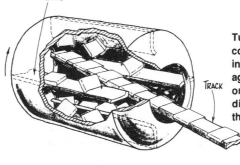

Tumbling barrel hopper handles complex forms like stampings. Vanes in rotating barrel continuously agitate parts. Those with proper orientation are caught on shaped discharge track, others fall back into the barrel.

be a case in which an out-of-tolerance condition did not affect feeding or orienting but was critical down line from the feeder in some later process.

Graham's point is that the twin feeder technique is useful but, he asks, why simply build a duplicate when you can go further and include other useful needed functions? And particularly so since typically you can do more things with an off-line feeder than it would be prudent to try with an on-line unit because jamming or poor feed rates can't be tolerated.

The off line feeder can run at some speed optimum for qualifying—and it can be run for a different number of shifts independent of, say, an assembly machine. Better, it might be located at the point a part is manufactured.

For many characteristics that need to be checked, tooling up a feeder as a qualifier is little different than tooling it up to orient parts. The designer uses his same bag of tricks to make use of differences in profiles, weight, and other physical features of parts. Examples of orienting-like tech-

niques useful in qualifying parts are shown in rough sketch form. Since the intent is to separate parts according to some feature, the techniques imply the use of some tracks and tooling outside a feeder bowl to allow take-away chutes to be used.

Many approaches

Despite the emphasis on vibratory bowl feeders, it should be understood that other types of feeders also lend themselves to qualifying operations. For example, elevating feeders can lift parts to a discharge chute where gravity takes over to move the parts through a series of go/reject stations. Various techniques used to orient and separate parts moving along vibratory feeder tracks can also be used on the feed tracks emerging from any type feeder. All types of feeders can discharge parts onto a pair of rotating rolls adjusted some precise angular distance apart; as parts move along the rolls, they drop through at the point the distance between rolls just exceeds their diameter. Another often used technique is to

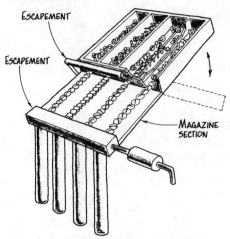

Oscillating box is useful for headed items, can be arranged to handle different sizes of parts in same hopper. Parts are tumbled as box oscillates and properly oriented parts pass through a selection gate onto holding tracks leading to escapement.

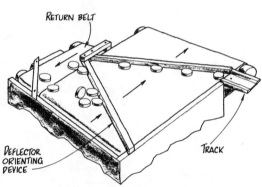

Horizontal belt feeder is designed to handle delicate parts when chipping, scratching, or nicking is to be avoided. Initial deflector passes correctly oriented parts which are then directed to a discharge track by a second deflector.

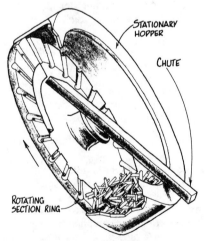

Rotating disc hopper has profiles machined or cast into face of a rotating ring. Parts picked up by the profiles are guided by a cam into a discharge chute as the ring rotates. Suitable for headed parts and cylindrical parts requiring no end selection.

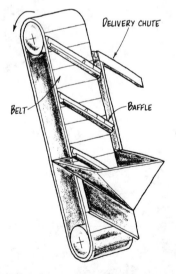

Elevator hopper handles large parts or large quantities of small parts. Flights on moving belt pick up some parts from supply hopper. Parts discharge into a delivery chute where orientation and selection can occur.

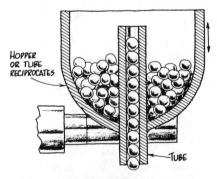

Reciprocating tube hopper picks up parts as either the hopper or tube reciprocates to present the mass of parts to the tube opening. Design of opening is such that parts straddling it are wiped away by reciprocating motion.

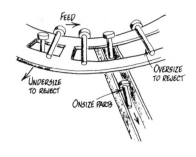

Oriented parts hanging by the head feed along the vibratory bowl's outer support track toward an undersize qualifying window cut into the track. If the length is undersize, it is not supported and falls through the window to a reject chute. Onsize and oversize parts are carried across the undersize window toward an onsize window. Those onsize lose support and fall through to an acceptable parts chute. Oversize parts continue to travel to a reject chute.

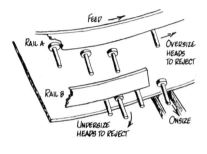

Flow of headed parts from feeder bowl travel with heads restrained by Rail A. At selection point along Rail A, both undersize and onsize headed parts pass under Rail A; oversize head diameters continue on Rail A to a reject chute. Undersize diameters pass under Rail B to a reject chute. Onsize-diameter parts continue on Rail B to an acceptable parts chute. This head gaging system can be combined with a shank length qualifying system.

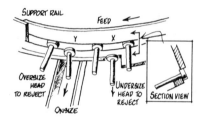

Parts hanging by head enter a selection station where support of shank ends has been removed. Parts are prevented from falling by a rail positioned parallel to and outside of the head-supporting track that contacts the side of the head of each part. This rail progressively rises with respect to the plane of the head-supporting track. Undersize heads fall out at Point X, onsize heads at Point Y. Parts with oversize heads travel to a reject chute. This selection technique can be used following head diameter selection.

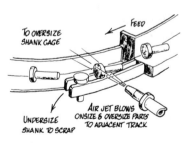

Parts oriented end-to-end feed onto parallel rails. The shank falls between the rails if it is undersize and the parts are rejected. Parts with both onsize and oversize shanks feed along the top of the rails to an air jet and are swept onto a track leading to a second set of parallel rails. At the second selection station, onsize shanks fall between the rails and the parts are directed to an acceptable parts chute. Oversize parts are swept off onto a track leading to a reject chute. Technique can check for rolled threads when threaded parts have larger shank diameter than blank part.

move parts across a screen or series of screens to remove debris, separate partial from whole parts, or segregate items into size classifications.

The strategy for classification by sizes is almost always the same: eliminate undersize items, including debris, from the flow then separate a desired size from oversize parts. However, Richard Straight, sales manager, Campbell Machines Co., stresses that users should not confuse qualification or sorting by feeder with precise part gaging. Qualification tolerances are relatively coarse compared to those obtainable in true gaging operations. Indeed, any gaging and inspection machine which is automatically fed should include a feeder tooled to qualify parts in order to eliminate downtime caused by gross oversize and deformed parts from jamming a close-tolerance gage.

The current association of automatic parts feeding only with large production quantities of a given design could change in the not too distant future. A significant clue is in some of the advanced manufacturing engineering occurring in Japan. Confronted with a number of conflicting needs—to improve productivity, react to uncertain markets, utilize facilities around the clock, maintain jobs, and satisfy workers with no burning desire to work second and third shifts—forward-looking manufacturing strategists envision a factory built around cells of integrated automatic production equipment that are set up during the day shift to run automatically with minimum human attention during the second and third shifts. That future factory will obviously depend upon flexible, trouble-free, automatic parts feeding which, in turn, requires a supply of orientable parts that consistently meet design and quality specifications. And in the sense that companies now strive for products and parts that are quick and easy to manufacture, the future will see more designs that are quick and easy to assemble as the real key to significant productivity gains.

High metal removal rates that cut the time a workpiece is on a machine lose their effectiveness unless there is a corresponding improvement in techniques used for loading and unloading the workpiece. Robotic devices will eventually reduce non-cutting time to a minimum, but meanwhile a crop of machine tool feeding devices and systems has emerged. These range from systems which are integral with the machine to stand-alone handling aids which can be wheeled into position whenever circumstances require.

Undoubtedly, the main trend in Britain is for integral workpiece handling systems which are being increasingly incorporated, in one form or another, in all types of machines. Such systems are now available not only on high-output, mass-production equipment—where even a small saving on each component will give spectacular results—but on heavy-duty machines used for removing large amounts of metal from very bulky workpieces in batch quantities, or even on workpieces run one at a time.

Typical developments

These trends can best be illustrated by recent machine designs. A typical development is the latest horizontal-table-type borer introduced by the Kearns-Richards Div, Staveley Machine Tools, Ltd, Birmingham, England.

This machine, the NC125 5-axis computer numerically controlled (CNC) borer (**Figure 1**) uses a pallet loading system particularly suitable for heavy machining operations. Equipped with a table 59″ square, the machine is capable of accepting a dynamic loading of 17,640 lb. A pallet system has been devised that will accept workpieces weighing up to 13,230 lb.

Pallets are located on the machine table with an accuracy of 0.0008″ by means of a tenon key and plunger. When machining is completed and the plunger has been automatically released, the pallet supporting the completed component is withdrawn under hydraulic power onto a carousel-type pallet changer.

By indexing the pallet changer through 180 degrees, a second pallet, supporting a fresh workpiece, is moved into position. This is then transferred hydraulically to the machine table and locked into position ready for the next machining cycle.

The first pallet can then be unloaded and immediately reloaded with another component while machining of the workpiece just transferred to the machine table continues uninterrupted. Changeover time is about one minute from cut to cut.

An example of pallet handling development comes from Kearney & Trecker Marwin, Ltd, Brighton, England, whose KTM 560 CNC machining center is available in three versions, each a development from a previous model.

WORKPIECE FEEDING SYSTEMS BOOST MACHINING OUTPUT

by Eric Ford, BSc (Econ), BComm
British Technical Writer

Reprinted from Tooling & Production, September 1980

First to be introduced was the standard model. This has a 40-tool magazine and a twin linear pallet shuttle system. It was followed by a 40-tool rotary pallet machine.

The latest version, an 80-tool rotary pallet type (**Figure 2**) has a 40-tool magazine located on both sides of the column. Each magazine has an associated tool transfer arm which moves the next tool required from the drum to the change position on the centerline of the column.

A rotary pallet system, as fitted to the most recent machines, is beneficial where there is a high ratio of loading to cutting time and a multi-pallet system is needed. Four pallets can be fixed to a rotating mechanism at the end of the Z bed and fed to the machine in sequence.

Many tools needed

When a different component is being produced on each of the four stations of the rotary pallet, particularly when these are complex items, a large number of tools is required—and this is the thinking behind the 80-tool machine.

Kearney & Trecker Marwin's market research showed that an unexpectedly large proportion of engineering products require four major components to make up one set of parts, needing more than 40 tools. An example is a typical electric motor, which has a body, two end covers and a connector cover box. All these require various combinations of milling, drilling, tapping and boring. The research also indicated that 80 tools would be sufficient to enable one set of components to be machined for each complete cycle of the machine.

Graham Felstead, the firm's technical director, says, ''This machine epitomizes our approach towards the ultimate goal of

1. To date, this system by Kearns-Richards Div, Staveley Machine Tools, has been applied to feeding a single machine, but it is envisioned that one pallet changer of this type could be located between two NC125 borers to serve them both in turn. This would enable them to form a part of a machining cell, carrying out first and second operations on workpieces. Such an arrangement would not only improve the work-feeding system of the cell but would enable the capital cost of the pallet changer to be spread over two machines.

2. *Many features of the KTM 560 by Kearney & Trecker Marwin contribute to its versatility as a multi-operation machine. Random pallet indexing of the 22" pallets enables the operator to transfer loaded pallets to the table on the Z-axis bed in any sequence. A system of coded pallets allows the machine to identify the component and select the appropriate part program from the memory stores of the CNC system—for which a wide range of software options is available. With a 15-hp DC drive, the spindle has a speed range of 20 to 3600 rpm, programmable in increments of 1 rpm. The working envelope is X axis 29.5", Y axis 25.6" and Z axis 19.7". For more information, circle E97.*

completely flexible manufacturing systems. Used as a standalone machine, the rotary pallet allows the user to accumulate practical experience in programming and loading families of components. This experience is an essential first step towards flexible manufacturing.''

The Genertron gear shaping machine, incorporating an advanced electronic control system, was introduced by W E Sykes, Ltd, Middlesex, England, about a year ago, and greatly speeded up gear production. Now the company has added an automatic turret loader consisting of a rotating plate supported on rollers and journalled in a needle bearing. Attached to it are turret plates, supplied in two halves to facilitate ease of changeover. The turret is rotated by a hydraulic motor. Angular location is achieved by a positive mechanical stop.

When loading extended boss and shaft type components into the work fixture, the whole turret assembly can be raised or lowered automatically to a maximum of 4.8". This avoids the need to raise and lower the work fixture and ensures a more

rigid and accurate location of the workpiece at all times. With a pitch circle diameter of 26", the turret will accept 18 components with a diameter of 3.5". For more information, circle E120.

As an indication of the importance attached to efficient machine loading, British manufacturers are bringing out new work-handling systems.

Between machines

Big improvements in floor-to-floor times can be achieved by reducing the time required to move workpieces from one machine to another, and transfer lines linking batteries of machines have been installed in the mass-producing industries for many years.

Developments continue with the design of such systems for a wider range of users in batch production, rather than mass production. They are made possible by recent control developments, such as the microprocessor.

The Drummond Div, Staveley Machine Tools, Ltd, Birmingham, England, has demonstrated how conventional machines can be offered as complete machining cells, tailored to a customer's particular needs. For example, the company's Maxiturn copying lathe and Maxicut gear shaper can be linked in any combination and planned to occupy a compact physical area to provide sequential machining on workpieces which are transferred between machines by fully automated handling units (**Figure 3**).

All units in the cell—there is no limit to the number of machines that are linked together—are controlled simultaneously to provide completely integrated cycling.

Costs recovered

Although all machine feeding developments are primarily aimed at reducing nonmachining time to a minimum, and success is measured in terms of increased productivity, there will often be a bonus in the form of space saving (an integral machine-loading system is generally more compact than the various handling devices around the machine it replaces) and operator safety.

Manual handling of workpieces can be responsible for many of the strains, sprains and other industrial injuries which result in inconvenience and suffering for the operator and loss of production for the employer.

A new machine tool with built-in work-loading arrangements as an optional extra may be more expensive than a model without these facilities, and it also costs money to fit such devices to an existing machine. But experience shows that the additional expense is quickly recovered. ∎

3. *A typical installation of the Maxiturn copying lathe and Maxicut gear shaper by the Drummond Div, Staveley Machine Tools, is a single automated cell producing a pinion shaft from the billet to the gear-cut form, ready for tooth shaving. Rough billets are loaded into the Maxiturn lathe and are fully turned at one setting over the shaft and gear profiles before being transferred directly on to a reservoir conveyor which feeds the Maxicut gear shaper. Workpieces from the conveyor are taken into a carousel wheel for feeding into the gear shaper.*

Reprinted from Tooling & Production, September 1981

Specifying automatic parts feeders

When planning for automation systems involving parts handling, selection of the most cost-effective orienting/feeding subsystem by a potential user depends on his knowledge of the various types available within each category. The functional characteristics and relative merits of each type are described here to help the user specify the kind that will efficiently serve specific needs.

Important to any automatic manufacturing or assembly machine handling small parts is the parts feeder that orients and feeds parts from bulk. No matter how efficient the downstream operations are, without the continuous movement of parts—precisely oriented and fed at an optimum rate—the entire system is ineffective.

by Cyril Williams
Managing Director
The Institute for Advanced Automation
Ogden, UT

Parts handling feeders or—more simply—feeders are basically custom-built combinations of various mechanisms and vibratory devices. They are used not only to feed or convey parts, but to present them in their most usable position and to maintain consistent orientation as they are being fed hour after hour.

Basic, commercially available feeder systems for bulk separation and parts orienting and feeding are broken down into categories of *mechanical, 60-cycle vibratory* and *120-cycle vibratory*. In ad-

dition to these types, there are other special systems used for conveying single parts such as pneumatic and those that use high-frequency electromagnetic fields. These methods are not in widespread commercial use and will not be dealt with here.

Types of mechanical feeders

Mechanical feeders can be divided into several general categories such as rotary, oscillating and linear. There are also various combinations of these. Other than

1. Sketch shows the elements of a centrifugal mechanical parts feeder. By adding a curved fixed fence or hook, the parts are aligned as the disc rotates and eventually reach the outer periphery, where they are oriented end-for-end and are tangentially discharged. Photo shows a mechanical feeder developed by Hoppman Corp, Springfield, VA, which uses a flexible rotating disc and clustered tapered rollers, termed Rollermat™ by the company, to feed metallized plastic caps at 350 ppm.

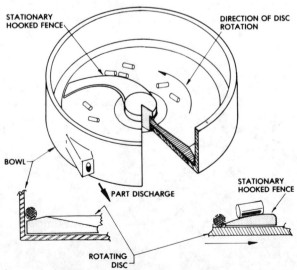

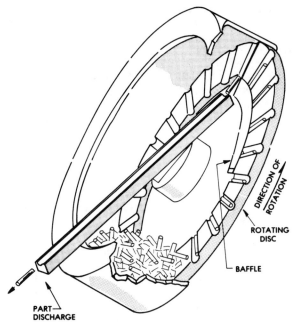

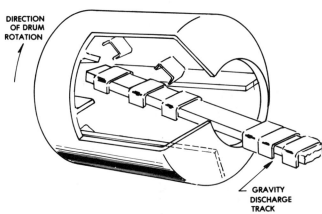

2. Another version of a rotating-disc parts feeder picks up parts in radial grooves of the disc as it is rotated through the bulk parts supply and carries each around the feeder's outer rim. At the 12 o'clock position the parts are dropped onto a fixed track, from where they leave the bowl by means of gravity.

3. As the drum of this mechanical parts feeder rotates, parts are carried to the uppermost position, where they drop randomly onto a fixed gravity track. Those parts that land on the track in the correct position are discharged by gravity. Those parts that do not fall properly onto the track are recirculated.

the types used to feed standard small screws, rivets and pins, all mechanical feeders are custom developed and constructed. They are developed on the bench as a one-on-one activity performed by a machine builder. The technique is strictly cut and try. While a concept may be followed, the development and final product is the result of whatever creative bent the builder possesses.

The disc or centrifugal type of mechanical feeder is bowl-shaped with a side wall and a spinning disc at the bottom. Generally, the disc rotates in a horizontal plane, although in some arrangements it may be tilted so that it is not level but skewed. In the spinning-disc arrangement, the parts are thrown by centrifugal force against the wall of the bowl or an outer track, **Figure 1.** At the periphery, if the parts are cylindrically shaped with a large length-to-diameter ratio, they will tend to line up against the wall as shown, and by the use of selectors* or wipers having different shapes, will follow in single file and be pushed out of the feeder tangentially through the discharge, end-for-end. Another type of rotary disc feeder is shown in **Figure 2.**

Barrel-type feeders are used for feeding parts that are U-shaped—similar to staples. Shown in **Figure 3** is a rotating outer drum with cleats inside, rotated around a horizontal or tilted axis. Storage for bulk parts is inside the barrel. In addition to rotating-type mechanical feeders, there are also oscillating types,

such as that shown in **Figure 4.** Oscillating boxes, which rock back and forth as shown in **Figure 5,** are used to feed headed parts such as nails, screws or pins. Linear types of mechanical feeders are manufactured in quantity, such as the ones shown in **Figure 6,** and are sometimes available off-the-shelf.

From an operational standpoint, mechanical feeders have good linear feeding control characteristics; vibratory feeders do not. For example, you can literally dribble out parts from a mechanical feeder in a small quantity and then, by turning the control progressively higher, increase feeding speed until reaching maximum. A vibratory feeder has only a small range of useful controllability at its maximum output. On the other hand, the mechanical feeder is sensitive to the amount of parts it can maintain in its storage and must be prefed from a separate bulk hopper to maintain its optimum level of operating efficiency.

Mechanical feeders are much quieter than vibratory feeders, and parts that are spongy, resilient, soft, oily or slick can be handled in a mechanical feeder where they could not be fed in a vibratory type. Mechanical feeders handle parts more gently with less recirculation.

Vibratory feeders

The basic vibratory feeder consists of a bowl which is "tooled" for a particular unique part or family of parts. The bowl is attached to a frame or plate,

which in turn is bolted to a cluster of leaf springs equispaced in a circle and in a skewed fashion around the center of the bowl. There can be 3, 4, 6 or 8 spring packs to support the bowl. The springs in turn are bolted to a heavy base which is supported by resilient rubber feet. The various elements are shown in **Figure 7.**

Construction of both the 60-cycle and 120-cycle vibratory feeder is similar in many respects and many users find it difficult to tell them apart. However, there are major differences. More significantly, most feeder builders specialize in only one, either the 60-cycle or the 120-cycle type. Operational differences between the two types depend on the manner in which the line current or input voltage is applied to the electromagnet. This determines behavior of the drive unit.

60-cycle drive unit. Alternating line current—at 60 Hz—is applied to the controller circuitry of the 60-cycle unit. The circuit receives full-wave line current and rectifies it by removing the negative portion of the wave form, leaving only positive pulses of current which energize the electromagnet in the drive unit in a pause-pull-pause sequence. The amplitude of the pulses is controlled with a voltage controller.

120-cycle drive unit. The 120-cycle system does not rectify the line current,

Selectors are special tooling devices designed to selectively accept, reject or correct the orientation of parts as they move past a point in the feed system prior to the point of discharge. Selectors accomplish this task by virtue of being designed to take advantage of the part's geometry, center of gravity etc.

but uses the alternating current just as it comes from the power line. Voltage, or amplitude, is controlled by a variable transformer. The electromagnet energizes every time the current rises from zero to maximum, both on the positive rise and the negative fall. In effect, the alternating current behaves like a switching circuit. As a consequence, the electromagnet pulses 120 times per second, as opposed to 60 times per second with the 60-cycle unit. These pulses represent the motion of the part during that one second (refer to **Figure 7**). At 60 cycles, the part makes 60 hops in 1 sec; the same part will make 120 hops in 1 sec on the 120-cycle unit. This does *not* mean that a part travels a greater distance in 1 sec on a 120-cycle unit than on a 60-cycle unit. The parts always travel the same distance in the identical period of time. However, the 120-cycle drive will cause the part to take *smaller* jumps or hops. This is the reason that when parts are feeding in a 120-cycle feeder, they appear to be moving without any bounce rather than the more pronounced motion of the 60-cycle unit. From a noise generation consideration, the 60-cycle unit is generally quieter than its 120-cycle counterpart.

Parts that have very distinctive geometrical features, such as slender cylinders, rectangles, prisms, spheres or other shaped parts that have strong natural resting features, can be readily oriented and fed in 60-cycle bowls. If aluminum bowls are used, they are less costly and easier to develop. On the other hand, parts with extremely small, delicate features that are used for orienting purposes might be run on a 60-cycle feeder, but the output would be severely penalized because of the relatively rough handling of the part in the bowl. In a 120-cycle feeder, these parts would be vibrated very smoothly, and selectors would be allowed to efficiently reject the parts or convert them to an acceptable orientation. The result would be a higher output in the 120-cycle system.

Feeder bowl options

Perhaps the most common feeder bowl configuration is an inverted "cascade" type made of cast aluminum. They are available in standard sizes from 5″ dia to 24″ dia and with clockwise or counterclockwise rotation. Some sizes are available with various multi-track discharges.

Tooling of cascade aluminum bowls is accomplished by attaching various devices—ramps, wipers, cutouts, rails etc—to the inside of the bowl track by screws or rivets. This type of bowl is gen-

erally used with 60-cycle base units for simple-shaped parts, and is the least costly.

The basic fabricated bowl construction consists of a rolled and welded band, and a domed and spiral track all integrally welded together. In some bowl sizes, double bottoms with sound-deadening material are used. Fabricated bowls are generally used on 120-cycle base units, but they can be driven with 60-cycle base units if the part warrants.

The linear distance available for tooling on a fabricated bowl is generally three times greater than with a cast aluminum bowl. A fabricated bowl can have tooling not only on the inside of the bowl, but around the outside as well. Fabricated bowls are built in standard 5″, 6″, 8″, 10″, 12″, 18″, 24″ and 36″ dia.

The fabricated bowl can have discharges in several places, such as 180 degrees apart, and can be fitted with multi-discharge tracks. They can be used to feed two different parts, such as bolts and washers, and have them meet at a common discharge and be assembled automatically. Quick-dump chutes can be provided on fabricated bowls for the purpose of quickly unloading parts from the bowl without having to feed them through one at a time or scoop them out.

4. *This type of mechanical feeder uses an oscillating centerboard which moves through bulk parts. Those that are oriented remain on the top edge of the blade and are nested on the upward stroke, where they will slide off due to gravity into a fixed discharge chute. Generally, this type of system is limited to cylindrical or slender parts.*

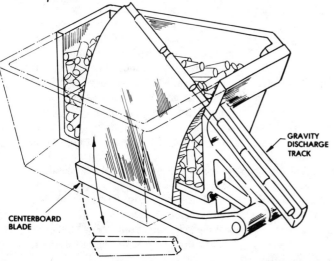

CENTERBOARD BLADE

GRAVITY DISCHARGE TRACK

5. *Feeders such as this are used to feed machines using large quantities of nails, screws or pins in multiple chutes. Oscillating of the compartmented box guides parts into slots to be escaped into their respective magazines.*

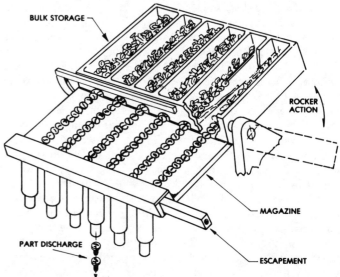

BULK STORAGE

ROCKER ACTION

PART DISCHARGE

MAGAZINE

ESCAPEMENT

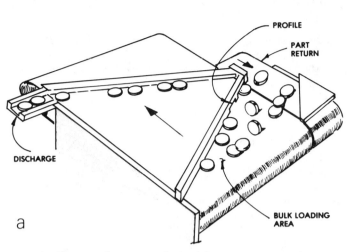

a

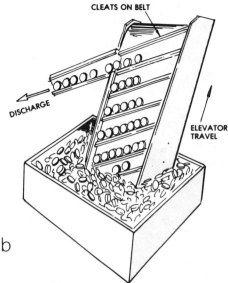

b

6. *Horizontal conveyor belts (a), moving in opposite directions adjacent to each other, make up a type of linear mechanical feeder. Properly oriented parts pass through a profile in a fixed fence. A second fixed fence guides the parts to the discharge chute. Misoriented parts continue along the first fence and are moved onto the first conveyor, where they are re-* *rected back to the main belt to begin a new cycle. Inclined linear feeders (b), also known as elevator feeders, both feed and orient parts. Correctly oriented parts reach the top where they drop through a profile opening and discharge by gravity.*

Specifying the correct feeder

Armed with the foregoing information on the differences between the various types of feeders, the user is better equipped to understand how to specify a feeder that will yield the most cost-effective service and operating life. As a general rule, these systems are best developed by feeder specialists, in the same context that the best bread is baked by a master baker and not the amateur chef.

Consider this fact: If the identical part was given to each of the three types of feeder manufacturing companies for a proposal for a feeder to orient and feed that part, each manufacturer would prepare a competent fixed-price quotation. Each company, in turn— using entirely different techniques—will state they can build this feeder to meet the required specifications. What *criteria* should be used to select the company to which the project would be awarded?

Should the award be based on price, construction, delivery, type, warranty, company background or construction materials?

The answer may be found in any one of several case history examples drawn from actual user experiences that we

7. *Two things that differ in the construction of vibratory feeders are the spring angle and the position of the electromagnet and its armature in the base unit. The relative position of the springs and the electromagnet coil determines clockwise or counterclockwise feed direction. With the upper and lower plates separated by angled flat springs, the upper plate revolves slightly when it is pulled down by the energized electromagnet attracting the armature attached to it. The upper plate, with the feeder bowl attached to it, rotates in one direction with the downward pull and reverses* *direction of rotation when the attraction is released. This appears as an oscillating, cyclic motion. Sketch (a) shows the electromagnet located in the center of the base. A tangential electromagnet arrangement (b) is used by many builders, particularly for 120-cycle operation. With this arrangement, if there are heavy loads to vibrate, several magnet assemblies can be used.*

a

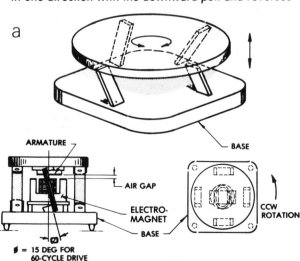

b

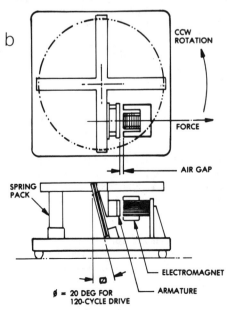

have documented. Essentially, these user experiences help point out some of the more critical considerations and specific application information needed before a reasonable evaluation of the feeder design can be made.

One case involved an electronics component manufacturer who selected a builder of premium printing (symbolizing) equipment to supply not only the printing head with which electronics components are marked, but an entire turnkey system, including parts feeder, escapement etc, all mounted in a single structure. Since the printing equipment builder does not build feeders, he secured fixed-price quotes from several feeder builders. All quoted the job, and each guaranteed to orient and feed the parts at the required rate.

Shortly after accepting the system, the component manufacturer experienced difficulty in feeding parts, eventually necessitating use of a full-time attendant to clear jams in the feeder. The difficulty stemmed from the fact that the user did not specify the type of feeder needed, and the printing device maker simply supplied the least expensive—in this case, a 60-cycle feeder, with tooling which wore imperceptibly over a period of time, causing jamming of misoriented parts. A 120-cycle feeder, though more expensive, would have saved the user the cost of manually attending the machine.

Other documented case studies have shown misapplication of feeders, incorrect tooling, disregard for the special nature or physical attributes of parts, or misunderstanding of the parts feeding

application, all of which are compelling reasons to specify the right feeder for the intended application the first time around. Feeder builders, if not aware of all the facts, cannot guide the user properly. If the responsibility is passed on to a turnkey system builder, he must also have all the facts.

The accompanying box includes a sample specification for a vibratory feed system listing the fundamental elements of information that should be supplied to a feeder builder.

This article is based on information contained in the text Automation Handbook, Copyright 1981, by the Institute for Advanced Automation. All rights reserved.

Sample specification

Vibratory feed system for Automatic Assembly Machine

Smith Company Part No 123—Plastic Widget

1.0 Technical Requirements:

1.1 *Scope:* This specification is for a vibratory feeder system consisting of the items listed.

1.2 Equipment shall consist of an 18″-dia straight-wall feeder bowl, fabricated of heli-arc-welded stainless steel mounted on an 18″, 120-cycle base unit with controller to operate on 115 VAC, 60 Hz, 10 A max.

1.3 Bowl shall be custom tooled to orient and feed Smith Co Part 123 and 1234 plastic widgets per samples furnished. Note: Parts package of both parts is identical. Smith Co will furnish part package drawings at vendor's request.

1.4 Rotation shall be counterclockwise.

1.5 Rate shall be 4000 parts/hr, min. When bowl has residue of less than 1000 parts, rate will not apply.

1.6 Part orientation shall be leads up, lens facing outward, in line, single row and all one way (see drawing).

1.7 The bowl shall discharge 6.0″ ± ¼″ from point of tangency unless otherwise specified in purchase order. Discharge may be "off" horizontal to obtain rate.

1.8 Vendor to quote a sound-deadening bottom.

1.9 Bowl shall not be required to operate under back pressure. However, parts in discharge shall be confined as necessary. Recirculation to be kept to an absolute minimum, as bowl will be operated on "demand." Quick-dump chute with provision for discharge of any debris shall be provided. Location is optional, although consideration must be given to operator convenience.

1.10 Interior surface of bowl shall be Teflon- or PVC-coated to prevent damage to parts. Exterior or nonfunctional surfaces may be coated at option of builder. Equipment supplier may omit coating on selecting or cam devices and discharges.

1.11 *Design flexibility:*
Vendor should not consider the above size or other requirements as limitations if changes or exceptions to the

above will produce a more efficient piece of equipment. Smith Co encourages innovations and suggestions to improve performance, reliability or maintainability of this equipment. Any such suggestions shall be identified in the equipment supplier's proposal.

1.12 *Acceptance criteria:*
Feeder shall be demonstrated to operate at the required rate for ten consecutive cycles of 5 min duration each, 10 to 60 sec off, and repeat without jamming. Vendor does not bear responsibility should parts entangle in bowl. The above acceptance tests may be performed at vendor's facility or at Smith Co by mutual agreement.

1.13 Air jets shall not be used without consent of Smith Co engineering.

1.14 Vendor shall fabricate equipment in a professional, workmanlike manner in accordance with accepted industry practices with regard to quality, reliability and safety. A 12-month warranty on workmanship and defective material is required.

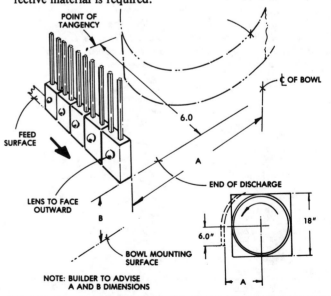

Reprinted from Tooling & Production, September 1980

QUIETLY FEEDING NOISY PARTS

by **Archie A MacIntyre**
President
Archie A MacIntyre & Sons Inc
Gladwin, MI

1. This 3"-dia steel stamping with a "castle" formed at one opening must be oriented and fed to an assembly machine. Because the part rings like a bell when struck against another object, it can be termed a noisy part.

Some metal parts, like the fuel pump valve housing illustrated in this article, ring like a bell when struck against other metal objects (especially each other). This attribute puts them in the category of noisy parts when they are processed through vibratory parts feeders to assembly machines.

OSHA and EPA regulations have underscored the importance of noise control to the extent that the term has become common in out everyday language. Uncontrolled noise is a hazard that both the US Government and industry as a whole have been attempting to regulate at the workplace.

197

2. *Picture on opposite page shows Mainline Feed System (right) that feeds fuel pump valve housings to dial index type assembly machine. Close-up photo of feeder (above) shows parts (top center) coming over top of magnetic belt elevator to descend to rotating table's conical surface where some drop to table below for return by cleated belt elevator (right) to storage hopper while others go through selector guides to in-line transfer conveyor (left).*

In the case of this fuel pump valve housing, the AC Spark Plug Div, General Motors Corp, Flint, MI, has applied one of our Mainline Feed Systems as the mechanical parts feeder used to serve a dial index type assembly machine.

The design concept of the Mainline Feed System is based on our philosophy that "parts handing" must replace the belief in "parts handling." Parts handling, in reference to orientation, accepts the idea that in any mass of parts a certain ratio will be in the correct attitude. Selector devices are positioned to accept those parts and reject the others. This requires mass movement to achieve the desired results. It is also the source of unnecessary noise.

The concept of parts handing reverses this process. Parts are placed in a single-file, unoriented manner and processed through the necessary selector devices to achieve the required attitude. Each part is treated like an individual.

With higher rates of feed of oriented parts requested by industry now and in the future, the parts handing approach points the trend in parts feeders. It is obvious that the less action and reaction necessary to process parts in a feeder, the less noise, violence and wear to parts and equipment will result.

In this application, the part, **Figure 1**, is about 3″ dia and is fed up out of a low-profile bulk-storage hopper by a magnetic belt elevator, **Figure 2**. The parts drop one at a time to the rotating table whose conical surface is covered with resilient material. This table turns clockwise under the selecting guides. Parts not right-side-up do not make this trip but fall through to another rotating

table that discharges them to a cleated elevating conveyor that takes them back to the bulk storage hopper.

Parts that pass through the selecting guides come off the conical surface onto the outer table surface and are delivered "castle up and leading" onto an inline transfer conveyor that takes them to the assembly machine.

This feeder arrangement delivers 25 ppm at under an 80 dB noise level. These results are achieved by what are inherently quiet mechanisms. Parts are carried and guided rather than pushed and "leap frogged" as in a vibratory feeder bowl. Care is taken in both the engineering and build stages to process parts with as little interface as possible, thus reducing the noise-generating areas.

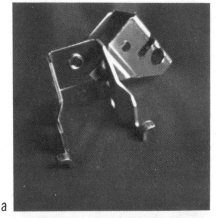

1. *Some sheet metal parts with tabs, projections, slots and holes, photos (a) and (b), may prove difficult to feed due to tendency to interlock. Photo (c) shows another problem encountered in automatic feeding of sheet metal parts; the U-shaped clip, in addition to having entangling tabs, may easily be distorted, causing the open end to be closed, thereby allowing one clip to wedge into another.*

Reprinted from Tooling & Production, September 1981

Parts feeding by design

The proper operation of assembly machines equipped with automatic parts feeders depends heavily on whether the parts are feedable. Basically, the design of the part must be such that it is capable of being oriented or selected by the tooling that has been custom designed to interact with certain geometric features of the part. Such interaction is brought about by causing the part to be in motion, either through a vibratory bowl feeder or some other mechanical device that separates or singles out parts from a bulk supply, and properly orients them for subsequent operations after discharge. Examples of both good and poor design features of parts, from the standpoint of automatic parts feeding, have been supplied by Feedmatic-Detroit, Inc, Southfield, MI, a subsidiary of Fusion, Inc, a builder of floor (elevator) feeders, and rotary-hopper, vibratory bowl and in-line feeders and parts placement equipment.

Spokesmen for the company say that the place to start is with the original design of the part. Giving consideration to the needs of automatic parts feeding at the design stage simplifies the eventual application of parts-feeding equipment and can help insure eventual trouble-free operation.

Of obvious concern in automatic feeding of parts is the possibility of damage to the parts being fed, caused by action of the feeder. Parts that tend to break, chip or in any way compromise part finish may not be candidates for automatic feeding, or at least may not be feedable at a rate which would make investment in a parts feeder economically feasible. Parts that are soft, or those made of pressed, granular material which is green or uncured, tend to break apart due to the tumbling action of the feeder. Such parts include those made of powdered

2. *This thin, flat sheet metal part with short legs or tabs at each end, photo (a), proves to be feedable because the tabs are designed in a way which makes it impossible for the parts to nest, one on top of the other. This permits layered parts to be separated via a wiper which acts as a "low bridge" to sweep away the top layered parts. An example of flat parts that will feed easily, but cannot be oriented, length for width, is seen in (b). This is because the feeder tooling sees these as square parts; the length and width are equal, even though the two notched sides give the appearance of a distinguishing feature. The feeder tooling would only recognize the parts' overall dimensions, without regard to the seemingly narrow leg.*

metal. Some parts tend to stick together, such as those that are magnetized or covered with grease or oil from previous operations, and are not able to interact with feeder tooling.

Assuming that a part design is suitable for automatic parts feeding, the part must maintain its design dimensions, from part to part. Unlike a human being, an automatic parts feeder cannot differentiate between part differences. Although some feeders attempt to include sorting of bad parts or foreign matter, often these attempts are marginal and require routine maintenance or cleaning out of the equipment to remove extraneous matter and malformed parts. A prime rule of parts feeding is that the parts must be uniform.

Long, thin parts are difficult to feed because of the problem of selecting or orienting the part. A profile, or cutout, in the feeder tooling, that matches the outline of the part, is about the only method that can be used. However, lamination or overlapping of long, thin parts often prevents parts from entering the tooling profile one at a time.

Assuming that a part is thick—not thin—and that it is rigid—not flexible—and that it will not nest, tangle, interlock or stick to other parts, then it is most likely a part that can be fed automatically. Moreover, a part should have one or more geometrical features that can interact with orientation tooling. Care must be taken to select the more predominant geometric feature, such as a step, chamfer, hole, groove or projection. Often, more than one tooling device may be necessary to gain the proper orientation of the part as it progresses through the feeder. These can be a combination of devices that selectively eliminate orientations that are not needed and subsequently return parts back to the bulk supply, keeping those that are properly oriented in a given position in the tooling sequence.

Parts stamped from sheet metal, **Figure 1**, are difficult to feed because of their tendency to nest, interlock and tangle. Such parts will not interact with the feeder tooling to gain proper orientation. It may be possible to feed stamped sheet metal parts that have only shallow depressions, ridges and reinforcing ribs, **Figure 2**. Preliminary testing of parts in prototype feeding equipment is often necessary to determine feedability.

3. An excellent example of nonrigid parts, which cannot be handled in automatic parts feeders, is seen here.

a

4. Many coil springs are difficult to feed automatically, due to a tendency to entangle. However, if they have closed, ground ends—such as those shown in photo (a)—entangling is minimized. Formed wire rings (b) whose ends are not closed also entangle easily. Feeding of these rings could be improved if the ends were angle cut instead of square, thereby allowing the opening to be squeezed close.

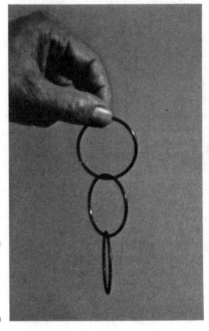

b

Parts must not be flexible, **Figure 3**; they must remain rigid in the feeding equipment to interact with feeder tooling which selects the parts. In addition to flexibility, wire parts present special problems such as entanglement, and may be difficult to feed, **Figure 4**.

Cylindrical parts with a reduced male diameter and a larger female diameter tend to telescope one into another. Similarly, cup or saucer-like parts tend to nest one into another and are difficult to separate in order to gain interaction with feeder tooling. However, on parts with different diameters, the difference in end diameters can be used for end-for-end selection, **Figure 5**.

If the difference in mass or weight from one end of a part to another is significant, gravity could be used for selection. Gravity is often used for threaded or nonthreaded studs with heads, **Figure 6**. The part is fed axis-horizontal, either end leading—the head or the stud—and dropped onto parallel rails. Because of gravity, the longer stud end will fall between the rails and the head will remain upright suspended on rails. However, in initial part design, care must be taken to control the ratio of the length to the head diameter so that the stud itself will always swing downward, keeping the head up, and not the opposite. Weight distribution, then, is important in using gravity for feeding.

It is sometimes possible to add tabs or projections to a part after it has been fabricated. These can be used for orientation purposes and removed later in a production process, or they could re-main in place provided they don't interfere with function. Holes or slots can also be added to a part and used for orientation purposes, again, provided such holes don't interfere with function. An example of radial orientation of a hollow cylinder with a sidewall hole using added slots for orientation is described in **Figure 7**.

Major factors to consider when designing a part for feedability are (1) the geometric shape of the feature, (2) location of the feature and (3) the size of the feature. If these considerations are kept in mind while designing the part, it will enhance the feasibility of feeding the part automatically as well as minimize the cost of the parts feeder.

5. End-for-end orientation of nut-like fittings, with a threaded hole at one end and a tapered diameter at the other, is accomplished by a feeder track designed with a scalloped, outer edge. The scallops serve to narrow the track, thereby permitting only parts with the smaller end down to pass over. If the part approaches the scalloped track with the threaded end (larger ID) down, it will fall back into the bulk supply.

6. Headed parts make use of gravity and parallel rails for orienting. The ratio of length of the part to head diameter should be at least 1.5:1. This permits the threaded portion to swing down between the rails for "heads-up" orientation. The center part shown here would have been wiped away before reaching the rails.

7. Radial orientation of this hollow cylinder with a sidewall hole was achieved by adding slots in the end of the piece. A projecting rail, added to the surface of the feed track, interacts with the two slots, permitting the part to pass over a section of track sloped toward the feeder bowl. If the nonslotted end is down, or the radial orientation incorrect, the part will fall off the sloped track into the bulk supply for recycling.

Basic Vari-Tran chassis for smaller assemblies, left, showing detail of lift drive shaft; and, above, fully tooled for automatic assembly of compound 6-way seat-adjuster motors at 1800/hr.

Reprinted from Tooling & Production, September 1981

High-speed transfer the nonsync way

Sometimes the best way to achieve high-speed transfer is to combine the best features of synchronous and nonsynchronous transfer methods. The Vari-Tran® transfer system developed by Rockford Automation Inc, Rockford, IL, is one way to speed up assembly operations.

Average total cycle time is generally in the range of 1.5 sec to 3 sec, with gross production rates from 1200 to 2400 assemblies per hour, single tooled. Higher production rates are possible with multiple tooling.

Unlike cyclical machines where the entire machine stops when one station is down, intermittent station downtimes are time-averaged since all other stations continue in productive operation. Slow operations can be intermixed without loss of overall production by using multiples of slow stations and performing the slow operations simultaneously on alternate pallets. On combinations of automatic and manual operations, the nonsynchronous machine produces at the average rate of the manual stations, while cylical machines have to be programmed for the slowest production at manual stations.

How it works

The drive system is the key to smooth acceleration and deceleration of the pallets. Each pallet has a drive wheel (double-roll cam follower) which rides in spring-loaded contact with the rotating main drive shaft. The wheel is located in a spring-loaded control arm which can pivot from the stop position (0 degrees with respect to a line perpendicular to the drive shaft) to the drive position (35 degrees), causing the pallet to transfer at the chosen fixed speed (100 fpm to 210 fpm) for the application.

When the control arm contacts the stop at the work station, the drive wheel rotates from 35 to 0 degrees, resulting in smooth deceleration to a complete stop in the work station. The pallet can be shot-pinned for positioning accuracy to ±0.002″. At 0 degrees, there is no transverse pressure against the pallet. When the station operation is completed, the stop is released and the pallet accelerates to its transfer speed.

To prevent contact between pallets, each pallet has a stop rod extending from the rear. As the following pallet approaches, it decelerates and comes to a stop when its control arm comes in contact with the stop rod on the preceding stationary pallet. A minimum space of 3″ is maintained between pallets, eliminating the possibility of physical injury to the operator.

A programmable controller can be provided for machine control reliability and the handling of part or process changes. In the turnaround area, an electromechanical memory device does not release a pallet to the other side of the turnaround until there is a space for it.

Competitive pros and cons

In the development of parts transfer devices in assembly machines, the principle choice was originally a synchronous (timed index, timed dwell) dial-type machine. For quality applications, barrelcam index units provided the smooth acceleration and deceleration necessary to keep loose parts in position until fastened together. Production rates from 600 to over 3000 assemblies per hour can be achieved with dial-type machines, single-tooled.

Barrel-cam-indexed, dial-type transfer chassis are still being applied successfully to modern assembly applications, but they do have some limitations. Only a limited number of operating stations can be applied to the dial because of space constraints. Even with large dials, the machine becomes crowded, inaccessible and difficult to maintain. In addition, tooling usually functions toward the center of the dial rather than in front where the parts placement and assembly operations can be easily observed. Finally, the cycle time is limited by the time required to perform the slowest operation on the machine.

Synchronous in-line machines use camcontrolled transfer bars to transfer parts or pallets. They maintain the advantages of the dial index chassis, but are slightly slower in production due to the heavier inertia loads in transfer. They also overcome two principal disadvantages—they provide more room and accessibility for stations and they allow better observation of tooling in operation.

However, as cylical machines, their index and dwell times are limited to the time required to perform the slowest operation. Another disadvantage, primarily related to capital investment, is the number of fixtures required. Fixtures must be provided for every index position over the length of the line.

Nonsynchronous machines were developed concurrently with in-line synchronous machines. They are driven by continuously moving chains to which pallets are clutched into and out of engagement, or upon which pallets are resting during transfer. Though considerably slower than in-line synchronous machines, they can offer time-averaging advantages.

The nonsynchronous machine is potentially faster than a synchronous machine for a combination of manual and automatic operations. An excellent example is nonsynchronous assembly of alternator stator and rectifier assemblies. The operator manually orients and loads one alternator stator to each fixture. The wire leads must be straightened and positioned during loading. Loading time will vary considerably, depending on the difficulty of straightening and positioning. On a nonsynchronous machine, loading time will be time-averaged. On a synchronous machine, the cycle time would have to be set at or near the longest time required to load the part or else suffer the loss in production of empty fixtures if run at a faster rate.

Chain-type nonsynchronous machines, even those with pallets riding on rollers, are inherently slower than synchronous machines because transfer speeds have to be low enough to dampen the abrupt stops when pallets come in contact with pallet stops or preceding pallets. Generally, chain speeds are limited to 50 fpm to 60 fpm, and production rates to 1200 assemblies per hour or less.

Reading clockwise from center top, typical products assembled on Vari-Tran machines include accumulator leak tester (900/hr), Super-8 film cartridge (2400/hr), alternator rectifier assembly (900/hr), shock absorber (1440/hr), fan-clutch assembly (1200/hr), lock-body assembly (2000/hr), strut piston and rod assembly (1200/hr), seat adjuster (1800/hr), AC clutch assembly (900/hr), isolator tube assemble and test (1200/hr), typewriter ribbon cartridge (1800/hr) and carbon canister assembly (2400/hr).

Pallet locomotion detail. In the upper view (looking down on three pallets), the two pallets on the right are at rest because their drive wheels are turning perpendicular to the rotating drive shaft. The left pallet, with its drive wheel 35 degrees from shaft perpendicular, is moving to the right about to hit the stop. Below is an end view showing control arm detail.

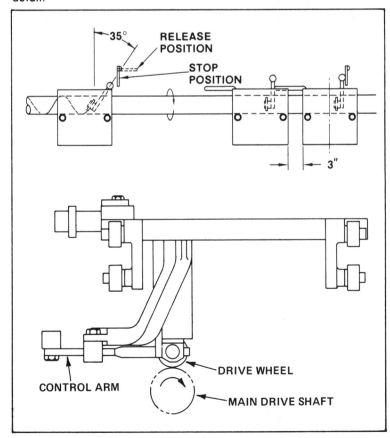

Selection criteria

As with any machine, the Vari-Tran is not the best machine for all assembly operations. Synchronous and nonsynchronous chain-type machines will continue to provide the best solution for many applications.

Key considerations in selecting the type of machine for a particular application include number of parts in assembly, number of stations, types of operations, part size and weight, production rate, use of operators or robots and capital investment.

Although all the above requirements are thoroughly considered for each application, Rockford uses the following general rules of thumb:

1) For 900 asemblies or more per hour and more than six operations, use the Vari-Tran.

2) For less than 900 per hour and more than six operations, use a nonsynchronous chain-type machine.

3) Where more economical and suitable for a specific application, use a synchronous dial.

Presented at RIA's Fifth International Symposium on Industrial Robots, September 1975

Handling Automation for Two Milling Machine Tools

By G. Spur, B. H. Auer and W. Weisser
Technical University of Berlin
German Federal Republic

Industrial robots open a new way for automation in low series production. The paper describes a practical test to implement an industrial robot into a production line of two programmable milling machine tools at the Siemens plant in Berlin. After dealing with the workpiece analyse, the requirement of the subsystem is derived. The development of workpiece magazines and chucking tools followed a laboratory test to eliminate all system troubles and accelerated the implementation into the shop. Finally the economics will be discussed.

2. Introduction

The production of smaller quantities of discrete parts characterizes the manufacturing at the Siemens AG, Berlin plant. Within the small batch production there are up to 48 % idle times and only 52 % operation time. The largest part of the idle time is the part handling time. It takes up to 60 %. The main part handling operations are transportation to the machine tool, pick up the part, positioning, inserting in the chucking tool, unloading and restoring. The percentage of the parts are characteristic for the plant and shown in figure 1.

In consequence of the lack of skilled operators the utilization rate of the machine tools could not increased. The plant is already working two shifts a day. Therefore remained for rationalisation only the following points:

1. Parallel machining on additional machine tools.

2. Substitude of the conventional machine tools because of numerical controlled machine tools.

3. Automatic universal chucking tools.

4. Handling automation with industrial robot in connexion with universal workpiece magazines.

The industrial robot opens a new economic way to save labour cost and increase the production rate. If the skilled operators only supervices the automatic handling devices and feed the part stores, a multi machine operation will be very economic. That was the reason for the Siemens plant to start in cooperation with the Institute for machine tools and manufacturing of the Technical University of Berlin a practical test with industrial robots.

The implementation of an automatic handling system in connexion with universal chucking tools and workpiece magazines followed four phases. First started a part analyse and the selection of the suitable machine tools. Then followed a study about the requirements of all subsystems, the development and the laboratory test. After a final trouble analysis chained the installation in the production shop.

3. System study and requirements

Part analysis

In a first step the parts within the small batch production has been analysed and all parts to be machined on programmable machine tools has been selected. Then the parts has been classified to part families suitable to similar gripper devices and workpiece magazine types. The following aspects for parts and machine tool selection has been observed:

- Operation time between two and five minutes,
- the first and the second cycle of operation must be on machine tools which are standing within the working area of an industrial robot,
- the manual operator must load and unload two machine tools,
- the machine tools should work two shifts a day,
- the machine tools should be balanced more than 25 % with the selected part family,
- work-piece weight less than 25 kg (55. 115 lbs).

The workpieces with similar size and shape have been selected for the automatic handling system. The weight of the parts was about 1,5 kg (3.307 lbs). The workpiece A will be milled in two operations. A first spot facing operation on a horizontal milling machine and the second milling operation on a vertical type machine tool. The whole operation time for both passes takes 2.3 min/part for manual loading and unloading. The first machining operation takes 0.7 min/part. The workpiece type B will be only machined in one operation on the horizontal milling machine. Operation and cutting time is equivalent to the first pass for workpiece A type. 15000 parts of type A and B will be machined every year in batches of 720 from each type.

Machine tools
The milling machine tools are plug board programmable with

30 program steps. The machine tools have a maximum feed
rate from 0.016 m/min to 2.5 m/min, a quick traverse of
3 m/min in the X and Y axis and of 1 m/min in the Z axis.
A cam surface for position measurement renders possible
for programming in each axis.

Material flow and system lay out

To outline the system lay out and the material flow for the
manufacturing system has been the next step. As the machining times of the two workpieces are quite different, material
flow charts have been discussed. It results, that the part
type A starts the spot facing operation on the horizontal
milling machine and goes then for the second operation to
the horizontal milling machine tool. During the second
operation of part A a workpiece type B can be machined,
because of the cutting time of part A in the second operation takes about two times longer as the cutting time of
part B. The time studies and the exact cycle diagramms will
be discussed later.

The system lay out has been outlined that the industrial
robot can rotate between the machine tools with its maximum speed and maximum working radius that the milling
machines need not much idle time for positioning in quick
traverse. In front of the industrial robot should stand the
workpiece magazines. Figure 2 shows the lay out of the
complete manufacturing system.

System requirements

After the workpieces have been selected and the material
is fixed. The machine tools could be stipulated the requirements of the industrial robot, the workpiece magazines and
the chucking tools. For the industrial robot resulted the
following demands:

- reach capability not less than 1.8 m,
- about 50 program steps,
- about 20 input and output signals,
- accuracy of position $\pm$ 1.5 mm,
- maximum speed greater than 1 m per second and
  greater than 120° per second,
- high reliability.

Because of the workpieces type A and B should be machined
at the same time, we must design two different workpiece
magazines and two chucking tools. The workpiece magazines
perform two functions. 1. The storing of the blanks and
the finished parts. 2. Separating the blanks in defined
position for the handling system. All functions must be

controlled, any faults must be indicated and should stop the robot. The store capacity of the workpiece magazine must be greater than 30 blanks and 30 finished parts. In order to have more than one hour for automatic working without manual recharge. A further demand has been a low local requirement and low direct labour.

The chucking tools must be operate automatically and must controll the defined part position. Any fault in clamping as well as chips being on the workpiece support must be also indicated and interlocked. The chucking tool for the horizontal milling machine must be suitable for clamping both workpieces.

4. System development

Industrial robot

We selected the industrial robot "Transferautomat E" produced by VFW–Fokker in Germany. The handling device is shown in figure 2. The industrial robot has five degrees of freedom and operates in a cylindrical working area. The X-axis has a reach capability of 1.8 m, the lift of the Z-axis is 0.35 m and the maximal rotation is 240°. The wrist yaw is 180° and the wrist swivel is 90°. Three axis are driven by electric DC discmotors and are closed loop servo controlled. The wrist yaw and swivel motion is pneumatically powered and has adjustable stop dogs for correct positioning. The gripper device is also pneumatic powered. The standard control system has a plug board program store for 30 program steps and 10 potentiometers for each servo axis. A 60 step plug board is as option available. Each potentiometer can be used at any program step. Five Input and five output commands are programmable. A slowly motion, program dwells and interlocks can be programmed. The positioning accuracy is ± 1 mm for the translation and ± 0.1° for the rotation axis. The maximum speed is 1 m per second for the linear motions and 100° per second for the main rotation.

The robot control has been enlarged up to 62 program steps and 15 input and 15 output commands. The C-axis (swinging about Z-axis) electric DC disc motor has been changed from a 500 W power to a 1000 W powered type. That the demanded rotation speed could be obtained.

The gripper device was built out as a simple clamping gripper with stiff fingers. A force measuring sensor has been developed and attached. The sensor signals a adjusted gripping force to the robot control.

Accessories

For a first laboratory test a simple gripper device, a slide magazine and a chucking tool has been developed. Of that followed an continuous working in these simple area. As an result of the test followed the construction of the gripper, part magazines and of the chucking tools.

A spiral chute (figure 4) has been developed as store for workpiece type A. The magazine has a load capacity for 54 blanks and 56 finished parts. The chute seperates the blanks automatically and fixes the part in defined position before the industrial robot starts the pick up operation. The finished parts will be layed down on the upper spiral chute and will be pushed by a pneumatic piston from this position. All operations will be controlled by limit switches and light barriers.

The workpiece type B is stored in a "shelf type" magazine (figure 5). The shelf has four stores, each store contains eight blanks. The blanks will be separated by a pneumatic piston and pushed in a defined pick up position. If a store is empty the shelf will be lifted so that the next store comes in loading position. The store for the finished parts is built up as an inclined push channel. It is mounted behind the shelf.

The automatic chucking tools work with hydraulic power and will be operated by the robot control. Figure 6 shows the device without the power installation. The parts will be clamped within the boring. Two limit switches control the defined position of the workpiece.

5. Laboratory test

Afterwards all devices has been proved, the whole manufacturing system has been installed for a two week laboratory test. In this installation all functions and interlocks were implemented except the two milling machine tools. The hydraulic chucking tools were setting up on two tables.

The robot has been programmed for the real time operation. The waiting period for the machining were simulated. All components has been proved in an endurance test for the two weeks, 16 hours on each day. All troubles has been notified, statistically analysed and removed. The handling program and the handling speed has been optimized. After that,the system has been worked over the whole two weeks, so we were sure, that the most troubles are repaired and only further troubles could occur in consequence of the cutting process.

Figure 7 shows the laboratory installation of the manufacturing system. At the left hand and in the front view are the tables with the hydraulic chucking tools. The spiral chute is in the background and at the right hand is located the shelf type magazine.

6. System implementation

Figure 8 shows the cycle diagram for the whole machining process. After the industrial robot has been started, it picks up a workpiece type A from the spiral chute and checks the correct gripping force before it swings to the horizontal milling machine. The robot puts the blank into the chucking tool and checks the correct position. The part will be clamped and the control master starts the machine tool during the robot swings to the horizontal milling machine. Here waits the robot for the end of the machining, the arm extends to the finished part, lifts it up, checks the gripping force and swings back to the spiral chute. After laying down the part on the chute, the industrial robot reaches back to the horizontal milling machine, which has meanwhile finished the first pass of part A, picks it up and feeds the vertical milling machine for the second pass. Because of the long machining time on this machine tool , the robot swings to the shelf magazine, picks up a workpiece type B and feeds the horizontal milling machine, which is still idle. The robot stayes for the end of the milling process and puts the finished part from type B on the inclined push channel.

Then it swings back to the spiral chute, picksup a workpiece type A, feeds the horizontal milling machine again and a new cycle can continue.

The handling operation covers 48 program steps and 28 set point potentiometers. Eight output commands and nine input interlocks were necessary. The robot control starts, interrupts and supervices the whole machining process. Figure 9 shows the input and output data flow between the industrial robot control and the peripheral devices.

For safety all swing motions with a workpiece in the gripping device were inerlocked to the force sensor signal. After a workpiece output the sensor signal "no force" also has been interlocked to the program stepping sequence. Further safety facilities has been a guard around the manufacturing system and a central emergency button.

7. Economics

With the discussed manufacturing system the attandant can operate three machine tools. In former times he was loading two program controlled milling machines. Now he operates one conventional machine tool and supervices the manufacturing system. So the manual work at the system could decrease of 90 %. Furthermore the idle time was reduced. The old operating time has been 0.7 min per part for workpiece type B and 2.3 min per part for type A. The cycle time of the manufacturing system takes for both parts 1.9 min. That has been a reducing of 37 % (without system down time).

For a first economics the following data has been calculated

Industrial robot	80.000,– DM	(34,335 US $)
Accessories and safety devices	40.000,– DM	(17,167 US $)
Development and laboratory test	10.000,– DM	(4,292 US $)
	130,000,– DM	(55,794 US $)

The payback period results with the equation [5]:

$$A = \frac{I}{L - E \pm q(L + C)}$$

were:

A = payback period in years
I = robot and accessories
L = total annual labour saving
E = expense of robot upkeep per year
C = capital value of associated equipment
q = production rate coefficient

The expense of robot upkeep was calculated with 6.500,– DM (2,790.– US $), the capital value was 36.000,– DM (15,451.– US $) and the production rate increased for 0,2.

$$A = \frac{55,794 \text{ US } \$}{18,026,– \text{US}\$ - 2,790.– \text{US}\$ + 0.2(18,026.– \text{US}\$ + 15.451 \$)}$$

A = 2.54 years.
================

8. Conclusion and future work

This paper described the linking of two programmable milling machine tools with an industrial robot associated with two workpiece magazines. The components of the system were introduced and explained. The material flow and the exchange of input and output commands between both machine tools and the master control has been discussed.

The apply of flexible handling systems for feeding machine tools for cutting shaping demonstrated us the demands to the industrial robot andthe accessories. The aim to develop a manufacturing system for middle series production has been fully performed. But long set up by the machine tools, the industrial robot and the accessories are limits. The set up time for the whole manufacturing system has been about eight hours.Therefore in this point has to start a further development.

A first step in this direction are numerical controlled machine tools in connexion with computer controlled industrial robots. The robot control should have a large program memory as well as branch, jump and subroutine facilities. Programming the handling devices should be done both by punching a tape as well as leading the robot arm manually. These are important conditions for working with paletts as workpiece magazines. With such facilities will the set up time decrease and the expenses for a plenty of magazine types in the small batch production could be abolished.

Today we are doing works in sensor developping, computer control systems for industrial robots and flexible accessories. The aim for the next years will be an automated flexible computer controlled manufacturing cell. That will be a machining and handling system for small batch production, which works a three shifts day. The system should work the whole night without manual operation.

9. References

1. Spur, G.; Auer, B.H: Die Steuerung von Industrie-Robotern. Zeitschrift für Wirtschaftliche Fertigung (ZwF) 68 (1973), H. 8, S. 381-387.

2. Spur, G.; Auer, B.H: Industrie Roboter zum Beschicken von numerisch gesteuerten Werkzeugmaschinen. Zeitschrift für Wirtschaftliche Fertigung (ZwF) 69 (1974), H. 1, S. 3-8.

3. Auer, B. H.: Industrial Robot Feeds Numerical Controlled Machine Tools. The Industrial Robot, June 1974, P. 173 - 177.

4. Spur, G.; Auer, B.H.: Die automatische Handhabung bei flexiblen Fertigungszellen. wt-Z. ind. Fertig. 65 (1975) S. 117-123.

5. Engelberger, J.F.: Four million hours of robot field experience. Proceedings of the 4th International Symposium on Industrial Robots. Tokyo 1974.

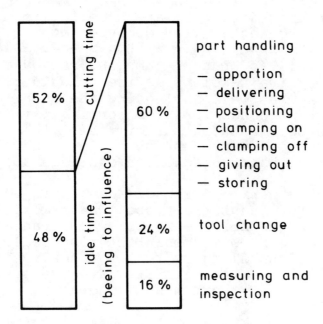

Fig. 1: Percentage of the operation time in the small batch production.

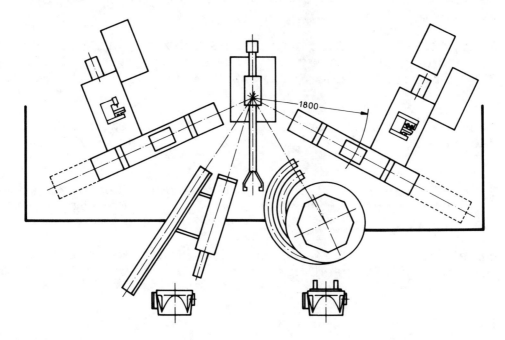

Fig. 2: Lay out of the complete manufacturing system. From left to right: Vertical milling machine tool, workpiece magazine type B, industrial robot, workpiece magazine type A, horizontal milling machine.

Fig. 3: Industrial robot "Transferautomat E", VFW Fokker GmbH

Fig. 4: Spiral chute for workpiece type A.

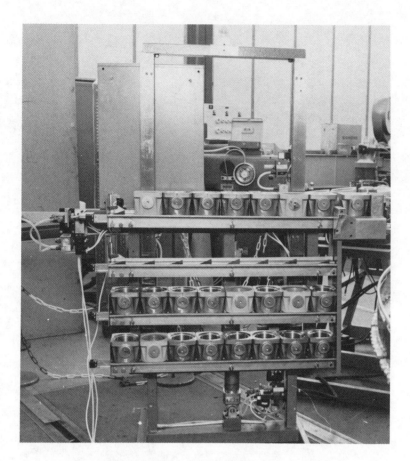

Fig. 5: Shelf type magazine for workpiece B.

Fig. 6: Chucking tool for both parts.

Fig. 7: Laboratory installation

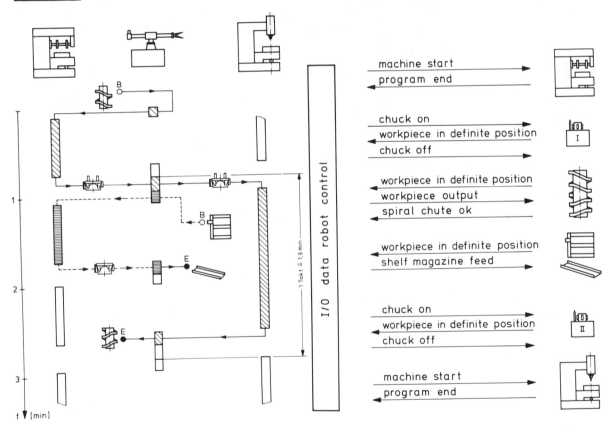

Fig. 8: cycle diagram for
the whole machining
process.

Fig. 9: Output commands and
input interlocks.

machine start
program end

chuck on
workpiece in definite position
chuck off

workpiece in definite position
workpiece output
spiral chute ok

workpiece in definite position
shelf magazine feed

chuck on
workpiece in definite position
chuck off

machine start
program end

I/O data robot control

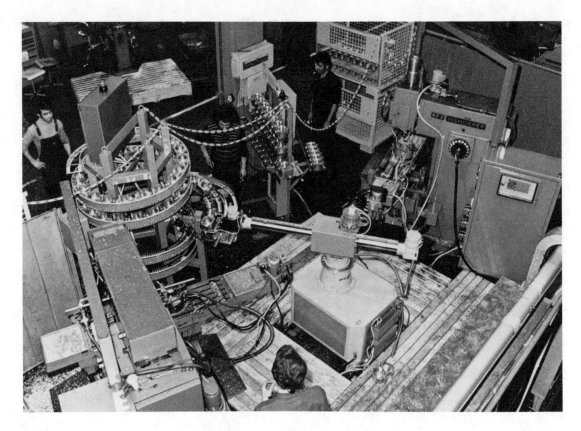

Fig. 10: Manufacturing system within small batch production.
From left to right: Horizontal milling machine,
spiral chute, shelf type magazine, industrial
robot, vertical milling machine tool.

Reprinted from Manufacturing Engineering, November 1979

Automated Product Evaluation and Regulation at the Assembly Line

Advancing electronics technology sparked the development of these powerful fourth-generation machines, whose capabilities now embrace component adjustment, quality control, and diagnostic testing

GARY GARCIA
Assistant Editor

AUTOMATED ASSEMBLY can increase process efficiency and improve product quality — this is well known. Indeed, automation has been utilized in manufacturing for many years, most notably in the automotive industry.

But assembly and fabrication are not the end-all applications of automation. With the powerful electronic controllers available today, automation can successfully be applied to such operations as component adjustment, quality control, and diagnostic testing. Such are the capabilities of four machines built for a major automobile manufacturer by the Automatic Production Systems Division of Ingersoll-Rand.

The primary function of the machines is to set the preload on 7¼" (184.15 mm) differential bearings and to set backlash between the ring and pinion gears, and the machines are called, appropriately, backlash/preload machines. Additionally, they perform numerous diagnostic tests on the part — and on themselves.

Among the part diagnostic tests are measuring initial pinion running torque to gage the accuracy of the initial pinion bearing preload setting, monitoring stretch of the differential carrier under load to verify its quality, and checking gear backlash at a number of points to determine if unacceptable gear run-out is present.

Self-diagnostics include verification of the accuracy of linear variable distance transducers (LVDT's), shunt testing of torque transducers, ascertaining that the pinion clutch drive is operative, and if d-c motors employed are stalled, for example.

Machine Configuration. This is not a newly automated process. Rather, it has been performed for a number of years by various machines from other manufacturers, and automation has yielded favorable results. Availability of improved technology and advances in the state of the art subsequently pointed the way for evolution of the system. The search for a sophisticated, fourth-generation device culminated in the purchase of four special Ingersoll-Rand machines for over $1 million.

Whereas the machines previously used employed analog electronics and were not "intelligent," the new machines utilize digital electronics, embodied in programmable controllers, and make many decisions during the course of the setting/testing sequence.

Modular in design, all sensors and tooling of the backlash/preload machines are grouped into various assemblies. This facilitates assembly and disassembly, incorporation of modifications, and maintenance. Moreover, all gaging is mounted on the front-column face of the machines for accessibility.

In keeping with the customer's wish to avoid use of pneumatics, all gaging is hydraulically controlled using a low-pressure system. The only exception is

◄ *BACKLASH PRELOAD machine and, at left, its control console.*
The main conveyor (not shown) runs along the front of the machine and parts are transferred to the machine tooling by a pop-up roller conveyor, shown supporting a 7¼" differential assembly and pallet. Safety guards removed for clarity.

MACHINE TOOLING with the angular resolver gagehead engaging the ring gear. The spindle nut drivers with protruding spring-loaded pins are completely backed off from the part. The small cylindrical items along the horizontal axis of the photo are the LVDT's. ▶

found in the spindle nut drivers (see drawing), which are powered by d-c motors. Pallet-stop cylinders mounted on the carousel are the only pneumatic devices associated with the machines.

The steel for the machines went up fast, but this accounted for only about 40% of the implementation time. The other 60% was needed for system debugging, revising setting and testing sequences, and, in this case, overcoming problems associated with newly discovered parts characteristics — a tribute to the machine's diagnostic capabilities.

During the course of the backlash/ preload setting and testing activities, dimensional data, the current sequence step, and reject codes are transmitted to a control panel for display. As long as parts are acceptable, they are automatically returned to the main conveyor and, generally speaking, monitoring of the control panel is not required. However, should a defect be detected, the test sequence is stopped and the operator, who can monitor all of the backlash/preload machines, is signaled with a red beacon.

Subsequent decision to accept or reject the part lies with the operator, and powerful diagnostic tools are available through the machine to provide information on which to base this decision.

To fully appreciate the backlash/ preload machine's abilities the complex setting/testing sequence and tooling should now be examined in detail, and these will now be described.

Part Transfer. As pallets enter the area of the four backlash/preload machines, they are apportioned among the stations by logic-controlled part-counting switches. When a pallet is in place in front of a machine a switch is tripped, signaling this condition to the machine. The pallet is subsequently transferred to the machine tooling by a pop-up roller conveyor. This device raises the pallet on a set of powered rollers, perpendicular to the main conveyor rollers, and drives it into the machine where it is latched in place. Power-accumulating rollers are employed here, and these are most forgiving in the event of a jam or crash. In this case, the pallet following the jammed one will simply roll against it and stop.

There are two types of differentials being set by the manufacturer, and these differ in ring-gear position by ⅜″

(9.525 mm). Moreover, these carriers are intermixed. This problem was overcome by incorporating a proximity switch in the machine that detects the ⅜″ variance and causes the machine tooling to shift by that amount when necessary to accommodate the differential currently in place.

Two clamping arms then raise the part, divorcing it from the pallet, and clamp the differential carrier's upper surface against a machined surface of the backlash/preload machine. Positional accuracy is ensured through the use of two locating pins, one round and one diamond, that engage two holes in the differential carrier. Another proximity switch is employed here to determine if the part is seated properly. If not, reject code 12, "Part Not Seated," is signaled.

Initial Gaging. The *A* nut and then the *B* nut adjusting spindles are next brought in at low torque. There are spring-loaded pins on the end of each spindle, and when these properly engage the adjusting nuts the spindles stall out. A fault may occur here: "Nut Stripped," signaled by defect code 15.

The parts enter the stations with no backlash and some preload, and at this point the preload must be removed. To accomplish this, two LVDT's are engaged on opposing sides of the part on the ends of the carrier tubes. The *A* nut is incrementally backed off until the LVDT readings remain unchanged for two consecutive nut positions, indicating the differential carrier has stopped shrinking and the preload has been removed. The *A* nut is then backed off one more increment to ensure a zero preload condition exists.

Interestingly, the dimension along the axis the LVDT's are gaging can vary

±⅛″ (3.175 mm) because it is not a closely held dimension. Since the working range of the LVDT's is so much smaller than this, they are automatically repositioned for each part. This was handily accomplished by mounting the LVDT's on a slide mechanism that is spring-centered about a gaging arm. To gage a part, the arms are brought in from a stand-off position until they contact the part. The LVDT slide moves along with the arm, and when the arm contacts the part the springs serve to position the transducers at the approximate center of their usable range. The slides are then clamped in place.

The nut-driving spindles are driven by d-c motors with controls that make them act like stepping motors in a closed-loop configuration. In combination with integrally mounted angular encoders, these motors can be signaled to turn a certain amount. The encoders supply positional feedback to ensure that rotation of the spindles corresponds with the command to the motors, shutting them off. Moreover, the motors are driven from the same pulse train and therefore can be moved synchronously. Overshoot is minimal and predictable.

Drag Torque Check. A dead weight is dropped against a shaft running through the *B* nut spindle to the differential. This serves the dual purpose of separating the differential ring gear and pinion, which are sometimes tightly engaged, and of aging the carrier, i.e., relieving the carrier of internal stresses. The weight will later be used to simulate the bearing preload force required.

Initial drag torque of the pinion is then checked to ascertain that pinion preload was properly set. This is done by engaging a pinion driver equipped with a torque transducer from beneath the part and running the pinion gear. Here, two defect codes may be signaled. They are 1 and 14, "Initial Pinion Torque Out of Limits" (high or low, and how much) and "Pinion Clutch Out" (differential assembly locked), respectively.

The torque transducer is also used to monitor for torque spikes under zero backlash conditions as the pinion is driven, and these are indicative of stalling and tight spots. To avoid noise in the form of unwanted torque spikes, a smooth-running spherical ball hydraulic motor is used to drive the pinion.

Preload Sequence. To avoid varying the preload setting when backlash is later set, the ring gear is run in to a zero backlash position, indicated by a torque spike from the pinion-driver transducer, then backed off a little. This permits the backlash to be set with a minimum of nut position adjustment, thus minimizing the effects on the preload setting.

The first step in the preload setting sequence is application of the previously used dead weight to the differential case. The magnitude of the weight is such that it results in the same amount of preload that is to be applied to the bearings. Consequently, the amount of carrier stretch resulting from application of the dead weight will be the same as that caused by proper preloading of the bearings. So, carrier stretch is monitored with the LVDT's with the weight in place. Then the weight is removed and the B nut is torqued until the same stretch dimension is achieved, indicating the proper amount of preload has been applied.

Two reasons for part rejection are associated with carrier stretch under load from the dead weight. They are "Stretch from Dead Weight Low" (less than 0.003"/0.076 mm) and "Stretch from Dead Weight High" (more than 0.006"/0.152 mm) identified by defect codes 2 and 3, respectively. Note that low and high carrier stretch acceptable to the

VERTICAL SECTION of the backlash/preload machine. Main conveyor flow is into the page.

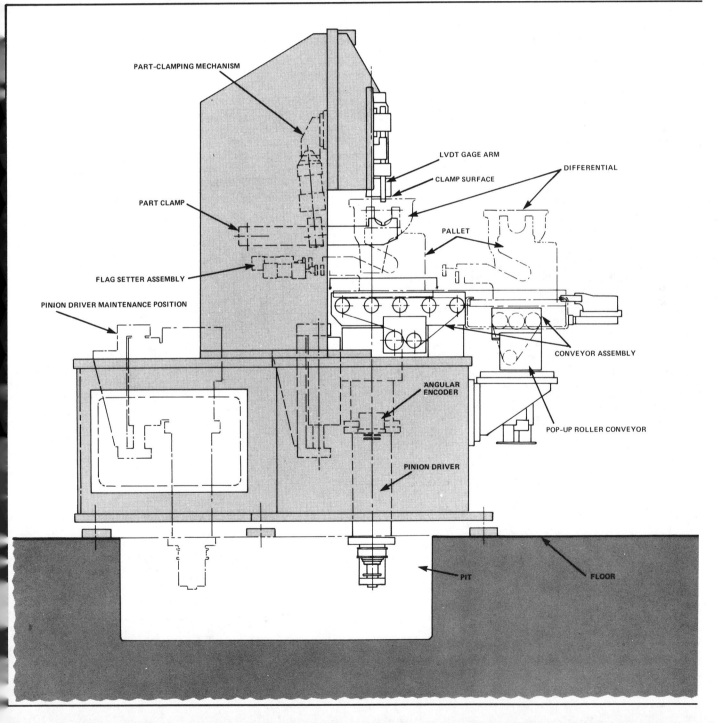

manufacturer are 0.002″ (0.051 mm) and 0.007″ (0.179 mm), bracketing the machine's acceptable stretch window. This variance compensates for possible machine inaccuracies. Another possible defect code is number 13, "Nut Motor Stalled."

Backlash Sequence. To set backlash between the pinion gear and ring gear, the latter is engaged with two wheels. One is a gage head for an angular resolver accurate to 0.01 degree, while the other is a drive wheel. The pinion gear is clamped in position with a Hydra-Loc chuck, and the ring gear is rocked back and forth by the drive wheel.

Angular readings are taken at the limits of the ring gear movement and delivered to the programmable controller in digital form. The programmable controller converts these angular readings into a backlash reading. The Hydra-Loc chuck then releases the pinion, which is indexed 180° and again clamped in place. Another backlash reading is then taken in the manner previously described.

A defect can occur here: "Backlash Variation High" (more than 0.003″/0.076 mm), defect code 4. If a defect does not occur but backlash adjustment is necessary, the machine will do this, centering the backlash window on the desired value. After this initial reading and adjustment, the machine will take another reading, attempt to adjust backlash again if necessary, then take a final reading. Defect code 5 will signal "Final Backlash Out of Limits" (0.004″/0.102 mm to 0.007″ /0.179 mm).

During this process the initial, intermediate, and final backlash values are displayed on the control panel. It is estimated there is no need for backlash adjustment 25 to 30% of the time, and that backlash is properly adjusted 50% of the time on the first try.

Final Tests. A final preload check is

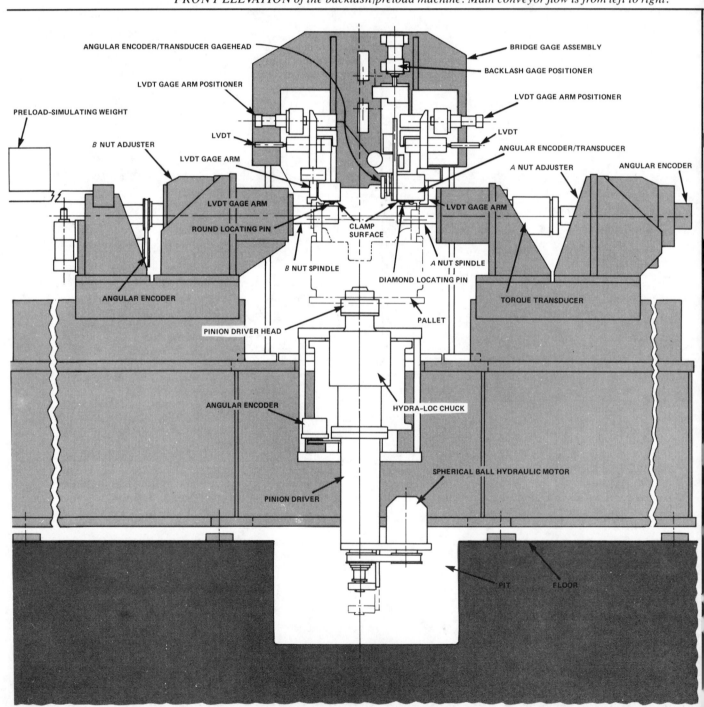

FRONT ELEVATION of the backlash/preload machine. Main conveyor flow is from left to right.

made by again measuring stretch of the part. Defect codes 6 and 7 can be invoked at this point. The former indicates "Carrier Stretch Low" (lost more than 0.001″/0.025 mm) while the latter indicates "Carrier Stretch High" (gained more than 0.002″/0.051 mm). If necessary, preload can be adjusted by varying the *B* nut setting by an amount small enough not to drastically alter the backlash setting but large enough to influence the preload.

Final pinion running torque is monitored as the pinion gear is again driven. Here defect code 8 will signal "Final Pinion Torque Out of Limits" (high or low, and how much). This completes the backlash and preload setting and testing sequence.

A flag is automatically set to indicate the part has been worked on and is not to be checked by any of the other stations. A good part is returned to its pallet and the pallet is sent back to the main conveyor.

Health Check. The backlash/preload machines do not sit idle between part sequences. Rather, after every cycle they perform a health check to ascertain that their transducers are operating properly.

The slides holding the LVDT's are moved against gage blocks, and the LVDT's are exercised against other gage blocks 0.005″ (0.127 mm) from the first. The LVDT readings are compared with this known dimension for accuracy. If an error is detected, defect code 9 will indicate an "LVDT Fault." Moreover, a visual dial gage is provided for comparison with the LVDT readings, but can be taken out of the machine sequence if so desired.

All torque transducers, too, are checked by passing a current through them and monitoring the signal returned. Defect codes 10 and 11 will signify "Pinion Transducer Fault" and "*A* Transducer Fault," respectively.

Importantly, if a machine fails a health check it shuts down and cannot be used again until it has been repaired and suc- cessfully passes a health check. And this shutdown cannot be overridden from the control console.

Cycle Time. Remarkably, one part can be cycled through the entire backlash/preload setting and testing sequence in an average cycle time of 45 seconds. Moreover, although there is a net production requirement of the four machines, each has a variable cycle time depending on part condition. For example, the part might arrive with the adjusting nuts backed off all the way. Since the nuts are adjusted at a low rate of,speed to facilitate detection of torque spikes, in this case it would take some time to run the nuts back in. The number of attempts needed to set backlash also influences machine cycle time.

Variable cycle time can be accommodated because the transfer line used is nonsynchronous. That is, parts are accepted and released as a machine becomes available or completes its sequence, rather than being tied to a fixed operational time period characteristic

EACH INGERSOLL-RAND backlash/ preload machine has a "brain" consisting of two Modicon 484 programmable controllers. The 484 CPU, semiconductor memory, and power supply are all contained in a 15 x 20½ x 6½″ (381 x 266.7 x 165.1 mm) housing weighing only 25 pounds (11.34 kg).

Communication with the outside world is accomplished through I/O modules, each providing four discrete inputs or outputs. Eight modules can be mounted in an I/O housing, and eight housings can be accommodated by an I/O channel,

THE MODICON 484 PROGRAMMABLE CONTROLLER

for a total of 128 discrete inputs and 128 discrete outputs in a 40 x 32 x 6½″ (1016 x 812.8 x 165.1 mm) assembly. With optional equipment this capacity can be doubled.

I/O modules are available with the following formats: 115 V a-c, 220 V a-c, 9-56 V d-c current source or sink/positive or negative logic, 90-150 V d-c, 5 V TTL (transistor-transistor logic), and proximity switch (input). Register I/O modules include a 16 x three-digit register multiplexer, plus A/D (analog to digital) input modules with four or eight inputs for 1-5, 0-10, and −10 to +10 V, and 4-20 mA. There is also a special purpose dual high-speed (up to 30 kHz) counter module.

The 484 is available with 256, 512, 1024, 2048, and 4096 words of 8-bit semiconductor memory. Memory battery power back-up is provided, and a battery condition indicator gives seven days warning of imminent battery failure. Typical scan time for a 4k 484 is said to be 19.5 ms.

The 484 is available with three different instruction sets. The basic instruction set provides for relay logic, timers (0.01, 0.1, and 1 second), three-digit counters, registers, and internal coils. The Enhanced I instruction set includes the basic instruction set plus eight 32-step sequencers, multiplication, division, addition, subtraction, and comparison of three-digit numbers, provision for transitional contacts with 1k-memory or more, 32 input and 32 output registers, I/O expansion capability to 256 inputs and 256 outputs, BCD (binary-coded decimal) convert (input-to-register, register-to-output), and built-in one-shots. The Enhanced II instruction set includes Enhanced I plus the SKIP function, MOVE (table-to-register, register-to-table), binary convert (input-to-register, register-to-output), and Modicon's MODBUS communication capability.

The P180 Programming Panel allows programming in simplified relay ladder logic using a CRT and pictorially encoded pushbuttons. Up to 77 contacts can be displayed at one time in an 11 x 7 format called a network, and many networks can be assembled, much like the pages in a book.

System, functional programming, and power-up errors are detected and displayed on the CRT, as is the amount of memory used. Register contents can also be displayed. With the P180, inputs and outputs can be disabled and outputs can be forced into a desired state. Moreover, the P180 is RS232C compatible.

The 484 is available for 115 or 220 V a-c, 50 or 60 Hz electrical systems and draws 100 watts of power. It is designed to withstand industrial environments, being rated for 0-60° C and 0-95% relative humidity. ∎

of synchronous transfer lines. Moreover, should one or more machines lock up as a result of a jam or detection of a reject, the duty cycle of the other machines is altered to take up the slack. This is possible through the use of limit-count switches and appropriate feedback logic.

Reject Mode. As previously mentioned, when any of the machines rejects a part, subsequent procedures must be initiated by the operator. These are accomplished by communicating with the machine through its control panel.

Each control panel incorporates six numeric light-emitting diode (LED) displays. Here are displayed the carrier stretch, running torque, initial backlash, intermediate backlash, final backlash, test sequence step number, and reject codes. When a reject is encountered the machine stops and that sequence step number and appropriate reject code are displayed. A list defining all reject codes is mounted on the control console.

The other five displays will show data available at this time. So, the cause for rejection is immediately apparent to the operator through the displays, and a number of alternate courses of action are possible.

► If rejection was caused by a marginal parameter magnitude, the operator can acknowledge the reject and continue the sequence. This can also be done for subsequently detected defects. Upon conclusion of the entire sequence all part defects will be sequentially flashed on the reject code display.

► If the condition of a part is such as to warrant immediate rejection, the test can be aborted and the defective part released from the machine.

► The part can be retested as many times as desired. This can be done as a complete automated sequence or in discrete sequence steps using the GET NEXT STEP control.

► By stepping through the test sequence and monitoring the data displayed, the cause for rejection can be traced to a specific part characteristic. This is invaluable for providing constructive feedback to assembly and fabrication process operators. Moreover, since measured parameters are temporarily stored by the programmable controller, a test history is available for inspection.

It should be noted that the bewildering assortment of pushbuttons and indicator lights on the lower part of the control console need not be used for these operations. These buttons and lights are provided for manual control of individual machine functions that make up each step in the sequence, and are primarily for use in maintenance and ma-

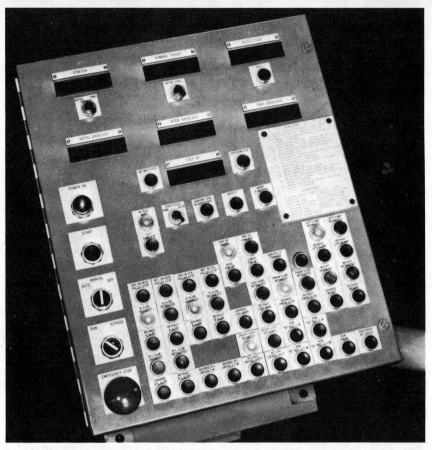

CONTROL PANEL of the backlash/preload machine showing the LED data displays at top and manual sequence controls at bottom. Normally, only the four pushbuttons in the center of the console need be used by the operator when the machine rejects a part. Note the defect code key and test step summary at right.

chine set up. Rather, the operator will typically use only four pushbutton contols located in the center of the panel: GET NEXT STEP, ACKNOWLEDGE REJECT & CONT., RETEST, and ABORT & RELEASE.

Electronic Control. Each backlash/preload machine is managed by two Modicon 484 programmable controllers. Why two? Original plans called for use of one 484 with small memory usage to provide as fast a scan time as possible, while the second 484 would do the bulk of the work. But Modicon subsequently introduced the Enhanced II instruction set for the 484, and this includes a SKIP function. This feature allows improved scan time for isolated high-speed machine sequences, which would seem to obviate the need for two 484's in this application.

But two 484's were still needed to provide the memory capacity required for the large number of software features incorporated by Ingersoll-Rand, as each 484 could support only 4k of 8-bit semiconductor memory. It can be expected that programmable controller memory capacities will be expanded in the future to support the software required for increasingly complex applications.

It is interesting to note that process revisions are easily implemented, thanks to the use of programmable controllers. Whereas use of a microprocessor-based system would have necessi-

tated the services of a software specialist, anyone familiar with relay ladder logic can program the Modicon 484.

Using a system built around hardwired analog boards, too, would have resulted in quite a predicament when the need arose to modify the setting/testing sequence. And extensive modifications were made, even on the factory floor. In fact, the process now used is totally different from that originally conceived.

Not only do the programmable controllers allow rapid process changes to be made, but with the addition of a CRT and a few software patches a veritable diagnostic laboratory is at hand. For example, the machine can be made to stop at critical points in the sequence, or can be made to repeatedly cycle through a particular function to facilitate data collection.

Moreover, the programmable controllers leave the door open for incorporation of future innovations. They could be tied into a host computer system for generation of part defect summaries, for example. Or, a hard-copy printer could be added for tagging defects with reject codes and associated diagnostic data. In short, provision has been made for incorporation of almost any conceivable modification, thanks to the versatility of programmable controllers. ∎

Reprinted from Manufacturing Engineering, July 1980

Parts Feeding: Some Classic Cases

Parts feeding and orientation belong to the realm of empirical engineering. Since there are no rules to go by, these examples of past successes may prove helpful

THERE ARE NO SET RULES or pat answers to the problems of feeding the wide variety of parts now being processed through automatic assembly machines. Many problems can be avoided by proper part design, but most parts require a comprehensive analysis and cut-and-try development for successful automatic assembly. This article presents examples of how some unusual and difficult parts feeding and orienting problems have been solved.

One surefire method of avoiding problems is to eliminate the need for part orientation. If this need can't be eliminated, the number of orientations required at the assembly machine should be minimized. This can be accomplished by retaining orientation from a previous manufacturing operation or an outside supplier, with the parts in stacks placed in or on racks, magazines, tapes, tubes, rods, arbors, or other transfer devices. This may be the only feasible method of handling fragile, easily damaged, or interlocking parts.

As an example, thin stampings are often produced in a continuous strip without separation. The strips are then wound into coils for placement on the assembly machine. Means must be provided on the assembly machine for separating the individual parts as required. Many parts from blanking dies can be inserted immediately into magazines or

1. PICK-AND-PLACE UNIT lifts chamfered washers from feeder system and places them into turnover jaws. Sensing switch determines whether washers are inverted.

2. TRANSFER DEVICE moves retainers (one shown in inset) from vibratory bowl (lower right) to a pickup device (top center), thus eliminating shingling and jamming.

transport racks by the punches.

Another way of eliminating the need for part orientation is to produce them on the assembly machine. This is an especially suitable method for simple parts such as shims, washers, gaskets, and springs. Formed-in-place gaskets can be produced on the assembly machine from hoppers of liquid silicone compounds or other materials. Many springs are being wound, and other parts that tend to tangle or are difficult to feed are being stamped from coil stock directly at the assembly machines.

Most feeding and orienting of small parts from bulk (random orientation) is handled by vibratory bowls or hoppers. Parts placed randomly in the bowl or hopper are induced by the vibration to climb a spiral track on the wall. Orienting devices, such as wiper blades, changes in track width, and bars, slots, or grooves to match part indentations and projections, are fitted to the tracks ahead of the outlets. These devices allow only parts in the proper position to pass. Those that are not correctly oriented fall back into the hopper for recycling.

Flat, thin, lightweight, and delicate parts may require nonvibratory feeders. These include centerboard and bladed wheel hoppers, stationary or revolving hook types, centrifugal units, rotary discs, and tumbling barrel hoppers.

BELLEVILLE WASHERS

The problem was that of feeding Belle-

ville-type bowed washers with their dished faces on one side only. This was accomplished simply by providing a bulge in the track contour. Washers bowed outward from the side of the track tumble

3. BLADE CAM (lower left) on vibratory bowl has a ridge projecting into feed track that untangles spring-type clamps and positions them in single file.

back into the vibratory bowl as they pass the bulged section because their centers of gravity are outside the washer trims. Washers that are bowed to the inside of the track continue to feed through the narrow section of the track in proper orientation.

CHAMFERED WASHERS

The problem was that of feeding washers — a routine enough problem that's normally handled by vibratory feeding. Complications were introduced by way of a 60° chamfer at one end of the bore. Thus the lack of perfect symmetry ruled out vibratory feeding, and other solutions were required. By adding a special escapement and probe mechanism, feeder manufacturers could easily determine whether washers were properly oriented. Those that were incorrectly oriented could be returned to the bowl for refeeding. This alternative would have been comparatively expensive, and would have required feeding the washers at twice the speed required by the assembly machine on the assumption that 50% of the washers would be misoriented.

The problem of automatically handling the washers was solved comparatively easily and inexpensively by Assembly Machines, Inc., Erie, PA. The solution involved use of standardized tooling components employed on the firm's basic AMi/1 assembly machines. With this method, the washers are fed

4. ESCAPEMENT MECHANISM separates and moves heavy ball studs from slotted inclined track (seen at lower right) into the placement jaws of a pick-and-place positioner.

flat with random orientation. At the end of the in-line feeder system, each washer is lifted by the outboard head of a double pick-and-place unit, *Figure* 1. Attached to this pickup head is a sensing switch which detects proper orientation. The washer is placed into turnover jaws located between the outboard and inboard heads of the double pick-and-place unit.

If the washer is correctly oriented, as determined by the sensing switch, it is not inverted. Tooling attached to the inboard head of the double unit removes the washer from the jaws in the intermediate position, and places it over the top of a shaft on the assembly machine. Simultaneously, tooling on the outboard head lifts another washer from the feed track.

When a washer is upside down — as determined by the sensing switch — the intermediate jaws turn it over, thus providing proper orientation prior to the washer's removal by tooling on the inboard head. Subsequently, it is placed on the assembly. This technique, which is completely reliable, uses 100% of the output from the feed system. Accordingly, the feed system only has to operate at half the speed that would be required

if each washer had to be probed, and misoriented parts returned to the vibratory bowl.

ELIMINATING SHINGLING

Shingling of parts was a major problem anticipated in designing a feeding unit for fragile bearing retainers. From past experience, it was known that the thin edges of the stampings, about 1″ (25 mm) in diameter, would cause the parts to shingle during feeding down a vibratory or gravity track. Additionally, the retainers are heat treated to spring hardness, a condition which contributes to part warpage. Both conditions would cause jamming within the track, and unreliable pickup, thus reducing the efficiency of the assembly machine.

A solution developed by Automated Process Inc., Milwaukee, WI, for reliable feeding was to eliminate the track. Instead, a transfer device is used to move the retainers from the bowl to a pickup device. This eliminates shingling or jamming, and isolates the part for reliable pickup.

The closeup view, *Figure* 2, shows a small section of the vibratory feeder bowl (lower right), the six-arm spider

transfer device, the pick-and-place unit (top center), the nest (left) at the loading station on the rotary dial table of the assembly machine, and a retainer (inset). Retainers are oriented in the bowl so that the fingers in their bores are projecting downward. Parts flow clockwise in the vibratory feeder bowl, and the spider transfer device rotates counterclockwise.

Each of the six arms on the spider transfer device has a pin that projects upward to hold the retainers. As each arm moves into the loading position under the bowl, a maximum of four retainers fall on its pin, the number being controlled by photocells. Nonsynchronous indexing of the spider carries the parts to an unloading position. At this point, a vacuum device in a pick-and-place mechanism lifts the retainers from the pins and moves them to the nest at the loading station on the assembly machine. This design eliminates the timing problems frequently encountered with one-on-one part feeding to indexing stations.

UNTANGLING CLAMPS

Assembly of spring-type clamps to the ends of tie rod tubes for steering linkages eluded automation for some time because of tangling and interlocking of the clamps. In the bulk stage, the clamps were often severely tangled, with up to four clamps interlocked. This resulted in jamming of the feed tracks and subsequent downtime of the assembly machine.

Close cooperation between the supplier of the vibratory bowl feeders, Midwest Feeder, Inc., Belvidere, IL, and the manufacturer of the assembly machines, Dixon Automatic Tool, Inc., Rockford, IL, led to an ingenious solution. It involved use of a blade cam, seen at the lower left in *Figure* 3, on each vibratory bowl. A ridge, projecting from the cam into the feed track, untangles the clamps and positions them in single file. Vibratory motion of the bowl causes the slots in the clamp to nest on a narrow, raised rail at the end of the track, thus providing proper orientation. Clamps not untangled and oriented fall back to the bottom of the bowl for recycling.

At the end of the track rail, the separated and oriented clamps are properly positioned for feeding to the assembly machine. At this point, a bolt is automatically positioned through each clamp, and driven into a nut escaped into position below the clamp. Then, clamp assemblies are pressed onto each end of the tie rod tube, thus effecting the assembly.

STUD FEEDING

The relatively high weight of ball studs

led to problems in their feeding, separating and positioning via inclined tracks. However, the problem was solved with the setup shown in *Figure* 4. The automatic assembly machine shown is a product of Dixon Automatic Tool.

Initially, ball studs are fed from a storage hopper up a moving lift conveyor. The studs are then oriented, with threaded ends down, in a slotted inclined track, hanging from their heads. With this inclined track, seen at the lower right, the parts can be easily fed. However, there is a force buildup at the parts-separation point because of the weight of the studs. Dixon designed a special escapement mechanism to easily separate and precisely move the parts one at a time into the mechanically actuated placement jaws of one of their standard pick-and-place positioners.

Another potential problem in the automatic assembly of these components was the variation in overall spring height. The springs are fed from a vibratory bowl feeder tooled to separate and position the springs in single file. An in-line vibratory track carries the springs to a standard Dixon pick-and-place positioner. All of these positioners include a sensing mechanism that inspects for proper placement. Inspection for placement was simplified by adjusting this built-in sensing unit to compensate for variations in overall spring height.

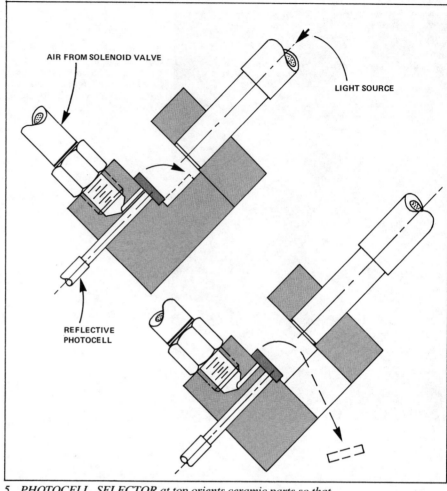

AIR FROM SOLENOID VALVE

LIGHT SOURCE

REFLECTIVE PHOTOCELL

5. *PHOTOCELL SELECTOR at top orients ceramic parts so that printed faces are all in one direction. Selector at bottom blows incorrectly oriented parts back into vibratory bowl.*

6. *KNIFE-EDGE SELECTOR uses bevel on one of longer edges on each ceramic substrate to properly orient the parts.*

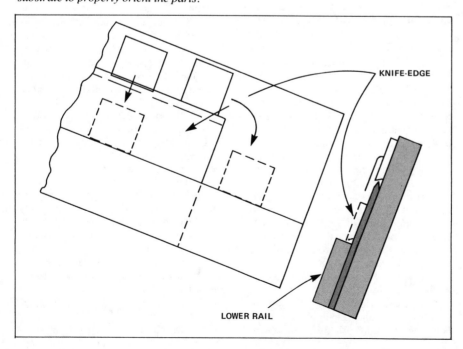

KNIFE-EDGE

LOWER RAIL

CERAMIC SUBSTRATES

Tiny white ceramic substrates present problems in that they have no external features to facilitate orientation mechanically. Measuring 0.68″ (17.3 mm) long by 0.16″ (4.1 mm) wide by 0.04″ (1.0 mm) thick, they do have a black stripe of electrically resistive material printed off-center on one face. The Anaheim, CA, operation of Swanson-Erie Corp. solved the orientation problems presented by these parts by using reflective photocells to identify the printed side of each substrate. Next, the offset direction of the printed stripe was identified for use on vibratory feeder bowls.

The first selector, shown at the top in *Figure* 5, orients the parts so that their printed faces are all in one direction. A reflective-type photocell is normally kept actuated by a light source. When a part passes in front of the photocell with its unprinted face towards the photocell, the white surface reflects the light emitted by the photocell, thus maintaining its actuation. However, when a part passes with its printed face toward the photocell, the reflected light is reduced, causing photocell deactuation. This energizes a solenoid valve that allows a blast of air to blow the part onto its opposite face, as shown.

Parts exiting from this selector are merged into a single feed line, with the printed faces of the substrates up. The parts are then mechanically inverted so that they enter a second selector, seen at the bottom in *Figure* 5, with their printed faces down. In this selector, the photocell will only see parts having a printed stripe that is below center. Such parts, being incorrectly oriented, will be blown from the track and into the vibratory bowl for recirculation.

Another vibratory bowl device developed by Swanson-Erie to select oriented ceramic substrates is shown in *Figure* 6. In this case, the parts, 0.34″ (8.6 mm) long by 0.31″ (7.9 mm) wide and 0.04″ (1.0 mm) thick, may or may not be printed at the time of feeding. The only identifiable feature on each part is a 20° bevel on one of the longer edges.

When a correctly oriented part is fed onto the knife-edge of this selector, the vibratory action of the feeder causes the part to slide from the knife-edge onto the lower rail because of its beveled edge. Misoriented parts feed the full length of the knife-edge, and fall from the end onto the lower track. Drop height is such that most of the parts rotate 90° during the fall. Those that do present a new edge to the next selector.

Two lines, each having four selectors, are used in the primary stage of the feeder. At 100% efficiency, the select-and-rotate action should result in orienting 50% of the parts fed. This reduces recirculation requirements because the number of parts correctly oriented in the first selector averages 12.5%.

The two lines of selected parts are merged into a single track, and the parts are passed through a final selector before discharge. Nonoriented parts are recirculated after passing through the fourth primary selector in each line. Using this system, oriented parts are fed at a rate of more than 180 per minute.

The vibratory bowl on which these selectors are used feeds the substrates through two separate printing operations. It can also be used to feed a high-speed sorter that inspects the parts for electrical resistance and continuity.

Swanson-Erie also developed a smaller vibratory bowl to feed this part to an automatic assembly machine. This bowl

7. *UNTANGLING/FEEDING SYSTEM automatically places 12 springs into the keys of a telephone dial assembly within five seconds.*

was equipped with a printed-side orienter, similar to the one shown at the top in *Figure* 5, and two selectors like the one at the bottom. This combination provides a suitable feedrate in a minimum space while limiting recirculation. Minimizing recirculation is important for ceramic substrates because they tend to pick up metal from the bowl surfaces. If they become excessively metalized, conductive paths can be established, a condition leading to defective assemblies.

FEEDING OF SPRINGS

Compression-type springs are among the most difficult components to feed automatically in high-volume applications. However, Automation Associates, Inc., Addison, IL, has developed a complete line of equipment capable of untangling and feeding single or multiple lines of oriented springs, including single spring escapement and placement mechanisms. Three basic units handle a range of spring diameters from 0.062 to 0.625″ (1.57 to 15.88 mm).

The systems utilize timed, pulsating air blasts to agitate and separate tangled springs that are bulk loaded into an upper bowl. The same air supply drives a number of springs through outlet nozzles and into plastic tubes. The tubes, which serve as magazines and are normally full, convey the springs, end-to-end, to manual operator pick-offs, portable pen escapements, or fully automatic escapement mechanisms. Periodic reversal of the air flow direction automatically clears the outlet nozzles of any tangled springs, driving them back into the bowl, thus eliminating downtime due to jamming.

A U.S. telephone manufacturer is using this system, *Figure* 7, to place 12 springs, 0.070″ (1.78 mm) in diameter by 0.500″ (12.70 mm) long, into the keys of each dial assembly. This is accomplished automatically within five seconds, using two six-line systems on five installations to feed springs to automatic escapements.

An operator hand loads each preassembled dial on a pneumatic slide nest. Hand contact with dual palm buttons initiates the automatic cycle, advancing the slide with its assembly to the spring loading station, locating the key spring holes, and positioning all 12 springs simultaneously. Prior to use of these installations, the operators would hand load the springs, which were fed from vibratory bowls.

In many applications where only one spring is required per assembly, two or three outlet systems have been tooled to feed springs to various locations, spaced as far apart as 15 ft (4.6 m). The units can be readily adapted to existing assembly machines, thus eliminating operator loading and control, increasing productivity, and reducing costs. These dispensers can also be used for conveying pins, small tubes, and similar parts.

Slight modifications of springs may be necessary to assure successful feeding. It also helps to have two coils close together at each end of the spring, or about three coils close together in the middle of the spring. On tapered diameter springs, the end of the wire at the larger diameter end should be drawn over the center of the spring axis to prevent tangling. Springs with two tapered ends should have at least three or four coils close together at both ends. It is also necessary that springs supplied to the bowl be dry and clean.

Another method of feeding springs is via a Hopperator incorporated in feeding systems made by the Vibromatic Co., Noblesville, IN. This unit, which serves as a bulk feeder and preliminary untangler, also maintains the level of parts in a vibratory bowl by means of a proximity control switch. Upon demand, the unit oscillates vertically to allow parts to pass through grids into the vibratory bowl, thus preventing the entry of large clumps of tangled springs. In cases of extreme interlocking problems, a mechanical separator can be incorporated in this system. Interlocked parts enter the separator and are ejected as single springs prior to feeding through the orienting and selecting mechanisms in the vibratory bowl.

HANDLING MICROCOMPONENTS

Microcomponents such as diodes, chips,

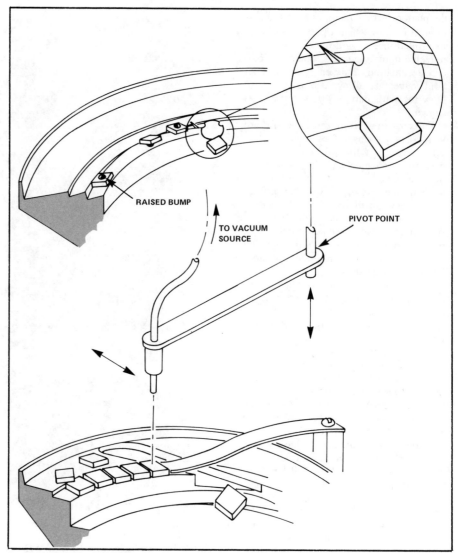

8. *ENLARGED VIEWS of sections of feeder bowls used for semiconductor microcomponents as small as 0.010″ square.*

dice, and ferrite memory cores used in the semiconductor industry present their own class of feeding/orienting problems. Some of these parts are as small as 0.010″ (0.25 mm) square, and many have to be oriented so that they leave the feeder proper side up. The Minimus Co., Venice, CA, specializes in the design and production of vibratory feeders and orienting devices for such parts.

Feeder bowls made by this firm are custom machined from aluminum bar stock. Precision machining of these compact bowls is difficult, but the surface finish produced is superior to that of conventional cast bowls. This provides gentler handling and less abrasion of delicate parts, especially semiconductor dice, integrated circuits, and similar items. Another advantage is that the bowls can

be machined to whatever track width, helix angle, exit configuration, or orienting feature required for a specific application. The bowls are normally plated or anodized to resist wear.

Feeder drives for the vibratory bowls are of cup core design, with fully enclosed coil and small magnetic gap. In addition, armature overhang provides shielding to eliminate stray magnetic fields. Feedrates are infinitely variable from less than one to over 100 parts per second. Models are available for clockwise or counterclockwise movement of the parts.

Enlarged views of sections of these feeder bowls are shown in *Figure* 8, with track configurations originally developed by Rudolf F. Zurcher of the Hughes Aircraft Co. In the top view, parts traveling

9. *WASHERS IN VIBRATORY BOWL seen at upper left drop over shanks of bolts as the components pass through assembly section (seen at center).*

clockwise have reached the end of the inclined portion of the track, and are about to slide onto the final, level part of the track. The parts being fed are dice having a solder bump on one side. It is necessary to orient them with the bumps up.

Near the end of the inclined track, there is a groove running parallel to the track and near its left-hand edge. Dice with their bump surface up will slide over this groove, and onto the level portion of the track. Those with their bumps facing down will get caught in the groove, and be dumped back into the bowl for refeeding when they reach that portion of the groove interrupted by a cutaway section. Debris such as pointed slivers of silicon are also disposed of in this manner.

The bottom view shows the dice after they have traveled to the end of the level track. They are prevented from advancing farther by an externally mounted stop. Properly oriented dice are commonly removed from the track, one at a time, by a vacuum lifting device that transfers them from the feeder to a test or assembly station. Any dice reaching the level track that have climbed each other and overlap slide off onto an overflow track. This overflow track, which is slanted toward the outer edge of the bowl, carries the dice to a chute for return to the bowl and refeeding.

ASSEMBLING WITHIN VIBRATORY BOWLS

There are applications where certain parts can be assembled within the vibratory

bowls themselves. When applicable, this approach can result in a considerable reduction in cost. One example, *Figure* 9, is a system made by the Vibromatic Co., Inc., Noblesville, IN, for assembling bolts and washers. Washers in the vibratory bowl seen at the upper left are fed clockwise, while the bolts in the other bowl are fed counterclockwise. As the bolts pass through the assembly section (center), washers drop over their shanks.

Built into this system are checking devices to make sure that a washer, and not more than one, has been placed on each bolt. Washers that do not fall on the bolts are ejected back into the proper bowl (upper left) for refeeding. Satisfactory assemblies are discharged to a placement station where they are positioned in nests on an assembly machine.

NOISE REDUCTION

Several methods are available to reduce the noise from vibratory parts feeders and handling systems. Coating the outside of the bowls with damping material can help, but acoustical enclosures of noise insulation and absorption material are generally more effective. Some commercial units consist of a transparent plastic hemispherical dome mounted on a cylindrical side enclosure fitted with sound-absorbing material. Quick-opening clamps, hinges, or air-actuated means are available for ready access.

Vibromatic Co. offers a mechanical-

10. *MECHANICAL FEED SYSTEM has central auxiliary hopper, and bottom of centrifugal feed system rotates, eliminating need for vibration and reducing noise.*

type feed system of centrifugal rotary-feed design. As shown in *Figure* 10, the system is equipped with a central auxiliary hopper; the bottom of the centrifugal feed system rotates. This eliminates the need for vibration to drive the parts, and offers other advantages in addition to noise reduction. For example, in the application illustrated, Oilite bearings can be fed faster because of the additional drive produced in the bottom of the system. Such parts usually do not feed well in vibratory systems because the oil causes the parts to adhere to the tracking surfaces.

CENTRIFUGAL FEEDERS/ ORIENTERS

Speeds faster than those attainable in vibratory types, gentler handling, and quieter operation are benefits of centrifugal feeder/orienter systems made by the Automation Div. of Hoppmann Corp., Springfield, VA. Parts entering the centrifugal feeder bowl first move onto a flexible, rotating disc which rotates on a fixed cam plate. A combination of gravity and centrifugal force moves the parts to the outer edge of the disc where they are transferred to an orienting rim that rotates independently around the disc.

The cam surface positions the disc at the same level as the rim on one side of the bowl, and slopes downward to form a trough area against the ID of the rim on the opposite side of the bowl. A fixed guide ring is positioned over the rim to direct the parts through the orientation process. Parts properly oriented pass through the qualification tooling and are fed out of the bowl. Those incorrectly positioned are wiped back into the trough area for orientation on subsequent revolutions.

One of these systems, developed to feed forged-steel ball-joint sockets, is shown schematically in *Figure* 11. Attempts to handle these parts in overhead mounted, vibratory parts feeders with gravity delivery tracks proved to be unreliable, orientation was difficult, and operation was very noisy. With the centrifugal feeder shown, orientation is simple, sound levels are maintained at or below 80 dB (A), and from 40 to 50 sockets are delivered per minute from random bulk.

The positions of five short projections on each socket flange with relation to a profile qualifier allow properly oriented parts to leave the bowl, while incorrectly positioned sockets fall back onto the disc. Guiderods at the feeder exit are positioned to pick up the flange lips and lower the sockets onto a powered delivery conveyor.

Another centrifugal feeder made by Hoppmann is being used to feed primed 5.56-mm cartridge cases from bulk, orient them open-end-up and diameter-to-diameter, and press them into clips on a continuously moving chain. This is being accomplished with a linear speed of 3800 ipm (96 520 mm/min) and a production rate of 1200 cases per minute, with an efficiency of 99.95%, eight hours a day, five days a week.

In addition to feeders, Hoppmann Corp. also makes complete parts handling systems and Continuous Motion (C/M) assembly machines. Some of these use continuous motion, input starwheel modules to escape the parts individually from the accumulation track positions in the hoppers, and transfer them to continuously rotating turrets of assembly systems equipped with cam actuated tooling stations. Continuously rotating transfer wheels can also be used to automatically unload completed assemblies. This eliminates the need for intermittent stopping or dwelling of the assembly machine at every cycle, thus increasing productivity. ∎

11. CENTRIFUGAL FEEDER for forged steel, ball joint sockets simplifies orientation and reduces the sound levels generated.

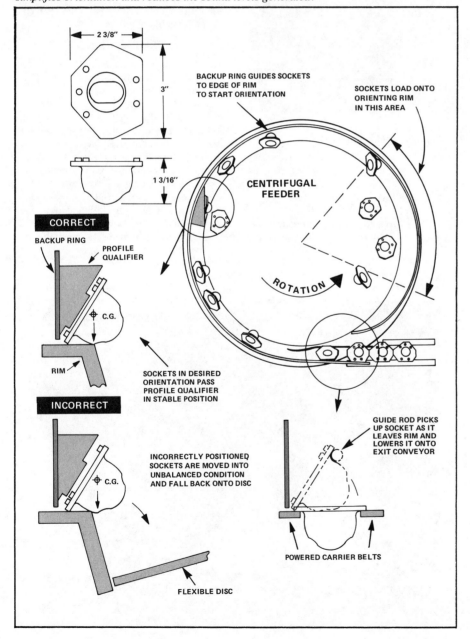

Presented at RIA's Robots Conference, October 1977

Workpiece Transportation by Robots Using Vision

By Robert K. Kelley, J. Birk, and V. Badami
University of Rhode Island

Workpieces transported to fixtures by robots may not have precise alignment with the robot hand. Practical methods to avoid obstacles without requiring the complexity of geometric modeling are presented. An algorithm has been developed which computes the arm joint values needed to compensate for this misalignment. The algorithm relates changes in the features of images from two TV cameras to changes in the position and orientation of a workpiece. During an instruction phase, known position and orientation perturbations are automatically applied to a sample workpiece to establish the correspondence between these changes. Also, safe ways to grasp the workpiece and collision free trajectories are defined. During production as each workpiece is handled, position and orientation variations are noted to permit the actual relation between the workpiece and robot hand to be computed. The workpiece can then be transported along a safe trajectory and brought to the fixture without a collision.

INTRODUCTION

Industry is looking to put robots to work at new jobs, especially in batch manufacturing and assembly. Computer-based manipulators (robots) have the desirable attributes which are needed for such tasks. They are easily re-programmed and of general purpose design. Their workpiece grasping systems include tongs, electromagnets and vacuum pickups. However, robot hands typically do not have the sensory capability of human hands. They cannot measure workpiece slip, for example, when acquiring a workpiece. For this and other reasons, workpieces which are transported by robots to fixtures or other goal sites may not be precisely aligned with the robot hand. The problem is to transport this kind of robot held workpiece to its goal site without a collision. That is, without regrasping the workpiece, compensate for the misalignment and avoid collisions enroute and with goal site structures as the robot places the workpiece.

WORKPIECE TRANSPORTATION ALGORITHM

An algorithm has been developed which computes the robot arm joint values needed to compensate for workpiece misalignment with the robot hand. In general, misalignment can be in any of three degrees of freedom in position and three degrees of freedom in orientation. Throughout this paper, the term "pose" will be used to refer to both position and orientation.

The workpiece transportation algorithm will be described by first considering the activities which take place during the production phase. Then the instruction phase activities which preface the production activities and make them possible are described.

Production Phase

During production the pose of each workpiece is checked as it is handled. Pose variations are calculated, corrections are computed and alignment is performed before the workpiece is transported along a safe trajectory to the fixture. A diagram of a typical production layout is shown in Fig. 1.

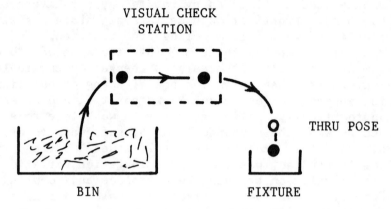

Fig. 1. Typical production layout with visual check station.

Workpiece pose check.--Each workpiece is brought to the visual check station. Initially, the robot assumes prespecified arm joint values to bring the workpiece to a predefined check pose.

Two television cameras are used to extract image feature values from the pair of workpiece images. For example, image features such as the center of gravity and direction of the minimum moment of inertia axis may be extracted from the binary image of a workpiece. These features are drawn on the binary image of a sample workpiece in Fig. 2. These feature values are compared with those expected for the workpiece having the check pose. If

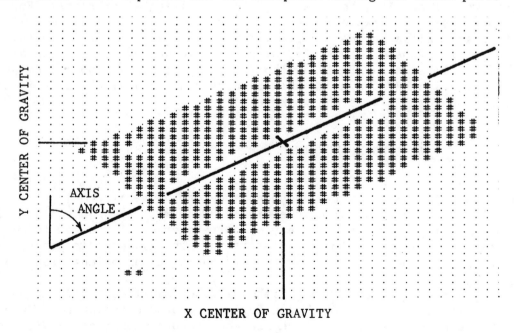

Fig. 2. Binary image of workpiece with features indicated.

the feature variations correspond to acceptably small misalignments in the workpiece pose, the workpiece can be transported to the fixture. If the feature variations indicate larger workpiece misalignments, the workpiece pose must be corrected before the workpiece can be brought to the fixture.

Workpiece pose misalignment is computed in terms of perturbations to the workpiece pose. The perturbations are calculated by weighting the image feature variations. In matrix form this calculation is expressed as

$$\overline{\Delta p} = [Q] \, \overline{\Delta f}$$

where: $\overline{\Delta p}$ = perturbation vector for the workpiece pose
$[Q]$ = weighting matrix relating feature variations to pose perturbations
$\overline{\Delta f}$ = variation vector for image features.

Thus, each constituent of workpiece misalignment can be obtained. Tests can be applied to each degree of freedom individually or in combination to determine whether or not the misalignment is acceptably small.

Using the results of the workpiece pose check, one of the following actions is appropriate:

1. Proceed to transport workpiece along a safe trajectory to the fixture when the pose is satisfactory.

2. Correct the workpiece pose and then
    a. transport along a safe trajectory if the effect of the correction does not need to be observed, or
    b. repeat the workpiece pose check if the effect of the correction needs to be determined.

3. Discard that workpiece and acquire another when the workpiece pose misalignment is too large.

Workpiece pose correction.--Workpiece pose perturbations define a correction which can be applied to the current estimate of the robot hand to workpiece relation (Ref. 1). Thus $\overline{\Delta p}$ defines a correction transformation matrix, T_c. The robot hand to workpiece relation is denoted by a transformation matrix, ${}^h T_w$. (The reader is directed to Ref. 2 where a detailed discussion of representing kinematic relations as transformation matrices is given.) The corrected estimate is computed by the matrix product (see diagram in Fig. 3)

$$({}^h T_w)_c = ({}^h T_w) \, (T_c) \; .$$

Since the robot base to workpiece relation for the check pose at the visual check station is known, the corrected pose for the robot hand can be obtained. Let $({}^O T_w)_{check}$ denote the workpiece check pose at the visual check station. Since

$$({}^O T_w)_{check} = ({}^O T_h)({}^h T_w), \quad {}^O T_h = ({}^O T_w)_{check} ({}^h T_w)^{-1}$$

defines the robot base to hand relation. This is illustrated in Fig. 4.

The robot arm joint values which correspond to this hand pose are found by solving equations which are specific to different robot kinematic configurations. (See Ref. 3, Appendix 7, for a typical solution.) When the arm joint values are adjusted to these new values, the workpiece pose should be closer to the nominal one.

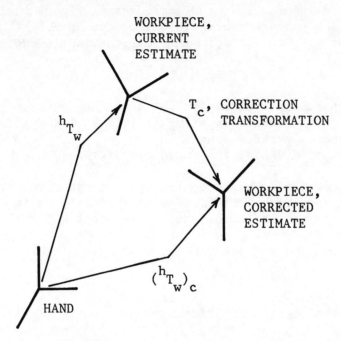

Fig. 3. Diagram for correction of ^hT_w estimate.

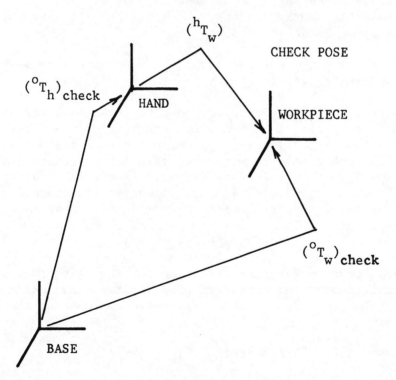

Fig. 4. Diagram for check pose.

Safe trajectory transport.--To transport a workpiece to a fixture, a selection is made from among the predefined safe trajectories. The selection is made on the basis of the deviation of $(^hT_w)_c$ from $(^hT_w)_j$ associated with a safe trajectory. If this deviation is small for the jth legal way to grasp the workpiece, then it is assumed that the robot arm structures will not collide with the workstation when the workpiece is moved along a safe, collision free trajectory.

The safe trajectory is specified as a sequence of poses (called thru poses) through which the workpiece progresses as it is transported from the visual check station to the fixture. The kth workpiece thru pose along the trajectory specifies a corrected robot hand pose. During instruction, $(^OT_w)_k$ is defined. During production it is required that the workpiece achieve the same pose. (See diagram in Fig. 5.)

INSTRUCTION

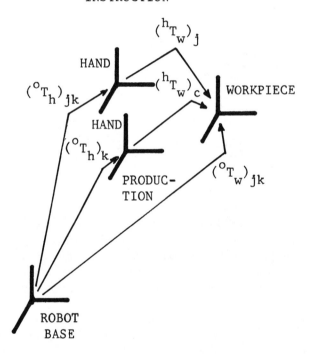

Fig. 5. Diagram for reestablishing k^{th} workpiece thru pose.

Thus

$$(^OT_w)_k = (^OT_h)_k \ (^hT_w)_c \ ,$$

where $(^hT_w)_c$ does not change as the workpiece moves along the trajectory. This is shown in the diagram in Fig. 6. Hence

$$(^OT_h)_k = (^OT_w)_k \ (^hT_w)_c^{-1}$$

specifies the corrected robot hand pose to achieve the desired workpiece thru pose on the trajectory. Notice $(^hT_w)_c^{-1}$ is computed only once for each trajectory and $(^OT_h)_k$ is solved for the robot arm joint values to achieve each thru pose on the trajectory.

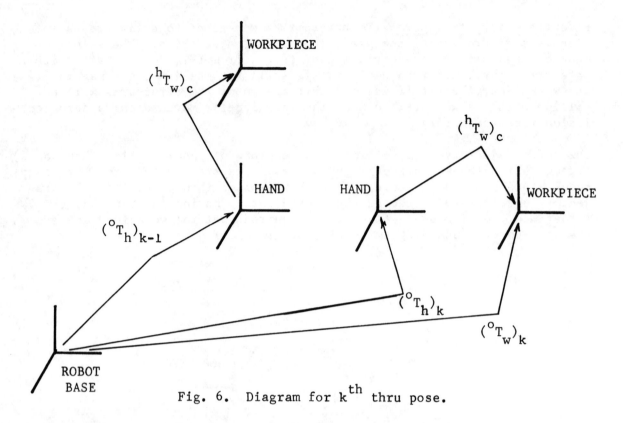

Fig. 6. Diagram for k^{th} thru pose.

Instruction Phase

During instruction a sample workpiece is employed to obtain the data which is needed to accomplish the production phase activities. In particular, legal ways to grasp the workpiece in the fixture are specified, safe, collision free trajectories between the fixture and the visual check station are defined by thru poses, workpiece check poses at the visual check station are defined, and the correspondence between workpiece pose perturbations and image feature changes is determined experimentally.

Legal ways to grasp workpiece.--To describe legal ways to hold a workpiece, a sample workpiece is placed in the fixture. Manually, the robot hand is brought to the workpiece and the workpiece is grasped without removing the workpiece from the fixture. The arm joint values are recorded.

The arm joint values are used to compute the robot base to hand relation $^{o}T_h$. For the first legal way to grasp the workpiece, the robot hand to workpiece relation $(^{h}T_w)_1$ is prespecified. For example, the workpiece coordinate system might be aligned with the hand coordinate system and displaced to a point near the center of the workpiece. Thus

$$(^{o}T_w)_{fixture} = (^{o}T_h)_1 \, (^{h}T_w)_1$$

describes the workpiece in the fixture.

To specify a second way to legally grasp the workpiece in the fixture without a collision, the robot hand is brought to the workpiece until the workpiece can be grasped, and then arm joint values are recorded. These arm joint values specify $(^{o}T_h)_2$. Since the workpiece is in the fixture (see diagram in Fig. 7), the robot hand to workpiece relation is obtained from

$$(^{h}T_w)_2 = (^{o}T_h)_2^{-1} \; (^{o}T_w)_{fixture} \; .$$

These steps are repeated for each $(^{h}T_w)_j$ which corresponds to a distinct way to grasp the workpiece.

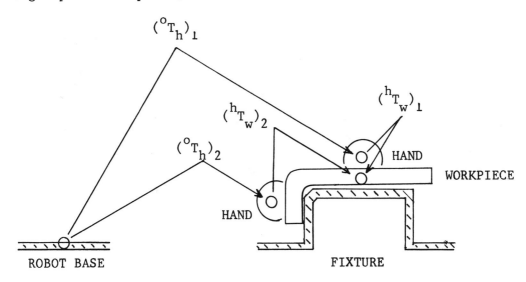

Fig. 7. Defining ways to grasp workpiece.

Specification of safe trajectories.--For each legal way to grasp a work-piece, a safe trajectory is specified. This trajectory is one through which the workpiece passes as it is transported between the visual check station and the fixture. (A common safe trajectory might be sufficient for all legal ways to grasp a workpiece.)

The safe trajectories are specified by grasping the workpiece while in the fixture in one of the legal ways, for example, the first way. The work-piece is then removed from the fixture to a thru pose by a programmer using manual controls to manipulate the robot hand. Typically the last thru pose is close to the fixture and differs only by a small displacement. Arm joint values are recorded. This procedure is repeated to define sufficient thru poses to guarantee a collision free path between the visual check sta-tion and the fixture. The programmer should select trajectories which maximize clearance between the arm and the goal site. Clearance between the workpiece and the goal site should also be as large as possible as this will not be controlled during production since the workpiece trajectory, not arm joint trajectory, is made to resemble trajectories defined during instruction.

The set of arm joint values at each thru pose along a trajectory define a workpiece pose. For example, referring to Fig. 8, at the kth pose along a trajectory (which begins with the pose nearest the visual check station) the arm joint values define $(^{o}T_h)_{jk}$ where j is used to indicate the way the workpiece is grasped. Then the workpiece pose at the kth thru pose is

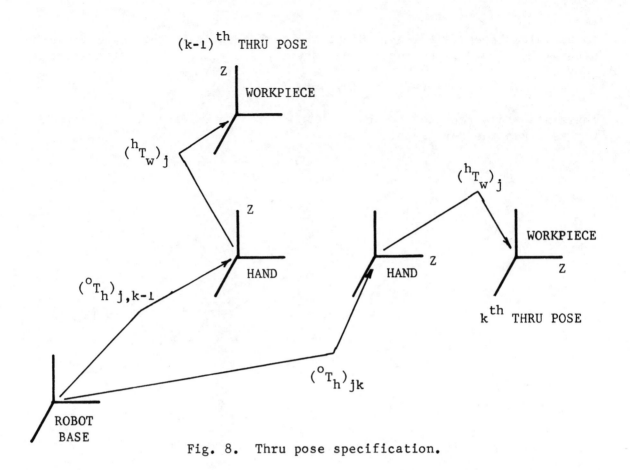

Fig. 8. Thru pose specification.

defined by the matrix product

$$(^{O}T_{w})_{jk} = (^{O}T_{h})_{jk}\ (^{h}T_{w})_{j}\ .$$

<u>Check pose</u>.--A check pose for the workpiece at the visual check station is defined for each legal way the robot can grasp the workpiece. (Ideally, a common pose would suffice.) Two properties which are desirable for the check pose are:

1. That the check pose have as many arm joint values as possible in common with those for the workpiece fixture pose to minimize the introduction of pose errors as a result of mechanical motions. This is particularly true of rotary joints for which small angular errors give rise to significant displacement errors when multiplied by the length of the arm links. (This consideration should also influence the selection of thru poses when specifying trajectories.)

2. That the feature values extracted from the pair of television images of the workpiece at the check pose possess sufficient sensitivity and range to characterize the pose perturbations of interest.

The second property can be facilitated by the proper placement of the pair of television cameras.

<u>Pose perturbations and feature changes</u>.--For each distinct check pose, the

correspondence between perturbations to the workpiece pose and the changes in the image feature values is determined. This is done experimentally by sequentially perturbing the workpiece coordinate system one degree of freedom at a time and computing the feature variations which result.

If the coordinate system is perturbed in position by $\Delta X, \Delta Y, \Delta Z$, and in orientation by $\Delta\alpha, \Delta\beta, \Delta\gamma$ (see Fig. 9 for definitions), then the correspondence

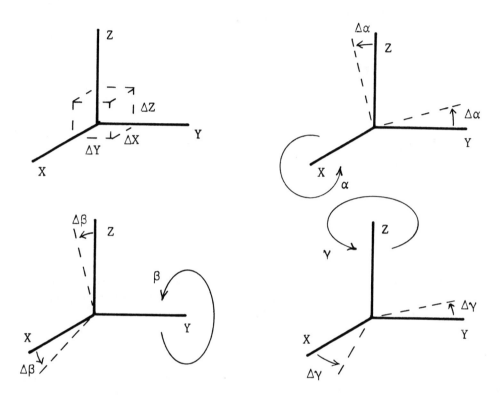

Fig. 9. Coordinate system perturbations.

can be represented as a weighting matrix $\lfloor R \rfloor$ so that

$$\overline{\Delta f} = \left[R \right] \overline{\Delta p}$$

where $\overline{\Delta f}$ and $\overline{\Delta p}$ are feature variation and pose perturbation vectors as defined previously, and $\overline{\Delta p} = (\Delta X, \Delta Y, \Delta Z, \Delta\alpha, \Delta\beta, \Delta\gamma)^t$. The columns of $\left[R \right]$ are obtained by sequentially applying

$$\overline{\Delta p_X} = (\Delta X, 0, \ldots, 0)^t, \ \overline{\Delta p_Y}, \ \ldots, \ \overline{\Delta p_\gamma} \ .$$

To be able to determine the pose perturbations which correspond to feature variations, a weighting matrix $\left[Q \right]$ as mentioned previously is needed to satisfy

$$\overline{\Delta p} = \left[Q \right] \overline{\Delta f}$$

where: $\left[Q \right] = \left[R^t R \right]^{-1} R^t$, pseudo matrix inverse (more than six features)

or $\left[Q \right] = \left[R^{-1} \right]$, matrix inverse (six features).

The configuration shown in Fig. 10 is being used to establish an experimental basis for applying the transportation algorithm described in the preceding section. During our experiments, image analysis and workpiece pose correction are performed alternatively as the ^hT_w estimate is refined.

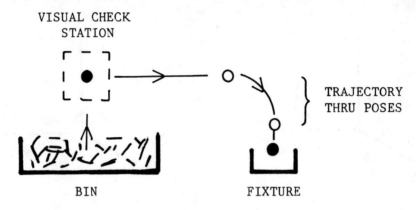

Fig. 10. Experiment configuration.

The investigation is being conducted at a single visual check station which features a compact visual zone. The visual zone is defined by the fields of view of the two television cameras. These cameras have a right angle relationship between the two principal rays as shown in Fig. 11.

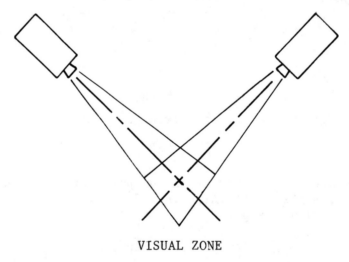

Fig. 11. Right angle camera geometry.

The URI Mark III robot which is used to perform the actual workpiece manipulations was designed and fabricated at the university. The robot has six degrees of freedom, three orthogonal linear joints, and three rotary joints. The robot and its design is discussed in Refs. 3, 4, and 5.

Our image analysis is presently based on closed circuit television camera images which are converted to binary format by applying a computer controlled intensity threshold to each analog picture element sample. Modules to automatically determine the perturbation weighting matrices [R] and [Q] have been written and are operational. This software includes feature extraction, robot kinematics, arm joint value solution, and mathematical computation modules. Almost all of the instruction phase software has been validated but has not yet been integrated. Specifically, the module for the specification of the legal ways to grasp a workpiece is working. The production phase software shares a large amount of the instruction phase software. The module for pose correction is being assembled with the largest portion of the production phase software awaiting integration and validation.

Example: Pose Correction

To show the nature of the pose correction algorithm, an artificial example will be presented which uses some data collected on the sample workpiece whose binary image was shown in Fig. 2. A single television camera was used. The camera had its principal optical axis collinear with the workpiece coordinate system Z axis. Thus the angles α and β are rotations about the axes in a plane parallel to the image plane and γ is an angle of rotation in the image plane.

The workpiece was rotated independently about each of the three workpiece coordinate axes. Data points for the change in a feature value (direction of the minimum moment of inertia axis) are plotted in Fig. 12.

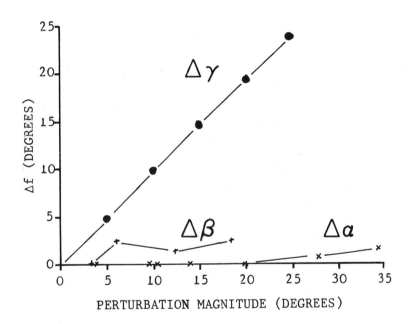

Fig. 12. Experimental data.

Notice that for angular perturbations within 20°, there is no correspondence between Δf and $\Delta\alpha$, only a very noisy correspondence between Δf and $\Delta\beta$, and a nearly one to one correspondence between Δf and $\Delta\gamma$.

The [R] matrix in this case is a row vector where:

$$R_1 = avg\ (\Delta f_1)/avg\ (\Delta\alpha)$$

$$R_2 = avg\ (\Delta f_2)/avg\ (\Delta\beta)$$

$$R_3 = avg\ (\Delta f_3)/avg\ (\Delta\gamma)$$

where avg is the arithmetic average of the quantities. For this data,

$$[R] = (0,\ 0.144,\ 0.973)$$

Because this is an underdetermined system, the pseudo inverse is given by

$$[Q] = R^t\ [R\ R^t]^{-1}\ .$$

The computation yields the column vector

$$Q = \begin{bmatrix} 0. \\ 0.149 \\ 1.006 \end{bmatrix}$$

Thus, the pose correction is based on

$$\begin{bmatrix} \Delta\alpha \\ \Delta\beta \\ \Delta\gamma \end{bmatrix} = \begin{bmatrix} 0. \\ 0.149 \\ 1.006 \end{bmatrix} (\Delta f)$$

Assume that the true workpiece pose can be represented by $[0,\ 0,\ 20^o]_{true}$ and it is desired to bring the pose to $[0,\ 0,\ 0]$. (Actually the angles must be ordered to represent orientation accurately for large angles.) Using the data which led to Fig. 12, $\Delta f = 19.29^o$ and the pose correction computed is $[\Delta\alpha,\ \Delta\beta,\ \Delta\gamma]_{computed} = [0,\ 2.9^o,\ 19.4^o]$. After applying this correction the true pose is $[0,\ -2.9^o,\ 0.6^o]_{true}$ which has a feature variation $\Delta f = 0.584^o$. The computed pose correction is $[\Delta\alpha,\ \Delta\beta,\ \Delta\gamma]_{computed} = [0,\ 0.087^o,\ 0.587^o]$. Performing the correction, the true perturbation is now $[\Delta\alpha,\ \Delta\beta,\ \Delta\gamma]_{true} = [0,\ -3.0^o,\ 0.003^o]$.

The correction procedure would terminate at this point since, for this example, the data indicates that $\Delta f = 0$ would be obtained from the workpiece image.

This example points out the sensitivity of this feature to rotations in the plane and the lack of sensitivity about axes in the plane for this particular workpiece. Additionally, this example demonstrates the hazards associated with an underdetermined correction system.

RESEARCH ISSUES

Through our experiments, the influence of factors such as the number of picture elements subtended by the workpiece, lens selection, the image features used, and the characteristics of the workpiece will be studied.

Some additional research issues to be resolved are the systematic selection of the check pose at the visual check station and the placement of the two television cameras, the characterization of ^hT_w estimation errors and their effect on the amount of detailed knowledge which is required about the structures in the workspace and which is needed to accomplish collision avoidance.

Other research issues are concerned with improving the utility of the transportation algorithm by finding ways to minimize the number of iterations of the ^hT_w estimation procedure and of generalizing from discrete to continuous specification of the legal ways to grasp a workpiece.

SUMMARY

An algorithm was described which permits workpieces which are not precisely aligned with the robot hand to be transported without collision to the fixture. Computation of arm joint values which compensate for this misalignment was explained. Practical methods to avoid obstacles without requiring the complexity of geometric modeling have been presented. That is, a method to specify collision free (legal) ways to grasp a workpiece in the fixture and a method to specify safe trajectories from the check station to the fixture were enumerated.

ACKNOWLEDGEMENT

This material is based upon research partially supported by the National Science Foundation under Grant No. APR74-13935. Any publications generated as a result of the activities supported by this grant may not be copyrighted without the specific written approval of the Grants Officer and the authors. Any opinions, findings, and conclusions or recommendations in this publication are those of the authors and do not necessarily reflect the views of the National Science Foundation.

REFERENCES

1. Kelley, R., et al., "Workpiece Orientation Correction with a Robot Arm Using Visual Information", Proceedings of 5th International Joint Conference on Artificial Intelligence, Cambridge, Mass., August 1977, p. 758.

2. Birk, J., et al., "Orientation of Workpieces by Robots Using the Tri-
    angle Method", presented at ROBOTS I Conference, Chicago (Rosemont),
    Ill., November 1976, SME Technical Paper MR76-612.

3. Birk, J., et al., "Robot Computations for Orienting Workpieces",
    Second Report - Grant APR74-13935, Department of Electrical Engineer-
    ing, University of Rhode Island, August 1976.

4. Birk, J., et al., "General Methods to Enable Robots with Vision to
    Acquire, Orient and Transport Workpieces", Third Report - Grant
    APR74-13935, Department of Electrical Engineering, University of Rhode
    Island, August 1977.

5. Seres, D., "Computer Controlled Robot with Visual Sensor for Advanced
    Industrial Automation", M.S. thesis, Department of Electrical Engineer-
    ing, University of Rhode Island, 1975.

Shown is a sampling of the parts that can be fed, properly oriented, the first time.

Feeding blind-hole parts *right* the first time

Tubular parts that are closed on one end, or parts with stepped bores, are difficult to feed properly oriented at a high rate of speed. The feeding concept shown here can increase output.

Reprinted from Tooling & Production, September 1981

Archie A MacIntyre has been involved in the parts feeding business for 28 years. In 1972, his company in Gladwin, MI, was awarded a patent for a mechanical feeding device that achieved noise abatement through vibrationless operation. The device, known as the Mainline Feed System, can be paired with his latest project to lift unoriented blind- or stepped-hole parts out of a specially constructed bulk supply hopper and deposit them on an in-line conveyor for further processing.

Until now, if a mechanical or vibratory feeder were used, some type of probing device would be needed to sense which end was being fed. Consequently, to achieve a feed rate of 30 to 40 parts/min, multiple units would be needed. Now, an ordinary elevator feeder can be modified to deliver these problem parts at much higher rates. In essence, the part is oriented to begin with.

The modifications necessary include the addition of a specially sized pick-up pin on the belt elevator, along with a custom trough or chute in the bottom of the bulk-storage hopper which concentrates the area of entrapment by allowing a group of parts to fall into the zone and be lifted, but not pushed away. By simply pulling a cleated belt through a mass of parts, the pickup success rate would not approach the 50 percent rate that is now attainable. The chute is only as wide as the special belt elevator and only slightly bigger than the diameter of the part; just enough for a part to fall into. In operation, the open end of the blind-hole part settles around the pickup pin and is lifted cleanly out of the hopper. With stepped holes, the pin only fits into one ID. The parts are lifted in phase and the need for orienting and checking around the periphery of a bowl is eliminated (unless the part must simply be flopped before proceeding). Any part that is long and cylindrical can be fed with this device; the only restriction is that the OD of the part must be less than the length.

To further increase the feed rate, it is possible to design a unit with multiple-belt elevators. The parts would then meet in the mechanical in-line or rotary feeder for further processing. Basically, the system consists of an electric motor, a power transmission train, the special belt and the fabricated hopper. This simplicity of design makes the device attractive from a maintenance standpoint. The system is not only applicable to assembly operations, but it can be paired with a centerless grinder or a similar machine tool to allow unmanned operation.

Stepped- or blind-hole parts settle around the pick-up pin and are lifted in phase.

Following pick up, parts are deposited into a chute or in-line feeder for further processing.

If the parts must be turned over, this can be efficiently accomplished using the Mainline Feed System.

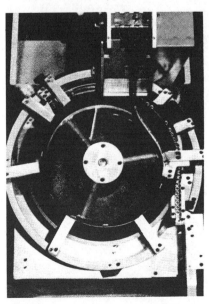

CHAPTER FIVE

MACHINING AND ASSEMBLY FIXTURES

Operated for the U.S. Department of Energy by the Bendix Corporation, Kansas City Division, under Contract No. DE-AC04-76DP00613

FIXTURING AND HANDLING OF MINIATURE PARTS

By

LaRoux K. Gillespie
Staff Engineer
Bendix Corporation
Kansas City, Missouri

The production of miniature components poses a number of unique fixturing and handling problems. In particular, paper-thin or precision-sculptured parts which are smaller than pencil erasers can require a great amount of fixturing and handling time and care in job shop situations. Thin and soft parts are easily damaged by the use of unsuitable clamping and handling methods. Because of their size, small parts are difficult to install in fixtures. Clamping is difficult because the clamps that provide the required holding force tend to cover a large portion of the part. Small-diameter locating pins in fixtures are easily bent by cutting or clamping forces. Parts which are small and thin are difficult to pick up and transport by conventional techniques.

INTRODUCTION

In the production of low-quantity batch-lots of small parts, the loading, unloading, and handling of the parts can require significantly more time than the actual machining time.

This is particularly true of precision parts which require more than one machining operation and which are smaller than 0.250 inch (6.35 mm) or thinner than 0.010 inch (254 μm). With parts such as these, which require milling, drilling, and grinding operations, the loading and unloading time constitutes up to 75 percent of the total production time. Thus, the minimization of cutting time, which is the traditional approach in improving productivity, is less effective than other approaches (Figure 1). The following observation recently was made by the chairman of one of the largest machine-tool builders in the United States.

> The average workpiece in a batch-type production shop spends only 5 percent of its time in production machines, and productive work is being done on the part only 30 percent of this actual time in the machines.[1]

A great amount of finger dexterity is required in the handling of minute parts. This alone significantly increases the handling time. Parts which have thin sections, precision finishes, or near-zero tolerances also necessitate extra handling care.

The combination of a small part size and close tolerances requires extra care and time in placing the parts in fixtures.

As an example, locating a part on a 0.016-inch diameter (0.406 mm) pin and assuring that it is seated flush with the bottom of the fixture is much like threading a needle with a thread which is only 0.0005 inch (12.7 μm) smaller than the eye of the needle.

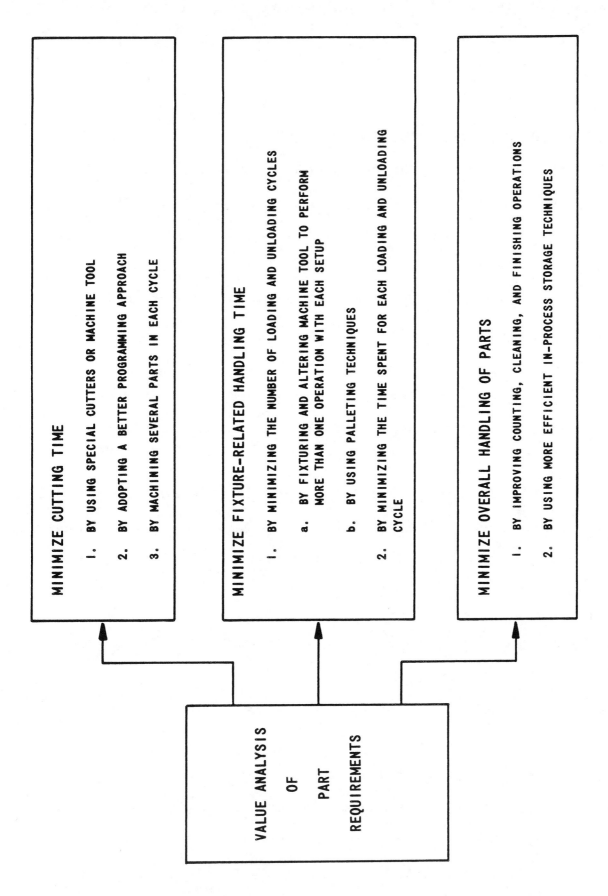

Figure 1. Approaches to the Minimization of Production Time

MINIMIZE CUTTING TIME

    1. BY USING SPECIAL CUTTERS OR MACHINE TOOL

    2. BY ADOPTING A BETTER PROGRAMMING APPROACH

    3. BY MACHINING SEVERAL PARTS IN EACH CYCLE

MINIMIZE FIXTURE-RELATED HANDLING TIME

    1. BY MINIMIZING THE NUMBER OF LOADING AND UNLOADING CYCLES

       a. BY FIXTURING AND ALTERING MACHINE TOOL TO PERFORM MORE THAN ONE OPERATION WITH EACH SETUP

       b. BY USING PALLETING TECHNIQUES

    2. BY MINIMIZING THE TIME SPENT FOR EACH LOADING AND UNLOADING CYCLE

MINIMIZE OVERALL HANDLING OF PARTS

    1. BY IMPROVING COUNTING, CLEANING, AND FINISHING OPERATIONS

    2. BY USING MORE EFFICIENT IN-PROCESS STORAGE TECHNIQUES

VALUE ANALYSIS OF PART REQUIREMENTS

A number of efficient techniques are available for high volume runs in which full-scale automation is feasible. This is frequently not the case for the production of low quantities of parts in batch lots.

INFLUENCES OF WORKPIECE CONFIGURATION

In searching for better approaches to the fixturing and handling of small parts, one of the first steps is to define the features that are common to many parts. As an example, typical handling and fixturing problems for thin, flat parts which, for example, are 0.001 inch (25.4 μm) in thickness, might include the following considerations.

- Picking up parts from in-process containers (any moisture present produces high-surface tension holding forces);

- Placing the part on tooling pins when the holes are small (perhaps 0.030 inch or 0.75 mm in diameter) and the tooling pins are within 0.0005 inch (12.7 μm) of the hole size;

- Handling the parts in a manner to prevent damage; and

- Separating the parts for cleaning and counting operations.

As an example of the difficulty imposed by thin parts, 130,000 of the 0.001 inch thick parts shown in Figure 2 will fit in the thimble shown. Merely picking up these parts can be a major problem, deburring them an even more challenging one. The following typical problems are encountered in the handling and fixturing of small milled parts.

- Designing a clamp to hold the part (because of their size, most strap- or toe-type clamps mask much of the part.);

- Providing clamping pressure that will prevent the movement of small precision parts without distorting them;

- Providing locating features that are not bent by cutting forces;

- Providing rapid clamping techniques; and

- Providing rapid part-insertion and part-removal techniques.

Figure 2. A Thimble Can Hold 130,000 Miniature Spacers

MAJOR HANDLING AND FIXTURING ELEMENTS

The following major elements must be considered in the handling and fixturing of small precision parts.

- A variety of possible techniques to transfer the parts from in-process shipping containers to the fixture and back;

- A fixture configuration to provide support for and rapid setup of the part, and to provide clearance for burrs (the use of multiple fixture stations and pallets must also be considered);

- Techniques to accurately locate the parts in the fixture through the use of tooling holes, external contour, or both; and

- A variety of possible clamping techniques.

Of these, the part-location techniques offer the least flexibility.

BASIC APPROACHES TO FIXTURING

An analysis of fixturing reveals four basic approaches for minimizing nonproductive handling time.

- Utilize single-purpose fixtures (e.g. drilling for a single pattern of holes);

- Utilize multipurpose fixtures (such as a traveling pallet);

- Utilize multiple part fixturing (load one part while another is being machined); and

- Design the process such that no fixturing is required.

The last approach in the above list is obviously the most desirable, when it can be accomplished. Chemical machining, single-stage blanking, and complete fabrication in a single NC operation are examples of the latter approach.

The use of duplicate fixtures allows an operator to load one part while the machine is cutting another. The machine-tool cutting rate then becomes the productivity limiting factor.

While for many parts this approach offers some advantages, it also has some noticeable disadvantages: tooling costs are nearly doubled since two fixtures are required; the approach is effective only on machines which do not require hand feeding and loss in dimensional repeatability occurs from the use of multiple tools because of dimensional differences between the fixtures.

For an initial approximation, the economic break-even point in the use of this approach is the point at which the saving in handling time equals the cost of the fixture. If a duplicate fixture can be made for $1,500, the maximum number of parts to be machined is 1000, and the fixture saves 90 seconds per part, the labor rate must be at least $60 per hour for this approach to be economical. Tables 1 and 2 provide additional data which is useful in evaluating the economics of the handling time.

For the manufacture of small parts, more than one part may be machined from a single strip of metal. If the parts remain attached to the strip until the final operation, only one load-unload operation is required for the entire group of parts on each strip. In addition, the strip often provides greater clamping accuracy because of the larger surface area. Figure 3 illustrates the application of this approach to the component shown in Figure 4.

Figure 3. Parts Produced From a 1/8-Inch High Strip

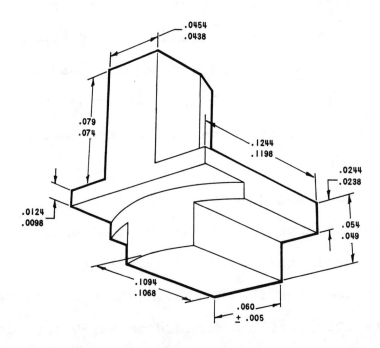

Figure 4. Miniature Sliding Detent Produced From a Strip

Seven separate operations are performed on these parts while they remain attached to the strip. This type of approach has been described in more detail elsewhere.[2] It is widely used in the fabrication of miniature electrical components by progressive dies.

If a fixture for the production of five parts in a strip were no more expensive than fixtures for the production of an individual part, the strip approach would eliminate 80 percent of the handling time. For most metals in a job shop application, the cost of material wasted by the strip approach to the production of miniature parts is insignificant.

LOCATING PARTS IN FIXTURES

One of the more time-consuming elements in the fixturing and handling of small parts involves the positioning of the parts on the locating devices of the fixture. Parts may be located on the fixture by any of the following four methods:

- External part contour;

- Datum holes and pins;

- Tooling holes which are external to the finished part; and

- Tooling bosses.

As datum-hole sizes and tolerances become smaller, placing the part over the locating feature becomes more difficult. This often becomes a major problem with thin aluminum parts because the thickness and shape of the part may not allow lead-in chamfers on tooling pins. In addition, thin parts can become distorted from forcing them over tooling pins.

When external contours must be produced with tolerances of only 0.001 inch (25.4 μm), the locating pins can be only 0.0002 inch (5.08 μm) smaller than the datum holes. When these holes are only 0.020 inch (0.50 mm) in diameter, unusual dexterity is required to load the workpieces. Furthermore, removing these parts from the fixture may require as much skill as loading them, since most of the parts are so small that cutouts for finger-holds cannot be employed. The cutouts often require the removal of material that is needed to support the parts.

TABLE 1: HANDLING TIME SAVED FOR PRODUCTION QUAN-
TITIES OF PARTS

Time Saved Per Part (Seconds)	Time Saved for Number of Parts Indicated (Hours)				
	100 Parts	500 Parts	1000 Parts	5000 Parts	10,000 Parts
10	0.28	1.39	2.8	13.9	28
60	1.67	8.35	16.7	83.5	167
90	2.50	12.50	25.0	125.0	250
120	3.33	16.65	33.3	166.5	333
180	5.00	25.00	50.0	250.0	500
240	6.67	33.30	66.7	333.0	667

Although many parts can be located from external contours, many more, because of tolerances, cannot be approached in this manner. If the approach can be used, it does have the advantage that part insertion is easier because the largest features of the part are used for part-location. Dropping a 0.250-inch diameter (6.35 mm) part into a 0.0002-inch larger (5.08 μm) cavity can be easier than placing it on a small pin, and this method eliminates the possibility of accidentally bending small pins. However, because parts can easily become locked in a

TABLE 2:   ANNUAL SAVINGS THROUGH ELIMINATION OF WASTED TIME

| Time Saved Per Day (Minutes) | Annual Savings at Indicated Hourly Labor Rate* | | | | | |
	$5 ($)	$6 ($)	$7 ($)	$8 ($)	$9 ($)	$10 ($)
5	108	130	151	173	194	216
10	217	260	304	347	391	434
20	434	521	608	694	781	868
40	868	1042	1215	1389	1562	1736
60	1301	1562	1821	2082	2342	2602

*values are based on 260 working days per year for direct labor only; no overhead is included.

tight-fitting cavity, the use of locating pins is often much easier.

In some cases, a blank much larger than that required for the part can be utilized during manufacture, and large tooling holes can be inserted in an area which lies outside the finished part (Figure 5). If closely controlled, the long spacing between the tooling holes provides more accurate orientation of the part than is possible through the use of closely spaced holes. When this approach is used, the small holes that are integral to the part are used only in the final manufacturing operation.

For many parts, the use of a tooling boss on the part is the most convenient approach (Figure 6). By this method, the boss, which can be of any convenient size, is used to position the part, and it also serves as a solid surface for clamping. After the part is finished, it is placed in a lathe and a cut-off tool is used to separate the workpiece from the boss. A grinding or facing operation then removes the small projection left from the cut-off operation. This approach is particularly advantageous for parts which are completely machined on N/C equipment, although it also can be easily used for multiple operation applications.

CLAMPING TECHNIQUES

The biggest single problem in the fixturing and handling of small parts is to find suitable clamping techniques. The clamps used must meet the following requirements:

- Be relatively quick to apply and remove;

- Not damage the part;

- Be small enough not to interfere with the cutters; and

- Be strong enough to prevent part movement.

Traditionally, simple **strap or toe** clamps have been used on tooling for short runs of parts because of their low cost (Figure 7). This kind of clamping, however, requires an operator to adjust one or two setscrews or thumbscrews which typically require from 10 to 30 seconds per part; for some parts, clamping may require considerably more time than this.

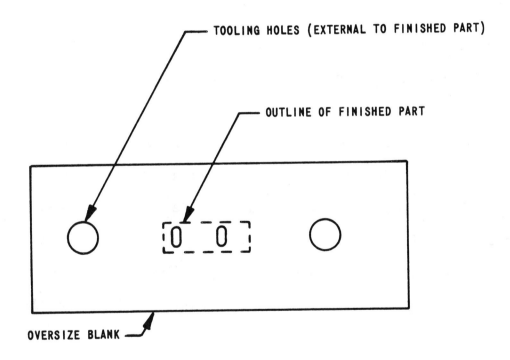

Figure 5.   Example Showing Tooling Holes External
            to Finished Part

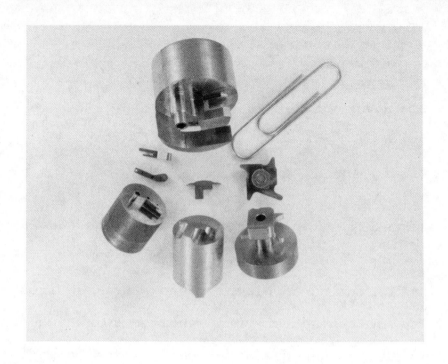

Figure 6. Parts Produced Using Tooling Bosses

From the standpoint of providing quick clamping and release, the first eight techniques shown in Table 3 offer nearly equal response time. Materials which retain magnetic fields, however, will require a somewhat longer time for removal from magnetic chucks. While magnetic, electrostatic, and vacuum chucks are generally used for flat parts, vacuum chucks have been used successfully to hold thin-flanged aluminum parts on the lathe. All of these approaches have the advantage of requiring no projections above the part which would interfere with the cutting tools.

The temporary bonding of parts to a fixture by the use of adhesives has been used successfully for both metals and ceramic plates. (*Eastman 910* epoxy, Argüeso's *Rigidax* and Devcon's *Zipgrip* have been used for holding metals.) Although this approach requires two additional manufacturing operations, it eliminates the problem of clamping brittle or easily distorted parts.[3] For some parts, there is an advantage in completely covering or filling the part with a low-melting temperature alloy. In general, the use of adhesives or low-melting temperature alloys increases the handling time for the parts; however, when groups of 20 or 30 parts are clamped in one unit by the use of these techniques, some saving in time can result.

TABLE 3: CLAMPING TECHNIQUES

Technique and Typical Maximum Holding Force or Pressure	Advantages	Limitations
1. Vacuum Chuck-- 13 psi (89.6 kPa)	Easily Accommodates thin parts with large surface area	Low holding force
2. Toggle-Action-Cam-Lock Clamp	Quick release	May distort thin parts; clamp may project too high for miniature tools
3. Magnetic Chuck-- 1125 psi (7.76 MPa)	No clamps in way of cutter	Parts must be magnetic and have reasonable surface area
4. Hydraulic Clamps	Fast actuation	Hydraulic supporting equipment is large
5. Pneumatic Clamps	Fast actuation	
6. Electrostatic Chuck	Can be used on nonmagnetic materials	Low holding force; use limited to metals
7. Collet	Quick release	Part geometry may not be suitable
8. Swivel Clamp*-- 68 pounds (302 N)	Quick release	
9. Strap Clamp**	Low cost	Slow; no control of force
10. Centers	No clamps in cutter path	Part geometry may limit application; requires additional machining operations
11. Thumb Screws**	Low cost	Slow; no control of force
12. Vise Jaws	Low cost	No control of force
13. Adhesive Bonding	Will hold any material	Requires subsequent operation for removal of adhesive
14. Potting in Low-Melting Point Alloys	Accommodates any shape; supports thin sections and hard-to-reach areas	Requires subsequent melting and cleaning operations; may load cutters

*Hydraulic or pneumatic clamps which release and rotate 180 degrees to provide easy removal of part (Stilson Corporation miniature air-powered Rota-Clamp).

**Torque-limiting screws can be purchased for these clamps, but the selection is limited.

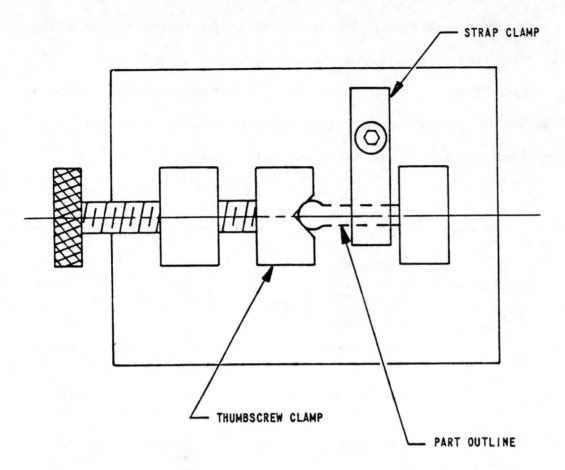

Figure 7. Typical Thumbscrew and Strap Clamps Used
in Tooling Short-Run Parts

Part deformation is possible when using toggle, strap, thumbscrew, or vise-jaw clamps since these devices will not accurately
limit the clamping force. Consider, for example, the part shown
in Figure 8. A 3/8-inch diameter (9.78 mm) four-flute end mill
is used to machine this part at a feed of 0.001 ipr (25.4 μm/rev)
and a speed of 2000 rpm. The predicted tangential cutting force
is 146 pounds (650 N) in 1045 hot-rolled steel (BHN 202) for a
1/8-inch (3.175 mm) depth-of-cut, cutting with the full diameter.
Thus, the clamping forces must resist at least the following amount
of force.

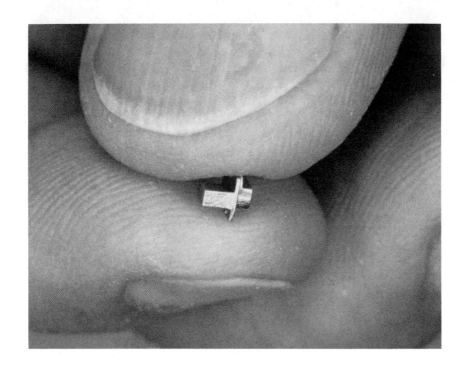

Figure 8.   Miniature Sliding Detent Which Is Susceptible to Deflection From Cutters and Damage by Clamping

$$F_{cut} = \frac{396,000 \; wnf^{1-b}d^{1-c}}{\pi CDV^{a}}$$

where

F_{cut} = tangential cutting force (pounds),

w   = radial depth-of-cut (inch,

n   = number of teeth on the cutter,

f   = feed per tooth per revolution,

b   = material constant (1/3),

d   = axial depth-of-cut (inch),

c   = material constant (1/6),

C   = material constant (3, for this example),

D   = cutter diameter (inch),

V   = cutting velocity (sfpm = $\frac{\pi DN}{12}$),

a   = material constant (0.125), and

N   = spindle speed (rpm).

The actual forces would be greater than indicated since the 1045 steel used in the example is not as hard as the 17-4PH stainless steel parts shown in Figure 8. In addition, a vertical force would exist which would pull the part upward against the clamp.

If only frictional forces were used to hold the part against the fixture, the clamping force would have to be greater by the amount shown in Equation 2.

$$F_{clamp} > \frac{F_{cut}}{\mu},$$

where

μ = static coefficient of friction.

Assuming that μ were approximately 0.75 for steel on steel, a clamping force of at least 194 pounds (865 N) would be required.

If the available clamping area were only 0.125 by 0.064 inch (3,175 by 1.626 mm), the clamping pressure must be approximately 25,000 psi (174.5 MPa). Using Shawki and Abdel-Aal's data[4] as a guide, the clamped area of the workpiece would be deformed 0.0002 inch (5 μm) beyond its theoretical elastic limit. Under a clamping force of this magnitude, the use of spherical locators would result in even greater deformation.

The use of tooling bosses, external holding features, and multiple parts in a strip provides a much larger clamping surface than is available on single parts alone.

In addition to making clamping easier, these methods significantly lower the pressure applied to the total part which, in turn, minimizes part damage. When tooling bosses are used, any damage which occurs is to the tooling boss and not to the part.

One of the major deterrents to the prevention of clamping damage has been a lack of knowledge as to how great the clamping forces should be. Similarly, little information has been published to provide engineers and tool designers with data on the forces which various types of clamping devices can exert on parts. Table 4 lists references for some of the more production-oriented sources of information from which cutting forces can be predicted reasonably close. Wilson[5] and Shawki and Abdel-Aal,[4] however, appear to be the only individuals who have published any readily available information on clamping forces and their effects.

As previously indicated, miniature parts require the use of miniature clamps. For profiling operations, it is highly desirable that clamps be arranged so that they need not be loosened and repositioned in order to complete the cut. Unfortunately, parts without tooling bosses or similar features frequently do require some reclamping during the machining cycle. This not only increases the clamping time, but it also introduces the risk of slightly altering the part location.

In addition to requiring potentially damaging pressures because of the small bearing surfaces involved, the clamps interfere with the movement of the machine spindle. The drilling of a 0.012-inch diameter (0.30 mm) precision hole, for example, requires the use of the shortest possible drill. A clamp which projects 0.030-inch (0.76 mm) above the workpiece requires that the drill length be at least 0.030-inch, plus an additional 0.010-inch (0.25 mm) for clearance, plus the thickness of the workpiece. Thus, the drill length required for clamp clearance alone is more than the total recommended length. (For maximum rigidity and accuracy, the drill length usually should not exceed three times the drill diameter.) This problem is even more critical when miniature end mills are used. A 0.026-inch diameter (0.6 mm) end mill with 0.62-inch (1.57 mm) flute length deflects 0.002-inch (50.8 μm) under a 2-pound (8.89 N) cutting force. An additional length of 0.036-inch (0.91 mm) for clamp clearance allows the tool to deflect 0.009-inch (0.23 mm) under the same load.[12] This results in a higher incidence of tool breakage and a decrease in part accuracy.

HANDLING OF SMALL PARTS

In one job shop, a typical part is handled individually at least 25 times throughout its fabrication cycle.

This assumes that each part is handled twice per operation, with eight machining, deburring, or finishing operations per part, two inspections, a passivation, and a final packaging. The part shown in Figures 3 and 4 is handled 12 times as a bar of five parts, 30 times as an individual part, and 34 times for inspection (assuming 100 percent inspection of all features).

In addition, the parts are handled 26 times as a group. Thus, each part is handled a total of 64 times. Each time, the operator must remove it from one location and carry it to another without damage or loss while maintaining some form of part orien-

TABLE 4: REFERENCES FOR SOURCES OF INFORMATION ON CUTTING FORCES

Operation	References*
Turning	9; 10; 11
Drilling	9; 10; 11; 12
End-Milling	9
Face-Milling	8; 9
Slab-Milling	9; 12; 13
Grinding	14
Broaching	8; 9

*Numbers shown refer to references listed at the end of this report.

tation. As the size of the parts decreases, the time required for handling increases greatly.

As shown in the following list of typical operations, most parts pass through a variety of production steps; some parts may be cleaned, inspected, and packaged several times:

- Machining;
- Deburring;
- Cleaning;
- Heat-treating;
- Passivating or plating;
- Inspection;
- Counting; and
- Packaging.

Because of the large amount of handling involved, the handling equipment must be carefully considered. As shown in Table 5, ten common handling techniques are available.

Fingers do not have enough dexterity to quickly pick up, orient, and place small parts such as those shown in earlier Figures. Tweezers may be used, and often are, but they can easily damage the threads on minute threaded parts, and the orienting of some parts can be very slow. Although the breaking of the bond between a magnet and a part has been a problem with many parts, pencil-type magnets having built-in release mechanisms are now available.

Vacuum probes provide one of the more versatile and rapid techniques for handling small parts with only the disadvantage of their inability to handle some shapes. Although a vacuum-probe unit is required for every operator (at a cost of $100 each), a saving of five minutes per day in handling will easily pay for such a unit in a year's time (Table 2). These units can lift up to 16 ounces (4.448 N) and have the added advantage of being usable for cleaning small particles from the surfaces of the parts.

The usefulness of vacuum probes is somewhat limited when parts must be pushed onto or pulled from fixtures. Tight-fitting pins may require the use of more than a 1-pound (4.448 N) force. As the part size decreases, the probe area must also decrease which, in turn, decreases the available lifting force.

As previously stated, the production of parts in strips eliminates many handling problems. Also, the fixturing of each part on a small pallet and leaving the part on the pallet until the final machining operation has been completed can be used to minimize handling problems and the time involved. Because of the need for multiple pallets, the tooling costs involved can be high; the addition of tooling bosses to the parts will accomplish the same objective at a much lower cost.

Vibrating bowl-feeders and the associated tracks for orienting parts (Figure 9) can be used effectively for large lots (from 3000 to 1,000,000 parts, for example), but the required set-up time will outweigh any saving in handling time for lots smaller than 500 parts. Units are available today which can handle electronic chips as small as 0.010-inch (0.25 mm) wide. Miniature pick-and-place robots, used in conjunction with bowl-feeders and cassette programming, also can be effectively employed for large lots. The use of these devices has been described in more detail elsewhere.[13], [14]

For large lots, air sometimes may be used to move parts from one station to another. One company blows small screw-machine parts through flexible pipelines from one process to the next.[15]

This approach improved the workflow between machines by replacing the batch-handling of parts in containers, thereby eliminating transit damage and handling time. Unfortunately, this approach is not applicable to small quantities of parts or to parts having widely varying geometries.

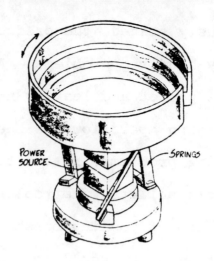

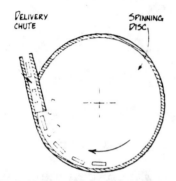

Centrifugal hopper throws parts to periphery of disc where they contact discharge point designed to accept items of proper shape or in desired orientation. Suitable for fragile parts which can't be tumbled.

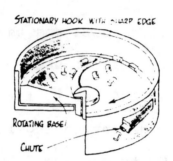

Stationary hook hopper separates mass of parts and orients and guides the parts upwards and outwards to edge of rotating base for feeding into a discharge chute. Suitable for delicate, simple forms.

Figure 9.   Vibrating Bowl Feeders Used for High Production Feeding and Orienting of Parts

One particular area which can require large amounts of handling time is the removing of miniature parts from loose-abrasive deburring media. Processes such as vibratory deburring or centrifugal-barrel finishing utilize random-shaped nuggets which often are approximately the same size as the workpieces (Figure 10). Thus, the screening of nonmagnetic parts from nonmagnetic media of the same size can require extraordinary amounts of time. As an example, one-hundred 0.020-inch diameter (0.5 mm) pins placed in two quarts of wet Number 24 media (approximately 0.030-inch or 0.75 mm in diameter) will require at least four hours to

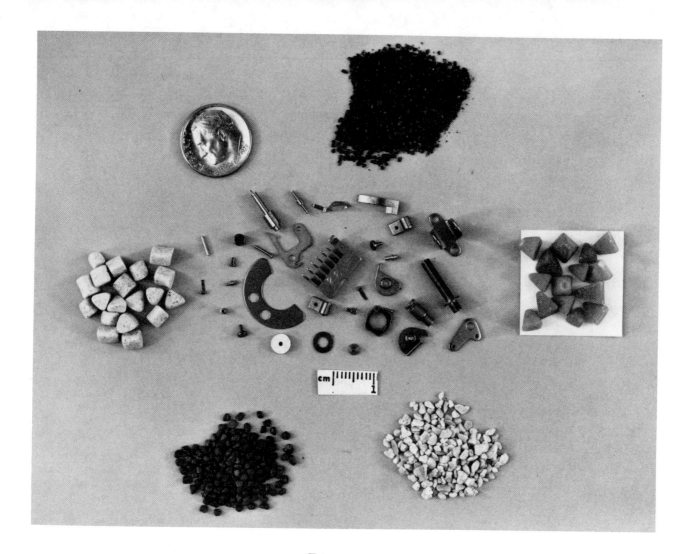

Figure 10.

Miniature Parts Compared to Small Deburring Media (Smallest media are N14s. Largest media are 1/4-inch plastic pyramids at right on photo)

sort by hand. In contrast, the deburring of such parts in a

centrifugal barrel tumbler requires only 40 minutes. Although

none of the existing commercial sorting equipment is capable of

sorting such a combination and still accommodating a wide variety

of other part configurations and sizes, the following three

approaches can be used to minimize this problem:

Reduce the amount of media used for deburring;

Purchase special screens in sizes that will sort out all

media except those which are exactly equal to the size

of the part; and

Approach	Advantages	Limitations
Use of Fingers	No tool cost	Cannot grasp minute parts
Hand Tweezers	Low tool cost	Relatively slow; can damage very soft parts
Hand Magnet	Low tool cost	Parts must be magnetic; residual magnetism must be overcome
Hand Vacuum Probes	Easily picks up and releases small, flat parts; not affected by material; no part damage	Not well suited for picking up surfaces having a radius
Parts in Strips	Convenient for handling with fingers	Some part-geometry limitations
Pallets	Convenient for handling by hand	Requires one fixture per part per lot
Bowl Feeders	Fast; low cost; feeder can orient as well as feed parts	Transporting from feeder to fixture requires separate tooling for each part; longer setup times than by hand; noticeable tool clearance or additional device to keep feeder track out of cutter's path
Pick-And-Place Robots	Fast	Not well suited for short runs of widely differing parts
Pneumatic Conveyors	Low cost	Not economically suitable for low-volume runs of widely varying part geometries
Tooling Bosses	No measurable material or tooling cost	Not suitable for all part geometries; requires two additional operations

Use vibratory feeders to sort the small volume of mixed parts and media that is left from the previous approach.

The first of these approaches permits the operator to search through less media to find the same number of parts; thus, handling is reduced in proportion to the amount of media used.

Although the separation of any proportion of media from the parts is advantageous, the standard screen sizes will not provide significant separation of the parts from some media. By utilizing special screens with openings which are graduated every 0.005-inch (127.0 μm), most of the media can be quickly removed. With only a small amount of media remaining, vibratory feeders with special separator gates can quickly provide the final separation.

IMPLEMENTATION OF FIXTURING AND HANDLING APPROACHES

While some techniques have unique applications, others can be used for many types of applications. Tooling bosses are now being used for many small parts because they provide easy handling and eliminate cumbersome, restrictive clamping. Furthermore, in most cases, they eliminate the need for special holding fixtures since standard colleting equipment is readily available.

The use of strips of parts and oversize blanks having tooling holes external to the parts are also used, but these approaches typically offer no advantage over tooling bosses when N/C equipment is used unless orientation of the parts between machining operations is required.

Adhesives are being used to hold ceramic-base integrated circuits for cutoff. They also have been used for holding metal parts, but not extensively. Vacuum chucks have been used to some extent, but the majority of small parts do not have sufficient surface area to provide an adequate holding force during machining; the same is true of electrostatic chucks.

The removal of thin parts from magnetic chucks is difficult because of residual magnetism in the parts.

Like vacuum and electrostatic chucks, magnetic chucks also require a certain amount of surface area, since several laminations in the magnetic chucks must be bridged to provide an adequate holding force.

Vacuum probes have been used for the handling of small parts, but, as previously indicated, they often do not provide adequate force for placing parts on fixtures.

REFERENCES

1. "News Digest," *Production*, June, 1972, p 7.

2. "Machining of Parts in Metal Strips," *Microtechnic*, Number 4, 1972.

3. "Plastic Tooling for Stainless Steel Parts," *Manufacturing Engineering and Management*, July, 1970, pp 26-27.

4. G. S. A. Shawki and M. M. Abdel-Aal, "Rigidity Considerations in Fixture Design--Contact Rigidity at Locating Elements," *International Journal of Machine Tool Design and Research*, Volume 6, 1966, pp 31-43.

5. Frank W. Wilson, *Handbook of Fixture Design*. New York: McGraw-Hill, 1962.

6. *Machining Data Handbook* (Second Edition). Cincinnati: Machinability Data Center, 1972, pp 869-881.

7. F. Wilson, editor, *Tool Engineer's Handbook* (Second Edition). Dearborn, Michigan: ASTME (Now SME), 1969.

8. Amitabha Bhattacharyya and Inyong Ham, *Design of Cutting Tools*. Dearborn, Michigan: ASTME (Now SME), 1969.

9. O. W. Boston and W. W. Gilbert, "The Torque and Thrust of Small Drills Operating in Various Metals," *ASME Transactions*, Volume 58, Number 1, 1936, pp 79-89.

10. Andor Horning, "Slab Milling Efficiency," *Tool and Manufacturing Engineer*, May, 1969, pp 59-62.

11. R. S. Hahn, "On the Nature of the Grinding Process," Paper Presented at Second International Production Conference, Birmingham, England, September, 1962. Also see "Grinding Fundamentals--II," *International Research in Production Engineering*. New York: ASME, 1963, p 207.

12. L. K. Gillespie, *Profile Milling* (Topical Report). UNCLASSIFIED. Bendix Kansas City: BDX-613-980 Rev., May, 1974 (Available from NTIS).

13. "Simple Automation Cuts Cost 92.5%," *American Machinist*, May 27, 1974, p. 41.

14. "Trends in Part Feeding," *Tooling and Production*, September, 1974, pp 34-47.

15. Richard G. Green, "In-Process Handling," *Automation*, November, 1974, pp 80-87.

PARTS PRESENTING UNIQUE HANDLING
AND FIXTURING PROBLEMS

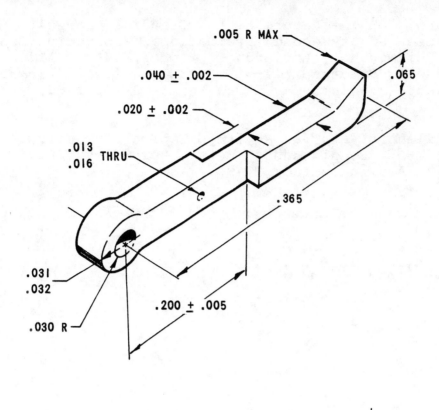

.005 R MAX

.040 ± .002

.020 ± .002

.013
.016 THRU

.365

.031
.032

.030 R

.200 ± .005

.065

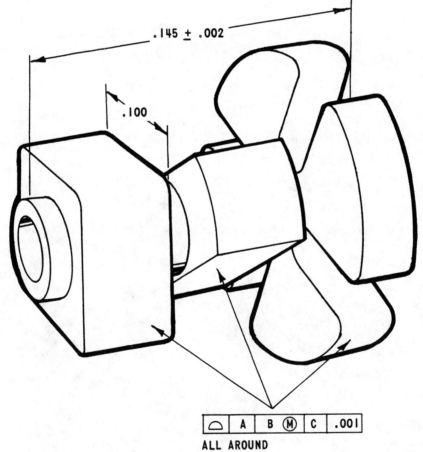

.145 ± .002

.100

| ⌓ | A | B | Ⓜ | C | .001 |

ALL AROUND

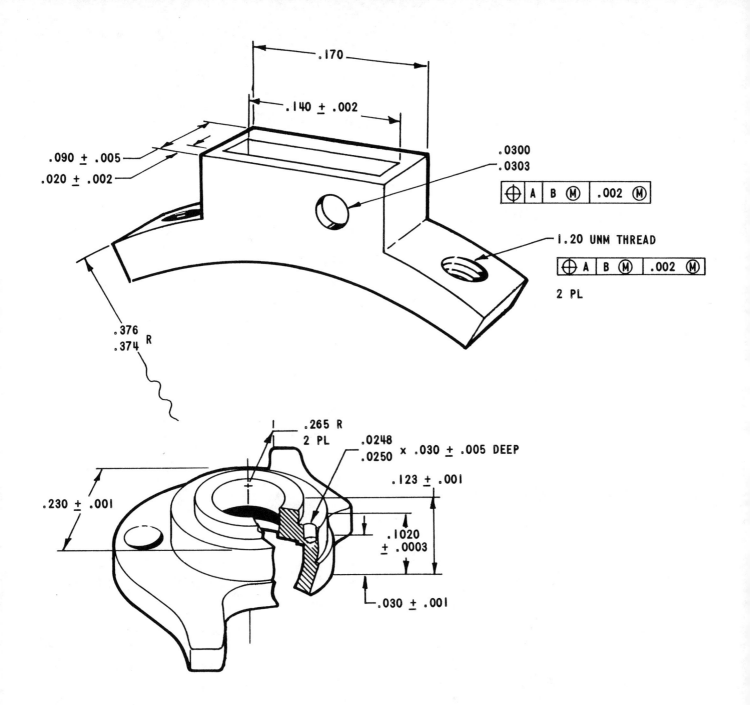

.170

.140 ± .002

.090 ± .005

.020 ± .002

.0300
.0303

| ⌖ | A | B Ⓜ | .002 Ⓜ |

1.20 UNM THREAD

| ⌖ | A | B Ⓜ | .002 Ⓜ |

2 PL

.376
.374 R

.265 R
2 PL

.0248
.0250 × .030 ± .005 DEEP

.123 ± .001

.230 ± .001

.1020
± .0003

.030 ± .001

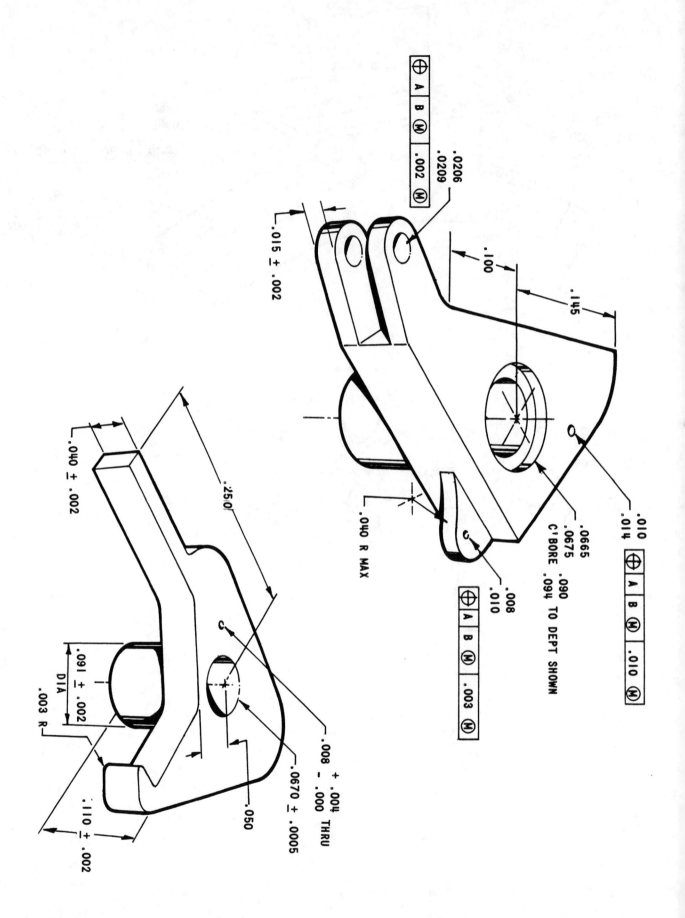

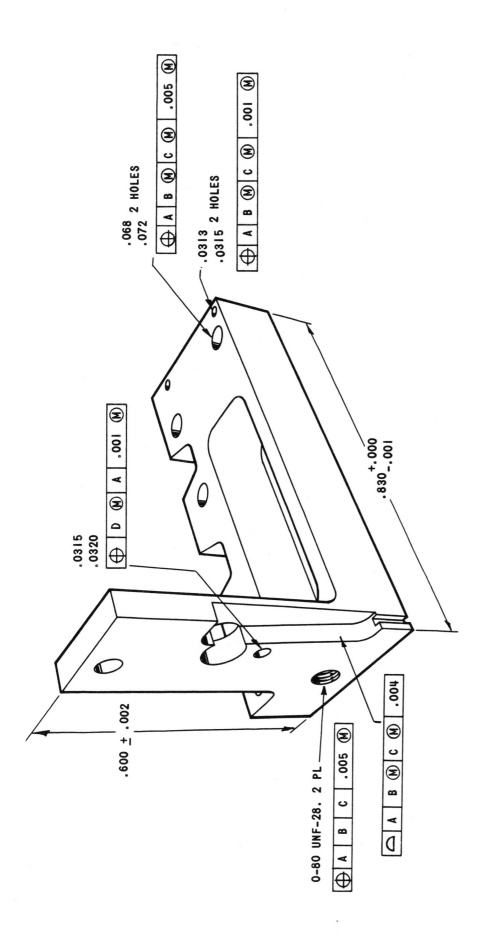

.068 2 HOLES
.072

⌖ | A | B Ⓜ | C Ⓜ | .005 Ⓜ

.0313 2 HOLES
.0315

⌖ | A | B Ⓜ | C Ⓜ | .001 Ⓜ

.830 $^{+.000}_{-.001}$

.0315
.0320

⌖ | D Ⓜ | A | .001 Ⓜ

.600 ± .002

0-80 UNF-28. 2 PL

⌖ | A | B | C | .005 Ⓜ

D | A | B Ⓜ | C Ⓜ | .004

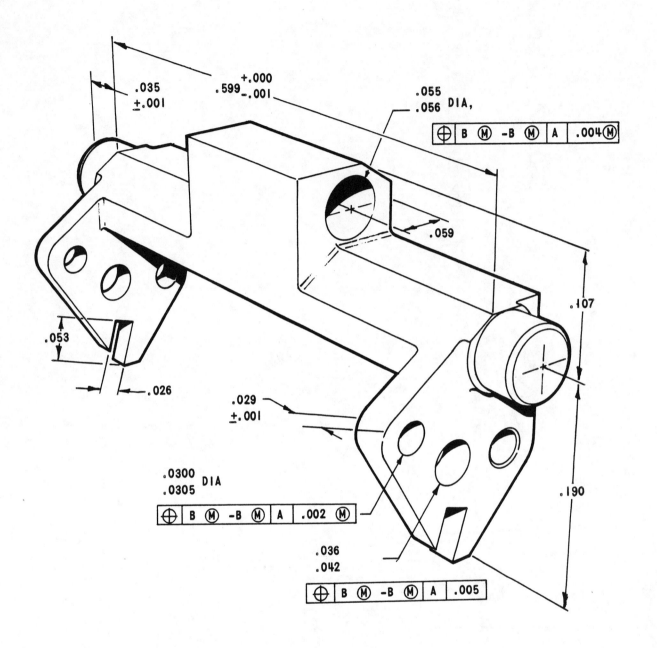

.035
±.001

.599 +.000 -.001

.055
.056 DIA,

| ⊕ | B Ⓜ -B Ⓜ | A | .004Ⓜ |

.059

.107

.053

.026

.029
±.001

.0300 DIA
.0305

| ⊕ | B Ⓜ -B Ⓜ | A | .002 Ⓜ |

.036
.042

| ⊕ | B Ⓜ -B Ⓜ | A | .005 |

.190

Presented at SME's Practical Welding and Brazing Technology Seminar, March 1977

Fixturing for Brazing Automation

By Raymond A. Hurlbut
Fusion Incorporated

Brazing has been a useful manufacturing process for a long time. New developments in mechanical systems, alloy and flux application, heat systems and post braze processing have made brazing automation more practical and economical than ever. The single most innovative and important contribution the manufacturing engineer can make is in the design and use of fixtures for brazing automation. This paper discusses the major considerations for optimum fixture design and utilization.

INTRODUCTION

Assuming that an economic analysis justifies the mechanization of the brazing operation and that equipment necessary to perform the mechanical operations can be available then the next consideration is fixturing. The optimum fixture design for a brazing operation results when the engineer has analyzed all of the conditions that will affect the operation and applies the best engineering data available into the basic design concept. The many variables that are present in the manual brazing operation are much more pertinent in a mechanized brazing operation. There is no textbook formula that will permit the substitution of values for the variables to arrive at an optimum solution.

Fixturing is probably the most significant manufacturing engineering contribution towards successful brazing automation. The fixture must hold the parts to be brazed in their proper relationship and at the correct position for the automatic application of flux, alloy, heat and coolant to the joint area. The major considerations for optimizing the fixture design are:

1. geometry of parts to be brazed

2. base metals and filler materials

3. desired production rate

4. number of joints to be brazed & heat effect

5. fixture material

6. heat sink effect of fixture

7. fixture cost

The purpose of this paper is to describe the practical approach to fixture design as it relates to mechanized brazing operations.

Geometry of parts to be brazed

Joint geometry will determine to what extent the parts to be brazed must be mechanically contained. There are two basic types of joints encountered in brazing operations, the lap joint and the butt joint, but there are many variations and combinations of these types. The lap joint offers the greatest possibility for control of the joint strength since the amount of overlap may be varied. One of the most common types of joint encountered is known as the modified lap joint shown in Figure 1. With the recommended joint clearance ranging from a press fit for copper brazing to 0.10mm (0.004 in.) for silver brazing the fixture designer must allow adequate containment of the parts in the fixture while not restricting alloy flow. In the case of round or tubular members the joint clearance is on the radius which means the hole can be 0.20mm (0.008 inch) larger than the rod or tube. Also remember that joint clearance must be maintained at brazing temperature which means that clearances at room temperature must take this into consideration.

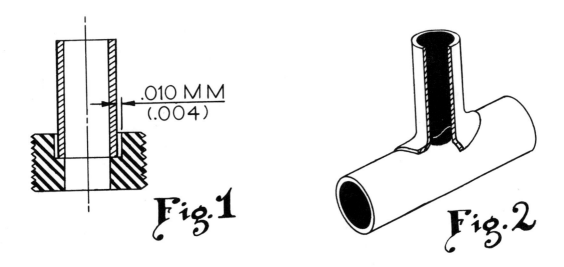

.010 M M
(.004)

Fig.1

Fig.2

Butt joints offer the advantage of a brazed assembly where the joint thickness is not increased. In order to obtain maximum strength in a butt join it is extremely important that the fixturing be designed to provide both proper alignment and also the optimum joint clearance for the particular base metals and filler materials being used. A common variation of the butt joint is the SADDLE TEE joint shown in Figure 2. The fixture used in this application while containing the part must not confine the vertical member to the extent that capillary attraction between the joint interfaces is hindered. As the base metals are heated there will be thermal expansion in more than one direction, if this condition is allowed to cause a mechanical interference than complete capillary action will not take place and the resulting joint will be less than fully satisfactory.

Base metals and filler materials.

The main fixturing consideration regarding the base metals to be brazed is must the fixture design include special provision because of the filler materials? If, for example, the joint is being made using a bronze filler material with an open flame heat source then the high temperature necessary can have a damaging effect on fixture parts. This might not only affect the fixture life but can also cause problems because of thermal expansion of the fixture itself.

The brazing of dissimilar base metals can present unique fixture design problems in that different rates of expansion must be dealt with. It may be necessary to build heat shielding devices into the fixture to prevent damage to one of the base metals. The type of materials to be brazed may dictate a specific brazing process and the designer must ascertain that the fixture will survive the process.

Desired production rate.

The production rate will not only effect the brazing process design but it will determine the ultimate fixture design. A brazing process involving two items to be assembled can usually be handled by manual operation up to approximately 600 parts per hour. Above 600 parts per hour or if more than two items are involved, then additional mechanical assistance for the operator may be required.

At production rates above 600 parts per hour the most common loading and unloading arrangement on a brazing machine uses an operator to load parts into the fixture and has automatic ejection after brazing and cooling is complete. This requires a fixture design that permits a mechanical fixture opening either preceding or in conjunction with ejection. Another possibility when high production rates are required is the automatic feeding or placement of one or more parts into the fixture. Here the designer must concern himself with not just the fixture concept, but the requirements of the various mechanical assists that will be utilized.

It is usually advisable to obtain input from a motion and time study specialist early in the fixture design phase of the project. A good practice is to use an isometric or pictorial sketch of the fixture and some loose parts to simulate the loading and unloading operation. The use of MTM (Methods Time Measurement) analysis is very advantageous since a study of the large time elements can direct the designer's attention to areas where the fixture may require changes or where mechanical assistance is advisable.

Number of joints to be brazed simultaneously.

When the brazing operations require the making of more than one joint simultaneously on an assembly consideration must be given to part orientation in the fixture. There are some successful brazing operations where filler material has been made to flow uphill by capillary attraction but utilization of gravitational force on the liquid filler material is more common. Fixtures can be designed so that all the joints are oriented in a compromise position that either favorably uses gravity or reduces its adverse pull on the filler material. Another technique is to design the fixture so that automatic repositioning can be accomplished on the brazing machine to obtain optimum part orientation.

A fixture designed to provide two different positions for brazing a fishing rod top guide is a good example. Two ends of a wire form are brazed to the side of a tube and the center portion of the wire form is brazed into the open end of the tube as shown in Figure 3. Because these 3 joints are in different planes on the tube it is difficult to get good brazed joints simultaneously at all three joints at one position.

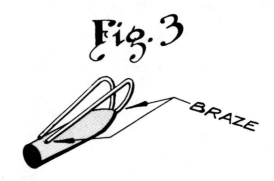

TOP LINE GUIDE

The fixture was designed with a hinge point so that the two wire ends can be brazed to the tube with the fixture laid back in an optimum position for these two joints, as shown in Figure 4. After the ends have been brazed, the fixture is engaged by a cam and shifted to the vertical position so that the center portion of the wire can be brazed into the end of the tube as shown in Figure 5.

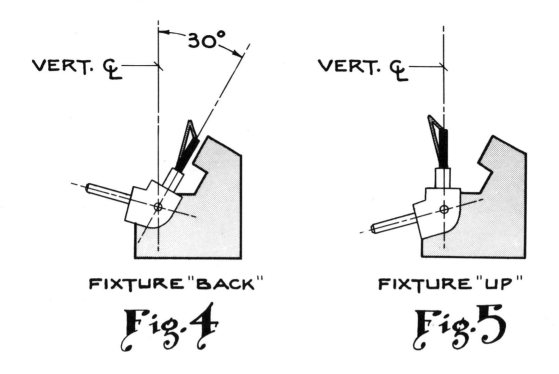

FIXTURE "BACK"

Fig.4

FIXTURE "UP"

Fig.5

When several joints must be made at one time the designer must thoroughly analyze the dimensional relations and tolerance of the component parts. Care must be taken in the fixture design that correct part location will result after the assembly is brazed. Most importantly the part must be removed from the fixture once all the components are brazed and become an inflexible assembly.

The need to braze more than one joint on an assembly not only must receive the considerations mentioned above but the proximity of the joints must be considered. Joints too close together may not permit the fixture design to use support members or locators in the normal fashion. Also there may be problems encountered with flux or filler material adhering to the fixture member so that loading or alignment is affected.

The heat required to bring the joint area and the filler material up to liquidus temperature can have considerable affect on the brazed assembly.

Not only is it necessary that proper joint clearance be maintained at the elevated temperature but the fixture must also compensate for expansion of both the base metals and the fixture components.

Often it is necessary to design the fixture so that thermal expansion will not be restricted during the initial heat cycle. This can present some dimensional problems especially if tolerances are critical. Another often used fixture feature is to design it in such a way that during the cooling cycle pressure is exerted by mechanical means to "force" the brazed assembly to certain dimensions. Some brazing fixtures are designed with sliding members that allow for thermal expansion and are manually returned to the starting point at the loading station by the operator.

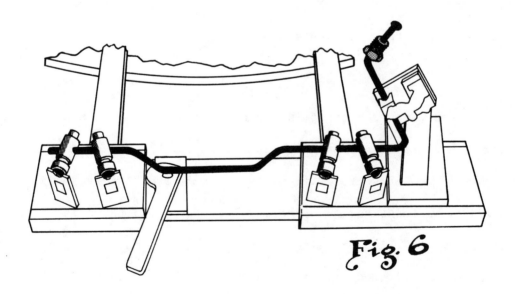

Fig. 6

The LEFT portion of this fixture is moveable and when the cam is released after heating and prior to cooling the fixture member moves with the contraction of the part.

Fixture material.

Just as the material of the parts being joined will have an effect on the process requirements, especially the heat requirement, the material used to construct the fixture must receive due consideration. The table below shows five stainless steels which provide properties needed for brazing fixtures. All five steels shown are standard commercial alloys but 304 and 410 are more readily available in various sizes.

TYPE NO.	MAXIMUM USE TEMP. °C	RELATIVE COST *	MILLING SPEED (SFM)	MILLING SPEED (CM/MIN)
304	895	4	40-60	1220-1830
309	1095	5	30-50	915-1525
410	675	1	40-60	1220-1830
430F	840	2	50-80	1525-2440
446	1095	3	40-60	1220-1830

* 1 = Lowest Cost

The corrosion resistance of the 300 series is much better than that of the 400 series. When water cooling is used the 400 series will show some rusting which is more a cosmetic than a functional problem. The machinability of 430F makes it a good choice for many complex fixture components.

When portions of the fixture are subjected to high temperatures as a result of flame impingement then the use of refractory metal or ceramic inserts may be used. Some of the new machinable ceramics offer interesting possibilities and will receive more attention when price and availability become attractive.

Heat sink effect of fixture.

The use of stainless steels in the fixture help to reduce the heat sink effect because of their poor thermal conductivity. This alone though is by no means sufficient and the fixture design must minimize the contact between the fixtures and the parts being brazed. The fixture will conduct heat away from the joint area which will increase the heat requirement or reduce the production rate. The fixture contact area can be kept to a minimum by altering the normally accepted geometry of various fixture components.

As an example the fixturing for a simple tube to fitting assembly would require some sort of plug to align the fitting on the fixture.

Instead of the normal round plug as shown in Figure 7 the plug can be cut away as shown in Figure 8. This design will result in reduced contact between the parts being brazed and the fixture.

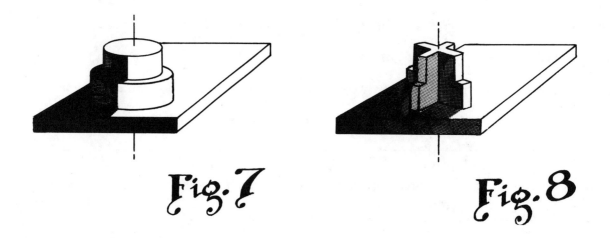

Fig. 7   Fig. 8

There are times when heat sinking is a desired condition, and here again good fixture design is essential. An example of this condition is the brazing of the diaphragm assembly to the capillary tube of a heat sensing control. Because the heat required for brazing can cause the diaphragm to be heated and perhaps affect its calibration, it is desirable to prevent the diaphragm from becoming heated. The fixture can be designed with a copper insert that contacts the diaphragm but not the stainless steel outer cover. In this manner heat is conducted away from the diaphragm and there is no deleterious effect as a result of brazing.

Fixture cost.

Now that it has been determined whether the fixture will be a simple "C" clamp or a complex assembly of exotic material it is appropiate to consider cost. A brazing process can require anywhere from 1 to 100 plus fixtures depending on the process. If the process uses a conveyor system to carry the part through the process stages then the fixture cost can be a significant part of total project cost due to number. A very common brazing machine in use today consists of a rotary turntable with fixtures attached to carry the parts through the various stations and back to the operator for unloading. If we assume as an example, the rotary machine uses twelve fixtures than the total fixture cost is the sum of:

1. Engineering design time including detail drawings

2. purchase or building of 12 fixtures

3. fixture set-up, debug and development time

Here are some examples of total cost of fixturing for brazing machines that have been completed recently.

EXAMPLE #1   FIGURE 9

This is a simple fixture to hold a stainless steel cup and copper tube in
in proper relationship for brazing. The fixture is designed to heat sink the
cup wall to prevent discoloration and the tube clamping is spring loaded.

Costs:

1.	Engineering (30.5 hrs.)	$ 579.50
2.	fixtures (12 purchased)	751.20
3.	install, debug, etc. (21.5 hrs.)	408.50
		$1,739.20

Cost per fixture = $144.93

EXAMPLE #2   FIGURE 10

This fixture has provisions for being rotated by the loading operator to adjust for the random position of the wire ends. Rotation is necessary to obtain optimum joint position.

Costs:

1.	Engineering (41 hrs.)	$ 779.00
2.	fixture (12 purchased)	1,900.42
3.	install, debug, etc. (61 hrs.)	1,159.00
		—————
		$3,838.42

Cost per fixture = $319.87

EXAMPLE #3   FIGURE 11

This fixture has a vertical sliding portion that is cam opened to permit removal of the brazed assembly from the fixture.

Costs:

1.	Engineering (48.5 hrs)	$ 921.50
2.	fixture (12 purchased)	2,475.00
3.	install, debug, etc. (70 hrs.)	1,330.00
		$4,726.50

Average cost per fixture = $393.90

CONCLUSION

The optimum fixture design will be the result of combining sound engineering and tooling practices with a review of the major considerations discussed here. It is also imperative that someone knowledgeable about brazing techniques be consulted prior to finalizing the fixture design and that the final design be reviewed before cutting any metal to construct the fixture. The equipment that the fixture will be attached to or used with must also be familiar to the designer because it can be critical to the final result.

A very worthwhile practice employed by one of the major suppliers of brazing equipment is the construction of a prototype fixture, usually out of cold rolled steel, which is then set-up on a lab machine. The heating pattern is developed as well as the loading, unloading and cooling techniques. After the lab machine has been set-up and the fixture debugged an actual production run is simulated and actual production rates established or verified. The finished brazed assemblies can be tested and submitted to whatever inspections are necessary to determine that the end result is the desired result.

Efficient fixturing for brazing automation can be achieved if the factors affecting the process are duly considered and if the manufacturing engineer utilizes the knowledge and experience available.

SUGGESTED REFERENCES

Brazing Manual, 3rd Edition, American Welding Society Inc.

Williams, G. N., "Increasing Productivity Through Brazing Automation",
Welding Journal, October 1973.

Williams, B. R., "Considerations in Automating a Brazing Operation",
Welding Engineer, March 1968.

Williams, B. R., "Selecting the Most Efficient Automatic Brazing Machine",
SME Paper #AD76-630.

Reprinted from Manufacturing Engineering, February 1975

Off-Spindle Fixturing Increases Productivity

*Providing duplicate fixtures so that
one can be loaded while a workpiece
in the other is being machined increases
productivity by reducing downtime*

RICHARD C. KNOX
*Manager,
Sales Engineering*
and
ALAN T. BAKER
Sales Engineer
*Hardinge Brothers, Inc.
Elmira, N.Y.*

WHEN AUTOMATIC second-operation machines are introduced into a plant, many fixturing methods that had been used for manually operated machines become inefficient. With a manual machine, the operator must concentrate on manipulating controls during machining. Then he must remove the part from the spindle-mounted fixture and fasten a new part. This fixture unloading and reloading is idle-spindle or nonproductive time. With an automatic machine, the operator has idle time while each part is being machined. If the workpieces are fixtured on the automatic machine, costly spindle idle time will be the same as for a manual machine. However, if duplicate fixtures are provided, nonproductive time can be reduced substantially.

Manual Method. Using an aluminum casting as an example, the workpiece is typically held on a fixture plate which is mounted on the spindle of a manual machine. The casting is center-ed, using a removable pin as a guide, and bolted to the fixture plate, as seen in *Figure 1*. Loading time is about one minute.

Precision tolerances (a total of 0.0003 inch on the bearing bore, and 0.0005 inch on the large diameter bore) are required on this casting, thus necessitating a rough and finish cut on both these surfaces. Total machining time is two minutes. After machining, the completed part is taken off the fixture plate by removing the four cap screws,

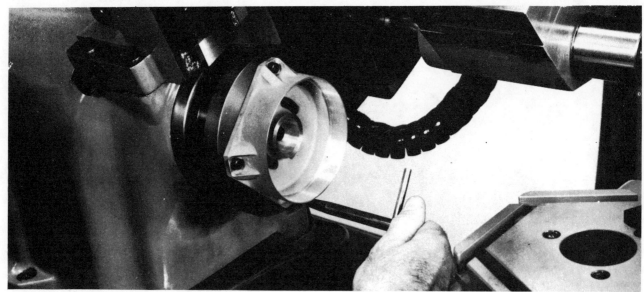

1. TYPICAL METHOD used on manually operated machines consists of bolting workpiece to fixture plate. Loading of this casting takes about 1 minute, and removal, 30 seconds.

which takes about 30 seconds. Total idle spindle time for loading and unloading each casting is 1 minute and 30 seconds, and floor-to-floor time would be 3 minutes and 30 seconds per part.

Off-Spindle Fixturing. Nonproductive time can be slashed on automatic machines by providing duplicate fixtures. While one casting is being machined, the operator can unload a completed part from the second fixture, *Figure 2,* and load an unmachined part. Each fixture is provided with a precision round hub which fits into a standard, hardened and precision ground step chuck on the machine spindle, *Figure 3.* Time required to mount a loaded fixture into the step chuck is about 7 seconds, and unloading time is 3 seconds.

Off-spindle fixturing thus reduces the total idle spindle time from 1 minute and 30 seconds to 10 seconds. Floor-to-floor time is decreased from 3 minutes and 30 seconds per part to 2 minutes and 10 second per part — an increase in the production rate of over 60%. Another important advantage of this method of fixturing is that there is no change in the chucking diameters, thus providing dead-length control.

Eccentric Fixturing. The duplicate fixtures used for off-spindle loading and unloading generally have round surfaces for gripping in a step chuck. However, they can have two different

diameters for chucking, and these surfaces may be eccentric to each other. The part shown being machined on the front cover of this issue is held in such an eccentric fixture, *Figure 4.* A Hardinge fully automatic chucking machine and two fixtures are used to produce

these parts.

The irregularly shaped workpieces are fastened inside the fixtures, using two set screws in each fixture for clamping. External, cylindrical gripping surfaces are easy and quick to load into the step chuck on the machine. After one side

2. OFF-SPINDLE FIXTURING reduces nonproductive time. Operator unloads and reloads one fixture while a workpiece on a second fixture is being machined.

3. *PRECISION ROUND HUB on fixture fits into a standard, hardened and precision ground step chuck mounted on spindle of automatic machine.*

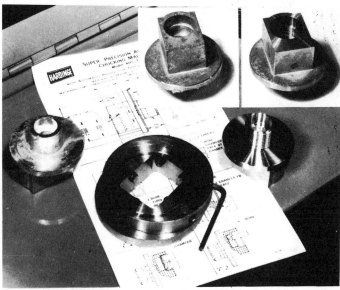

4. *ECCENTRIC FIXTURE for holding irregular shaped workpieces. Opposite end of part before and after machining is shown in inset (upper right).*

of the workpiece has been completely machined, the fixture is end-for-ended to grip on the second chucking surface, which is eccentric to the first by the amount of off-center required on the workpiece. Thus, both sides of each part are machined in a single fixturing, holding tolerances and concentricity as close as 0.0002 inch.

Other Applications. Fixtures may also be reversed in the step chuck to perform the same or similar operations on opposite ends of a workpiece. For example, the same operations are performed on both ends of an aircraft fitting with the setup shown in *Figure 5.* Again, two fixtures are provided so that one can be unloaded and reloaded while a part in the other is being machined.

Square or rectangular fixtures may be required for holding some workpieces. Such fixtures would be mounted in a step chuck having a square or rectangular hole. This permits performing operations at 90 degrees to each other, as well as on different center lines.

Many other variations of these techniques are possible to suit specific production requirements. In some cases, production rates on precision automatic, second-operation machines can be increased more than the 60% for the example previously discussed in this article. ∎

5. *FIXTURE REVERSAL in the step chuck permits performing the same operations on both ends of an aircraft fitting. Two fixtures are provided to minimize downtime.*

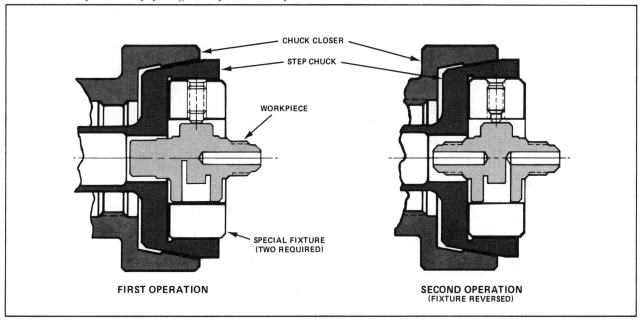

CHUCK CLOSER

STEP CHUCK

WORKPIECE

SPECIAL FIXTURE (TWO REQUIRED)

FIRST OPERATION

SECOND OPERATION (FIXTURE REVERSED)

Presented at SME's 1974 International Engineering Conference, April 1974

Tooling for the Extrude Hone Process

By David E. Siwert
Extrude Hone Corp.

The Extrude Hone abrasive-flow finishing process was originally developed to meet extreme quality requirements of deburring microscopically inspected aerospace components. Due to its basic simplicity, flexibility, and controllability, Extrude Honing has developed into a cost effective method for commercial deburring, polishing, radius generation, sizing, recast removal, and surface finish improvement. Advances in media formulation, machine design, and fixture design have made it a feasible process for finishing metal, plastic, and ceramic parts.

INTRODUCTION

The Extrude Hone process is a proprietary abrasive flow machining technique based upon the controlled extrusion of a semi-solid abrasive laden media. The media formulation is carefully prepared to produce the desired rheological effect and abrading action with a high degree of consistency and control. The typical media formulation exhibits rheopetic characteristics as it flows and encounters any form of restriction causing a rise in shearing stress. The viscosity (kinematic) increases and a "stiffening" effect occurs. At this point, energy is imparted to the abrasive grain to accomplish the abrading action. Process pressure, media temperature, and volume flow rate are carefully controlled to produce laminar flow of media through the part.

Currently the process is still an "infant" in comparison to most other "conventional" finishing processes. But its high degree of flexibility and control have been recognized by many manufacturers and its potential has only been touched.

Basic fundamentals involved in the tooling for the process are presented to enable manufacturers to understand its capabilities and potential applications.

TOOLING

The Extrude Hone process involves three types of tooling that can be altered independently to obtain the desired finishing action at a specified production rate. These include: (1) The machine tool--to accomplish and control media flow; (2) The fixturing--to confine and direct media flow; and (3) The proprietary abrasive laden media (perishable tooling with average machine life of 500 hours).

Successful application of Extrude Honing is dependent upon understanding the basic fundamentals of these three tools.

MACHINE TOOLS

Extrude Hone machines (Figure 1) have been designed and built in numerous sizes and with many optional features as required by manufacturers to meet their specific applications. All machines, regardless of size, are positive displacement systems designed to force the abrasive laden media through a fixtured workpiece at a selected pressure and flow rate.

The basic elements of the machine tools are as follows:

1. Hydraulic Power Unit

2. Media Chamber Assemblies

3. Hydraulic Clamping Cylinders

4. Support Structure

5. Control System

A hydraulic unit develops the fluid pressure required by the hydraulic clamping cylinders and media extrusion rams. Standard units operate within the
0-1600 p.s.i. (0-112.5 Kgs/Sq. cm) pressure range with flow rates up to 30
gallons per minute (113.6 liters/minute). Units developing greater pressures
and volume flow rates are forecast as manufacturers press for machines with
greater production capacity.

Current machines are designed with two vertically opposed media chamber assemblies. The assemblies consist of a honed cylinder, a specially designed piston
and cup seal subassembly, and a hydraulic ram. Diameters and lengths of the
honed cylinders determine the volumetric capacity of a given machine. Units
commercially available range from 100-1300 cubic inches (1639-21307 cc). One
media chamber assembly is mounted in the base structure of the machine while
the second is mounted in the head structure that moves vertically on a set of
guide rails. Hydraulic clamping cylinders are mounted in the base and attached
to the head structure. The head is moved down for clamping fixtures between
the media chamber assemblies and up for unclamping. Clamping force can be
adjusted to meet specific requirements for a given fixture design. Clamping
forces of more than 200,000 pounds (90718 kg) are possible on larger machines.
Overall dimensions for standard units range from 36" wx X 36" dx X 80"h to

72" w X 48" d X 130" h (91.4 cm x 91.4 cm x 203 cm to 182.8 cm x 122 cm x 330 cm). (Figures 2, 3).

A control system is required to operate the machine to obtain the desired process parameters. The degree of control required by a manufacturer will determine the type and complexity of the system. All units are extremely flexible to permit wide application of the process. Basically all units are offered with controls on:

1. Hydraulic System Pressure (clamping media)

2. Clamping and Unclamping

3. Volume Flow Rate of Media

4. Advance and Retract of Media Pistons

These controls are offered either manually or electrically operated.

The more sophisticated control systems provide an even greater degree of flexibility for the manufacturer but are not required for successful application of the process in most cases and as a result are offered as options on the standard units. These include:

1. Automatic Flow Timers

2. Automatic Cycle Counters

3. Pressure and Temperature Compensated Flow Control Valves

4. Independent Pressure Relief Valves for both Media Chambers

5. Volumetric Displacement Systems

Optional equipment often utilized in "high" production applications include media heat exchangers, temperature control units, and media refeed units. Heat exchangers and temperature control units are used for precise control over media temperature during machine cycles. Various size units are available to meet all cooling requirements. Refeed units automatically fill the media cylinders and keep them filled. Media is sometimes carried off in fixtures or is lost due to fixture leakage and must be replaced. The refeed units do this automatically to permit the operator to avoid manual reloading. Higher production rates often justify the additional cost of these options.

MEDIA FORMULATIONS

The key to successful application of the process lies in the proprietary media formulations. Several years of research, development, testing, and production utilization have lead to unique controllable formulations. Simple laboratory adjustment of formulations can produce a vast range of finishing characteristics. Formulations vary in aggressiveness, radius forming capability, and rheology. Average machine life is 500 hours for most standard formulations. The three components basic to all media formulations are: (1) Media base material--or vehicle, (2) The abrasive grain, and (3) proprietary additives.

1. Base Materials

Several different base materials are currently being utilized in the proprietary media formulations. The carriers differ mainly in their rheological properties. The most widely used carrier is a high-viscosity rheopetic fluid (i.e., its apparent viscosity increases with time to some maximum value at any constant rate of shear). Other carriers include newtonian fluids, pseudo-plastics, and dilatents. The base material is selected to meet specific finishing requirements of a given application.

The base material supports and separates the abrasive grain as it flows through the workpiece. Its inherent cohesion and tenacity are important in media formulations. The ideal base material for most applications is one that will flow virtually like a "solid slug" as it passes through a restriction. The base material must have enough tenacity to drag the abrasive grain along with it as it contracts, expands, and changes direction.

2. Abrasive Grain

Different types, sizes and quantities of abrasive grain are mixed with the media base material to meet specific finishing requirements. The four main types used are: (1) Aluminum oxide, (2) Silicon carbide, (3) Boron carbide, and (4) Diamond. Silicon carbide is the most commonly used grain due to its good cutting characteristics and economic price. Boron carbide and diamond are used in limited applications (such as polishing carbide dies) due to their high cost.

Abrasive grain sizes range from very fine #500 grit (small hole applications) to coarse *8 grit (roughing and stock removal applications). The finer abrasive grain generate extremely fine finishes and small radii while the coarser abrasive grain removes more material and generates larger radii. Even the largest grain size produces a finish much finer than would be expected because the "flexible" base material cushions grain as it is forced against the workpiece surface.

Abrasive grain is mixed in varying ratios with the base material. Abrasive grain to base material ratios vary from 1/4:1 to 4:1 with a 2:1 ratio most common for production applications. Generally speaking the higher ratios generate larger edge radii and increase stock removal rate. Maximum effective grit ratio is dependent upon the tenacity of the base material and its "saturation" point. Beyond the "saturation" point, the base material can no longer support and separate the abrasive grain in a uniform manner.

3. Proprietary Additives

Numerous proprietary additives are utilized to modify the base materials to obtain desired changes in their rheological properties, temperature stability, and solubility. Other additives are used to adjust cohesion, plasticity, viscosity, and lubricity. All additives are carefully blended in predetermined quantities to obtain consistent formulations.

FIXTURING

Fixture design is often a very important factor in achieving the desired effects from the abrasive-flow finishing process. Basic functions of fixtures include:

1. Holding the part/or parts in proper position between the two opposed media chambers.

2. Directing media flow to and from the areas of the part to be worked on during the process cycle.

3. Protecting edges or surfaces from abrasion due to media flow by acting as a mechanical mask.

4. Providing a restriction/or restrictions in the media flow path to control the media action in selected areas.

5. Containing the media and completing the closed-loop system required for multiple machine cycle operation without loss of media.

6. Assisting, loading, unloading, or cleaning operations.

Regardless of the ultimate capacity of the production tooling, the process feasibility and fixture design concept is usually developed and testing using "one-up" trial fixtures. This gives manufacturers an inexpensive way to evaluate the process, establish designs, forecast production rates, and define costs.

In most applications, a very simple fixture is all that is required. As production rates increase, more sophisticated fixturing is likely to become desirable in order to process several parts simultaneously. In all cases material selection is critical due to the extremely abrasive nature of the media and the pressures utilized in the process. Wear prone areas of the fixtures are designed with replaceable inserts to reduce operating costs. Rotary arms and tables (Figure 4) are used in "high" production systems with multiple fixtures to permit loading and unloading one fixture while another is being Extrude Honed. Production rates as high as 1,000 parts per hour have been achieved with speed of loading and unloading fixtures usually being the limiting factor. Improvements in fixture designs and material handling techniques would significantly increase the production rate possible.

For the purpose of basic fixture design, any part can be classified as a sleeve, tube, plug within a tube, or a combination of the aforesaid classifications.

Hydraulic manifolds and housings, (Figure 5) hollow turbine vanes and blades with air cooling holes, master brake cylinders, fuel spray nozzles, (Figure 6) and hydraulic valve sleeves (Figure 7) are a few of the countless parts that are included in the "sleeve" classification. "Sleeves" are characterized by a cavity with significantly smaller intersecting passages and proper media flow path is into the main cavity and out one or more of the smaller intersecting passages. Normally no stock removal occurs in the main cavity except at the point of intersection because the media does not become significantly abrasive until it is restricted. The key to fixturing in this case is to calculate total cross-sectional area of the main cavity and compare this to the cross-sectional areas of the intersecting passages. In all cases, the exit area must be less than entrance area if one desires to work on the areas of intersection. Selectively sealing off intersecting holes is required in some cases in order to obtain the desired results.

Extruding dies, compacting dies, cold heading dies, and drawing dies (Figures 8, 9) are prime examples of parts that can be classified as "tubes" for Extrude Honing. Glass bottle molds, internally splined gears, printed circuit boards, thread plates, and similar parts are all "tubes" because their configuration provides an enclosed flow path. Fixturing is used to direct the flow to the "tube" and the parts' internal configuration is the point of restriction for media flow.

External surfaces of almost any contour or configuration can be Extrude Honed by using fixturing to direct and restrict media flow around the outside surface. In this case, the fixture is subject to as much abrading action as the part and material selection is critical. These applications are classed as plugs within a tube because the part is the plug and the fixturing is the tube. Hydraulic pump rotors are good examples of parts that are processed as plugs within a tube. (Figure 11).

CONCLUSION

In summary, proper application of the machine tool, fixturing, and perishable tooling (media) can provide a cost effective method for manufacturers. The process offers greater flexibility and control than is available with other finishing processes and is limited only by the ingenuity of the process engineer.

The Extrude Hone process, media formulations, and machines are available, under U.S. and foreign patents issued or pending, exclusively from Extrude Hone Corporation, Irwin, Pennsylvania.

Figure 1

Figure 2

Figure 3

Figure 4

Figure 5

Figure 6

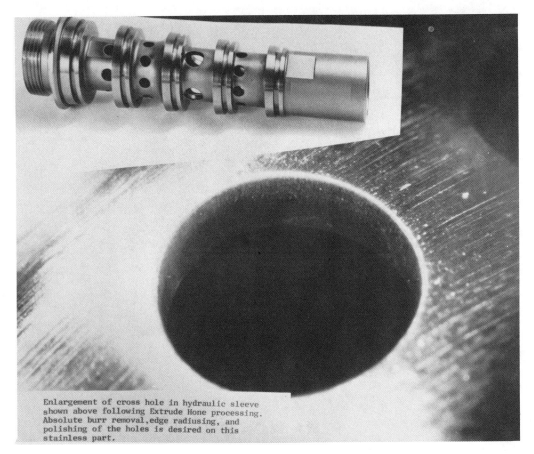

Enlargement of cross hole in hydraulic sleeve shown above following Extrude Hone processing. Absolute burr removal, edge radiusing, and polishing of the holes is desired on this stainless part.

Figure 7

Figure 8

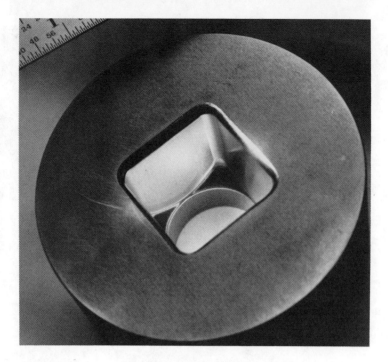

Figure 9

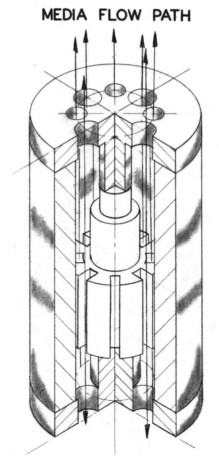

MEDIA FLOW PATH

HYDRAULIC PUMP ROTOR

Figure 10

Presented at SME's Nontrad '76, October 1976

Abrasive Flow Machining—Concepts of Fixture Design for the Dynaflow Process

By John Stackhouse
Dynetics Corp.

Basic design concepts of fixturing for the DYNAFLOW PROCESS are discussed. Methods of proving tool designs by using "soft-tooling," and how production fixture design parameters can be extrapolated from tests using such soft-tooling, are explained. Design considerations relating to work-piece size and configuration, machine capacity, media velocity and properties are also discussed.

INTRODUCTION

Abrasive Flow Machining (AFM) utilizes a flowable, abrasive laden media which is forced through orifices and over edges on many types of components. Edge-treatment such as burr-removal and radius-generation results, often in conjunction with surface-treatment such as thermal re-cast-removal and polishing.

Unlike many of the Free Abrasive Machining (FAM) methods, which are used for "overall" deburring, flash-removal or polishing, AFM usually concentrates on finishing particular areas of a component. Economic considerations can prevent parts that can be adequately processed by bulk-finishing free-abrasive methods from being processed by AFM, but components with inaccessible areas to be finished or with appreciable stock removal requirements can often be processed successfully with AFM.

The abrasive-laden media is used in specially-designed machine tools to process parts held in fixtures and provides an action comparable to that of a "flowable file," with capabilities ranging from a light buff to coarse stock removal.

The actual machines used for AFM have been described in a previous paper[1] but all basically consist of the same elements. A media chamber is situated either side, one above and one below, a fixture aperture. A mechanism provides fixture clamping and a hydraulic power supply forces the media from one chamber to the other and back again, through the part/fixture arrangement. Figure 1 shows such a machine.

Model	Distance Between Pillars	Maximum Daylight	Maximum Media Pumping Capacity	Main Motor Drive
HL60	37"	29.5"	60 gpm	30 HP

317

Figure 1
Model HL60
DYNAFLOW Machine

The media[2] is flowed at pressures of 150-475 psi and contains an abrasive, most commonly silicon carbide, aluminum oxide or boron carbide. A "DYNAFLOW Media System"[3] has been established to enable users to maximize their application of a particular media and allow mixing or modification of media, from prepared "bases" and the addition of abrasive grain, to give the greatest possible flexibility to the user.

Figure 2
AFM Media

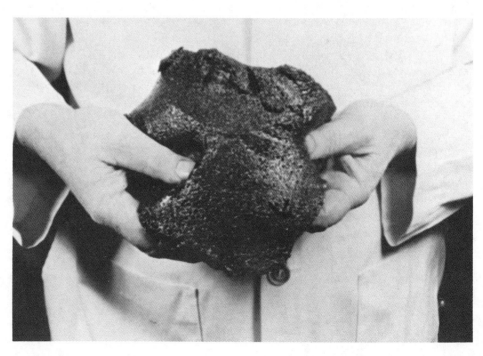

Another paper[4] has been written which discusses this systematic classification, the effect of using different types of abrasive grain and the effects of using different operating pressures.

A serious attempt has been made to impart science into the "art" of AFM, by documenting these details of Abrasive Flow Media, but little has been done to date in detailing the other element used in the process, which is fixturing.

The following discussion will serve to introduce those not familiar with AFM to fixture design concepts, and the case-histories discussed will detail the progression from experimental, or "soft-tooling," to production fixturing.

FIXTURES

Unlike Free Abrasive Machining (FAM) methods, each workpiece processed by AFM required its own fixture, but in some cases similar parts can be accommodated in "shuttle" fixtures, using various replaceable inserts.

Production fixtures can be single or multi-cavity, depending upon such factors as workpiece size or configuration and total production quantities. The materials and concepts for such production fixtures are discussed later.

Experimental test-fixtures, or "soft-tooling" as it is called, is normally a single-cavity configuration. These are used to establish one or more of the following:

o Feasibility of AFM, thereby demonstrating to a potential user the effect of AFM on particular parts.

o Optimum media formulation.

o Final production-fixture configuration.

o A cycle time, and from this extrapolate production rates and facility requirements for a particular potential user.

FUNCTION OF FIXTURING

The function of fixtures used in AFM is the same as that for conventional machining i.e. to locate and clamp the workpiece and to provide a guide for the cutting tool. The cutting tool in the case of AFM is the abrasive media.

Normal media flow is a reciprocating one, from the lower machine chamber to the upper machine chamber and back again which is normally repeated a number of times. A most important characteristic of AFM is that the abrasive media cuts most where the velocity increases the most, and any other cutting, or machining, is a function of the relative velocity.

A simple rule of thumb is:

<u>Greatest</u> AFM action @ <u>Greatest</u> Restriction.

<u>Least</u>   AFM action @ <u>Least</u>   Restriction.

From this rule of thumb, if typical part/fixture arrangement is considered, as seen in figure 3, it can be seen that, providing the fixture porting is of sufficient area, no abrasion should take place on the upper fixture plate.

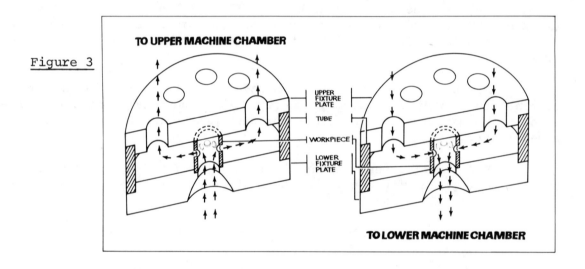

Figure 3

This figure shows a tubular workpiece with holes drilled through the sidewall which must be deburred. This configuration is typical of many hydraulic components, bearing retainers, etc. The left view shows arrows denoting media flow from the lower machine chamber to the upper, at which stage primary action is deburring <u>internal</u> intersections, as this is where <u>maximum velocity increase</u> is occurring. The right view shows flow reversal and primary action is deburring <u>external</u> intersections as this is where <u>maximum velocity increase</u> is occurring.

Workpiece configuration and size also affect the outcome of AFM processing with a particular fixture design. In the example above, processing would be successful only if the total area of the sidewall holes was sufficiently smaller than the area of the internal bore of the workpiece. On a workpiece with a bore of small dimension, and with sidewall ports of relatively large total area, the fixture concept more appropriate would be as shown in Figure 4.

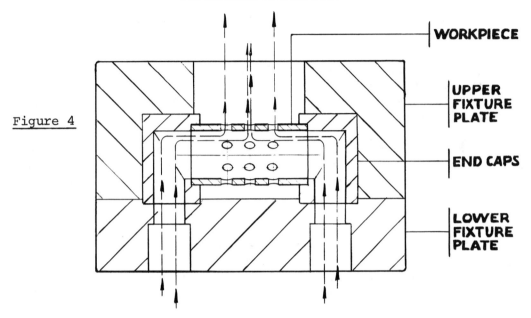

MEDIA FLOW TO UPPER MACHINE CHAMBER

WORKPIECE

UPPER FIXTURE PLATE

END CAPS

LOWER FIXTURE PLATE

Figure 4

It can be seen that, by using both ends of the workpiece to flow media to the area to be processed, the actual inlet area has been doubled, thereby doubling the ratio of $A_1 : A_2$ where A_1 = Total Inlet Area and A_2 = Total area of holes to be processed.

It should be noted that no hard and fast rules have been established to date, as factors other than the above also can affect the result.

Some of these are:

o   Total length of workpiece.

o   Location of holes and grouping.

o   Wall Thickness.

o   Tolerance of workpiece bore.

o   Tolerance allowed on edgebreak (in some cases total deburring can be effected but because of flow-patterns preferential edge breaking can occur).

o   Tolerance of sidewall holes.

Unless the workpiece is very similar to one which has been previously fixtured, the decision of which fixturing concept to adopt is largely empirical, that is, the most straightforward method of flowing in one end is tried first, using "soft-tooling," additional refinements and modifications being made during development.

FIXTURE DESIGN CONSIDERATIONS

As with any machining process, there are several considerations when designing fixtures for AFM. They include:

o Workpiece and fixtures are subjected to pressures up to 500 psi during the process cycle. As AFM normally utilizes reciprocating flow, each side of the assembly is subjected to these forces many times whenever a workpiece is processed.

o AFM Media contains abrasive grain and this must be considered. Fixture components with sliding fits should have sufficient clearance for ease of assembly and disassembly. For example, compression and tension springs may become clogged with abrasive, thus preventing their effective use.

o Areas contacting the workpiece should be coated with a resilient material to prevent abrasive grain impingement into the fixture or work-piece. In certain cases where production requirements or other factors do not allow this resilient coating, areas should be easy to wipe clean, and clamping should positively hold the workpiece to prevent separation during processing and subsequent ingress of abrasive.

o Internal seal areas must be designed for optimum effectiveness, as minor leakage will erode the fixture and workpiece. If a leak-condition persists without rectification, the leakage will become more and more severe, affecting process efficiency as less media passes through the area to be processed.

o The number of workpieces processed in each fixture will depend on the size of AFM machine and viscosity of media to be used. Often this is determined by reviewing the results obtained in preliminary tests using single cavity "soft-tooling."

o The material used for the fixture body will depend upon overall size and weight. Steel, aluminum and nylon are used, nylon generally being used for fixtures 10.00 inch diameter or less. On larger fixtures steel or aluminum are used, often with nylon end-rings to minimize abrasion where the fixture contacts machine fixture-plates. The material used as a resilient coating is polyurethane, which is normally bonded to aluminum inserts which are then secured to the fixture by socket head screws.

EXAMPLES OF FIXTURES FOR A.F.M.

It is not possible to cover all the various possibilities of fixture design because of the inherent flexibility of AFM, and the following case histories reflect that flexibility.

Example No. 1 ..... Stainless Steel Valve Block (See Figure 5) - deburr and radius internal intersections of holes and slots with main bore.

Figure 5

Initial discussion with the manufacturer resulted in "soft-tooling" being fabricated to show the effect of AFM. The parts had previously been deburred by hand, a very tedious, time-consuming and non-uniform procedure. This valve block is used in an aircraft servo control system and has to be absolutely free of burrs. Any burr present when the part is assembled could cause jamming of the system if it became dislodged in service. "Soft-tooling" was fabricated from nylon, to allow flow into both ends of the main bore. The manufacturer could not tolerate preferential edge-breaking and hence the decision to flow into both ends. For purposes of fixture design discussion the part can be considered "tubular," and media flow could be identical to that shown in Figure 4.

Discussion with the manufacturer about the processed parts indicated that some burrs remained on two internal slots. After reviewing this condition it was found that the two "scalloped" slots in question were substantially larger in area than the holes that emanated from them to the outside of the part. Further tests were run, with parts prepared by removing the large burrs at these two slots and successful deburring resulted. The "action" at the two slots, although not as agressive as at the hole entrance, was sufficient to remove the small burrs remaining and generate a small, uniform radius around the edge.

Final fixture design was identical to the "soft-tooling", except that areas contacting the part were coated with polyurethane, 0.125 inches thick.

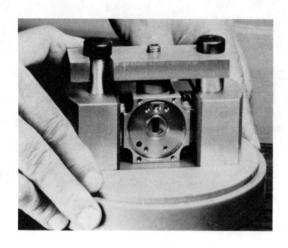

Figure 6   Figure 7

Figure 6 shows a part loaded into the fixture (6.00 inches diameter), note the polyurethane coated blocks and swing-clamp bar. Figure
7 shows the assembled fixture.

Single-cavity fixturing was determined to be most cost-effective,
because of relatively low production quantities. On similar parts with
higher production quantities the same configuration with multi-cavities
could be used.

Example No. 2 ..... Aluminum Metering Sleeve (See Figure 8) -
deburr and radius requirement 0.001 - 0.003 inches, at intersection of
sixteen 0.033 inch diameter holes with 0.315 inch diameter honed bore.

Figure 8

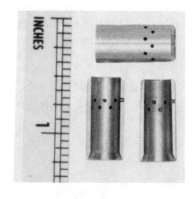

BEFORE   AFTER

Because of the extremely critical radius tolerance, these parts also were processed in "soft-tooling" flowing in from both ends of main bore, as shown in Figure 4. Also because of the extremely critical radius tolerance the parts were prepared by passing a pin through each hole, to weaken the burr remaining in the holes.

Production fixturing was designed, using polyurethane coated end-blocks to prevent damage to parts by the abrasive grain. Total process time for seven parts was twelve seconds, at 300 psi, using an easily flow-able No. 220 silicon carbide media.

Example No. 3 ..... Jet Engine Fuel Swirler (See Figure 9) - Reduce core-flash, polish surfaces and radius edges through vane section to increase air flow readings.

Figure 9

Figure 9 shows two parts, one processed and one unprocessed, in front of the assembled production fixture.

Single cavity "soft-tooling," was made from nylon, clamping on outer flange and incorporating a central plug attached to a cross-bar, to allow flow only through vane section. It was found that an anti-rotation pin near lugs was required to prevent the "turbine-effect" rotation as the media flowed through the vane section. While testing media it was found that Boron Carbide was significantly more effective than Silicon Carbide, as these parts are made from investment - cast Stellite, a parti-cularly hard material.

When designing production fixtures, a three-cavity configuration incorporating blank-off central plugs and an inside diameter to contact lugs and prevent rotation evolved. Fixture bodies, 8.00 inches diameter by 4.00 inches high, were made from nylon, with mild steel insert rings where parts are clamped around flanges. The stellite is such a hard material that abrasive impingement does not occur, even though media containing No. 20 mesh and No. 36 mesh abrasive grain is used.

Figure 10

Figure 10 shows the fixture in the open position, with one part in position. Note the nylon bodies and mild steel insert rings and blank-off central plugs. Also shown is a dowel for radial location of the two fixture halves.

CONCLUSION

Although many of the fixture possibilities of AFM have not been explored in this paper, a general introduction and design guidelines have been presented.

Each user may have different manufacturing techniques, burr-control and finishing requirements, but by using fairly inexpensive "soft-tooling," each can find if AFM can be an economic solution to his finishing problems.

Production fixturing can vary from simple, single-cavity to multi-cavity. No discussion has been presented on fixture-loading devices, but these are based on standard sliding and rotary-loading concepts which are attached to the machines. Fixture unload/reload stations can be positioned at the front or rear of the machines. Unloading and reloading of fixtures can be performed while a duplicate fixture load is being processed, if production quantities warrant such a set-up.

The DYNAFLOW PROCESS is covered by Patents and patents pending and is available from Dynetics Corp., Woburn, Massachusetts, U.S.A.

REFERENCES

1. J. Stackhouse, "Deburring by Dynaflow" SME Paper MR75-484, 1975.

2. U.S. Patent 3909217, September 13, 1975.

3. "Dynaflow Media System," Technical bulletin, available from Dynetics Corp., 1976.

4. W. B. Perry, "Properties and Capabilities of Low Pressure Abrasive Flow Mediae" SME Paper M75-831, 1975.

CHAPTER SIX
FIXTURING FOR AUTOMATION AND NC

Presented at SME's Jigs and Fixtures Clinic, March 1979

Principles and Concepts of Fixturing for NC Machining Centers

By Colin J. Gouldson
Canadian Institute of Metalworking

1. <u>N.C. MACHINING CENTRES</u>

   An N.C. machining centre has the ability to encompass many machine tools into <u>one</u>. With this fact in mind the obvious extension to this newfound capability is to design and build fixtures that will allow as many of these different operations to be performed in the least number of setups. How NC effects the design of fixtures may be derived from the principle of operation of these machines.

   Briefly, an NC machine is positioned by accurate ball screws coupled to servo motors. Feedback systems attached to these ball screws can measure the exact position to within one tenth of one thousandth of an inch(.0001). Therefore, the accuracy of a machined part is not reliant upon the operator but in the machine tool, the cutters and the fixture.

   In NC machining, the part remains fixed while positioning and motion is carried out by the tool. The only purpose of a fixture in NC machining is to locate accurately and hold the part rigid while the cutting tool is engaged in the workpiece.

   There are two major types of machining centres which need to be categorized, one having a horizontal spindle and a rotating work table, the other a vertical spindle. Both types of machines incorporate automatic tool changing.

2. FIXTURE DESIGNING FOR NC

Programming NC machines in many cases is accomplished using the computer but it has not yet become a reality in fixture design although successful implementation of CAD/CAM systems have produced automated drafting, process planning, scheduling and part design. Fixture designing for conventional machines relies on a thorough knowledge of basic machining practice, previous hands-on experience and familiarity with machine tools.

Tooling up an NC machine with regard to fixtures requires the tool designer's normal requirements together with teamwork on the part of the process planner, programmer, and tool designer. Each must know what the other has in mind and before final plans are made a sequence of events should be formulated as outlined in the following procedure.

1. The process planner decides the basic method of manufacture and writes his process sheets accordingly.

2. The NC parts programmer reviews these plans and indicates the size of material required, pre machined surfaces, stock allowance and his tooling requirements.

3. A basic layout of the proposed fixture is given to the tool designer by the programmer which may show the locating surfaces, part clamping regions, clearance areas required, and location of the fixture with respect to the machine tool. The tool designer should have reference material showing the physical elements of the machine such as slide travel and limits, table configuration and tool changer constraints.

4. The preliminary drawing should then be shown to the programmer for his approval and for him to obtain basic dimensions instrumental in his program. The information of concern to the programmer is as follows:

(a) dimensions from part datums to reference positions on the fixture.

(b) the dimension of fixture reference points to machine table locations.

(c) the exact physical location of clamping devices and similar obstructions.

The latter point is important because the path of cutting tools is predetermined and must be taken into consideration when the program is written.

One positive way to obtain the optimum fixture design is to have the tool designer learn the full capabilities of the NC machine from the programming aspect and from the operating view point. The designer is experienced in conventional machines and fixtures, the new insight should lead to a better understanding of NC and how to fully exploit its potential.

The tool designer may help the programmer with his task. One way is to supply a full size view of the fixture and located part as it would present itself to the spindle of the machine. In this way the programmer would use the prints as an overlay to duplicate the cutter paths for his reference and also to remind him where the clamps and obstructions are located.

3. ECONOMICS OF FIXTURE DESIGN

Reports have shown that with batch sized lots varying from 10 to 500 parts, an NC machine may produce an economical return in the production of parts. Although NC will not remove metal any faster than conventional machines, it will do it more efficiently. Where the advantage lies is due partly to the reduction in multiple setups, simplication of fixturing, reduction of lead time and reduced inventory.

It has also been reported that of the machining cycle time of one average part only about 30% is spent as productive time where the tool is engaged in removing material.[1] The remaining 70% is used up by setup, loading, unloading, inspection, idle time and miscellaneous items such as tool replacement and operator intervention. (See Fig. 1)

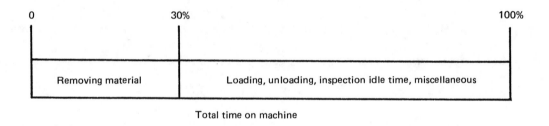

Fig. 1 Percentage of actual metal removal during the cycle of 1 part.

People engaged in tooling are concerned with setup, loading and unloading. These areas can have a much greater impact on the final part costs than that of the increased efficiency of the NC machine versus the conventional machine. This is especially true for shorter machining cycles where removal of parts, chip clearing and loading far outlast the cutting time. The ability to load and unload parts quickly leads to the first criteria of NC fixturing.

[1] NCS N/C CAM Journal, July, 1977

The fixture should be designed to reduce the part setup time.

One economical advantage of NC machining lies with the fact that smaller batch lots may be produced thereby reducing inventory. To accomplish this it means that the fixture is going to be set up on the machine more frequently, hence the second criteria for NC fixturing.

The design should enable all fixtures to be located and attached easily and accurately.

If one has an investment of hundreds of thousands of dollars in one machine, shouldn't a small percentage be spent on designing as near perfect a fixture as possible? The support personnel backing up that machine represents a substantial investment, namely programmers, tool setters, and maintenance people. Any error in design might require the needless services of any one of them. The overall effect of a soundly designed and manufactured fixture will lead to many advantages.

1. With an increase in quantity there will be a decrease in the cost per part.

2. Part accuracy will be maintained.

3. Inspection time will be reduced

4. Rework will be minimized.

5. Tape prove out will be accomplished faster

In summarizing the economical aspects of fixture design, there are two points worth considering. First, the accuracy of the part is directly dependent on the holding device no matter how well the part is programmed and processed. And, secondly, it is justifiably and financially prudent to examine all aspects of holding and locating the part firmly and accurately.

4. ACCURACY OF FIXTURES

The accuracy with which an NC machine produces parts is not wholly reliant on an operator but on the information fed to the machine. The final assembly of a fixture and its accumulative dimensions to locating surfaces and reference points in most cases is of little importance. Once the fixture has been manufactured and assembled it can be measured and the resulting new dimensions used to compensate for irregularities or deviation from the original drawing. The compensation may be performed by various means.

The programmer may use the new dimensions as a base for his program. With computer assistance it may mean changing only one statement which would produce a new set of coordinates or if manual programming were used the programmer may elect to only re-calculate the critical areas that may effect the final part. Another method for compensating is to simply instruct the operator to offset the zero coordinate system on the machine by the amount needed.

Certain 4 axes machines become more complicated to compensate for misalignment when machining is being carried out while the rotary axis is being used. In this case the machine control can be utilized together with an option that automatically keeps track of the fixture position as the table rotates.

5. STANDARDIZING FIXTURES

Generally fixtures are not consumed or used up but are an expense item. They are not capitalized but are subject to obsolescence once the parts have been manufactured. To become a candidate for cost recovery they must be either used for large batch runs or adapted to different parts. The adaption referred to is that of attaching the fixture to the machine and the placement of the fixture within the machine working zone.

Standardized fixtures are probably the most beneficial to any machine shop. They can be used over and over again and there is no lead time involved in designing and manufacturing the fixtures once they have been made. The concept involves common locations and fastening patterns for interface between all machines and all fixtures.

The basic item from which to build from is the base plate. Usually, this is the size of the actual machine table having accurately positioned dowel holes machined in a standard pattern. Also included is a pattern of tapped holes to accept studs and clamps. This plate is fastened to the machine table and located with either dowels or datum edges so that it will not move if the cutter and part are involved in a collision.

Other items may become the interface between the base plate and the actual fixture. This eliminates the bulkiness required when certain fixture heights must be maintained. These items may take the form of angle plates or riser blocks. (See Fig. 2).

The manufactured fixture components are made to conform to the common hole pattern used in the basic items and can, therefore, be mounted in any position on a predetermined grid pattern of holes. The advantage of this arrangement is that all components will be square and parallel to each other and to the machine spindle. Further, the built up fixture can be removed and replaced in exactly the same position on one or more machines.

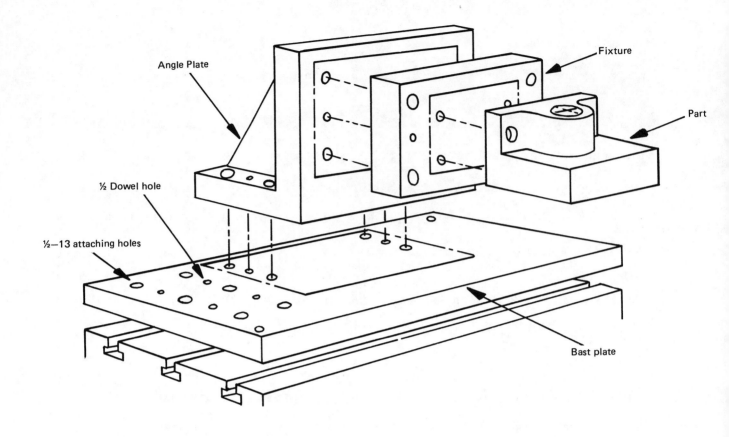

Fig. 2   Assembly of standard fixtures

6.   TYPES OF NC MACHINES AND FIXTURE LOCATIONS

NC machining centers as mentioned earlier can be categorized into two
major types, the horizontal spindle and the vertical spindle.   The more
complex is the horizontal spindle machine which usually includes a fourth
rotary "B" axis.   Supplied with this machine are two edge locators accurately and equally positioned from the center of rotation.   These form the
datums for any fixtures placed on the table.   (See Fig. 3).

The vertical spindle machine can be further split into two types, one
having a fixed zero, the other a floating zero.   The fixed zero machine

can be more difficult for fixture manufacture because there is no compensating feature for their inaccuracy. Further, the fixture must be placed at a fixed pre-determined position making programming a more difficult task.

There is more flexibility in the floating zero machine because compensation may be used to digitally correct the fixture placement to its new position squarely on the table. As the machine becomes more complex by adding another axis, fixturing becomes simpler. With a 5 axis machine any cutting tool can reach within one half of a sphere of the part. Or, considering a squared block, 5 of the 6 sides may be machined without multiple setups. The fixture in this case would remain attached to the movable table while the machine movement would replace the necessity of another setup.

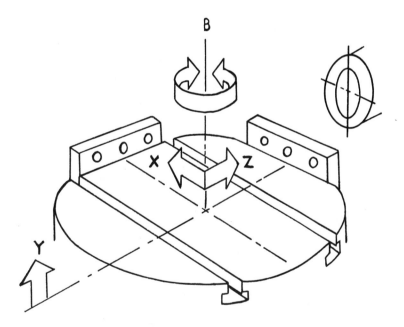

Fig. 3 Edge locators on a horizontal machining center with rotary table.

7. BASIC CONCEPTS OF FIXTURING

The manner in which the basic concept of fixturing applies to NC is no different from that of its counterpart, conventional equipment. Three principals apply, the first involves support of the work piece.

Any three points determine a plane in space. These points form the first basic location. Once established, two additional points form another plane which is a traverse location and one final point the third plane. These complete all the location requirements. (See Fig. 4).

Variations of these planes in reality will exist but the basic concept should remain unchanged. One modification to this idea is to retain the first support plane and incorporate tooling holes in it. This procedure will make the other two planes redundant and actually give a greater positive location to the part.

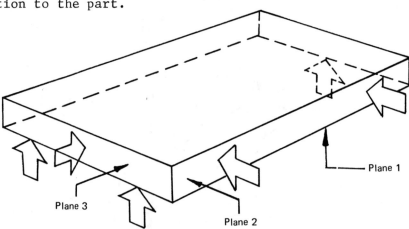

Fig. 4 3 planes locate any part.
Plane 1 uses 3 points, plane 2 uses 2 points and plane 3 uses 1 point.

In addition to locating the parts some thought must be given to where the part and fixture is to be positioned with respect to the machine table. The following points must be observed to find out where the fixture should be placed:

1. Machine Slide Capacity

   Does the fixture position the part within the machining zone?

2. Will all the tools reach?

   By consulting the gauge lengths of all tools, the maximum Z axis machine motion, and checking the shortest tool in conjunction with the deepest portion of the part to be cut, the optimum position of the fixture in one axis can be established. (See Fig. 5).

3. Are the Tools Too Long?

   If the tools cannot be moved out of the way for certain machine operations like rotating the table or tool changing then the fixtures will have to be positioned to alleviate this problem.

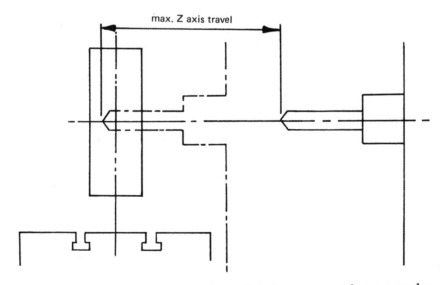

Fig. 5 Position fixture so that all machining operations can be performed.

4. <u>Has there been some consideration given to Programming?</u>

Where computer assisted part programming is used the part may be oriented in any convenient position with very little extra work and no loss of accuracy. When manual programming is used, the manner in which a part is oriented on the machine is important because it may generate excessive calculations. To eliminate these, the part should be fixtured so that the majority of datums and dimensions on the print are square to the axes of the NC machine.

5. <u>Rotary Work Table</u>

If the machine has a rotary table with a fourth "B" axis then the part should be positioned so that each face is an equal distance, if possible, to the centre of rotation. This again is for simplicity of programming as well as to enable all tools to be at an equal distance to all machined faces of the part.

6. <u>Vertical Spindle</u>

If the machine has a vertical spindle with a floating zero, the part and fixture may be located at any position on the table. If the machine has a vertical spindle with a fixed zero then the part should be located so that the dimensions from M/C zero to part zero are easily worked with. As an example, let us say the part were placed on the machine such that the reference edges shown on the print were at arbitrary dimensions with respect to the machine zero. The result would be that all coordinate data in the programmers tape would need these arbitrary dimensions included leading to the possibility of error. (See Fig. 6)

7. Minimum Setups

The last point to observe is that the part should be positioned to machine as many surfaces in the least number of setups. The part should be located such that it may be removed in the middle of an operation and then replaced without machining mis-matches occurring in subsequent operations relating to previous ones.

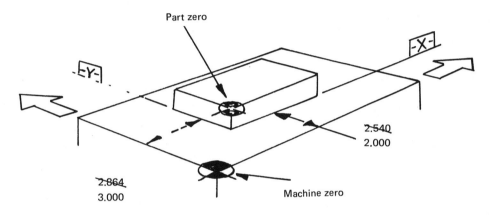

Fig. 6 Position the part on the table using nominal dimensions.

8. FIRST OPERATION FIXTURES

Most machining operations may be broadly categorized into two types. The first operation fixture where no work has been previously performed and the second operation fixture where some type of machining has taken place.

With the first operation fixture, two types of parts have to be considered. Is the part a casting or forging, or is it bar stock? If it is the casting type then inherent in these parts are variations in shape which must be accommodated for. If it is bar stock then allowance has to be made for irregularities in cutting and for manufacturers production tolerances.

Depending on the part, a premachining operation should be carried out if possible to establish a datum surface to aid initial fixturing. If this is not feasible, then by using adjustable supports and locators in conjunction with a pre-determined target point or line, variations in the part may be recognized. It should be pointed out that the NC machine has a much higher operating cost than a conventional machine so that any adjustable fixtures should be limited. As a solution, have the conventional machine cut initial datums or references.

9. SECOND OPERATION FIXTURE

The second category is one which relies on the first machining operation to establish reference surfaces or locating holes for accurately positioning the part. It is from this point on in the machining sequence that accurate parts are made by not allowing the part to move when the cutting tool is in contact with the part.

This type achieves the definition of a true fixture in that it holds the work but has no special arrangements for guiding the tools. Therefore, they may be manufactured as an inexpensive steel plate or fabrication containing nothing more than clamps, blocks and locating holes. The blocks or holes position the part with respect to the fixture and the fixture with respect to the machine table.

In considering the part orientation in the fixture, there are two possibilities which are derived from the type of milling operation that may be performed. We can either use the side of the cutter in a profiling mode or the end of the cutter using a facing mode. The latter cutting technique is preferred because it eliminates cutter deflection producing a more accurate part. As a result, when accurate angles must be machined, the part should be positioned to align the angled face perpendicular to the spindle.

10. CLAMPING

After positioning and locating the part, it must be held. Ideally, clamping should take place directly over the supported areas. Where the clamping pressure cannot be directed through the part onto the supports, it is essential that they fall within an area encompassed by them otherwise distortion will result.

The programmed cutter path usually dictates the regions left for clamping while the part shape and surfaces to be machined suggest the supporting areas. The final design of the fixture is the best combination of clamping and supporting. Automation of clamping should be investigated where quantities are warranted. If hydraulic or pneumatic clamping is a consideration, care must be taken where they are to be used in conjunction with a rotary table. In this case, all equipment must be an integral part of the fixture and protected from the environment of metal cutting operations.

11. LOADING AND UNLOADING

Loading and unloading constitute a considerable portion of the non productive cycle time of each part. By simplifying this operation more parts per hour can be produced. Two types of location should be employed to facilitate this operation. One is a primary location and the other is a positive one. The part should be roughly guided by the primary location to the positive location points. If the operator were to virtually throw the part onto the fixture there should be some means whereby the part should settle onto the locating devices almost unaided. Also, all clamps should be clear to allow ease of loading and unloading.

As with any repetitive work involving human intervention, a point is reached where carelessness may result. By designing into the fixture methods to prevent incorrect part orientation we can help to eliminate scrap parts. To foolproof the operation, there should be only one way that the part can be located and clamped. Once in position, positive stops should be employed so that the part will not move even if a collision occurs. It is much safer to have a cutter break than have a part break loose from the fixture.

NC machines by comparison with conventional equipment generally produce a far greater amount of chips in the same time frame. Even though devices such as scrap conveyors remove the material from the machine, there still can remain a problem of removing the metal from the fixture area. This is especially true on locating faces and under the part. For this reason, it is highly desirable to use rest buttons, minimize flat areas under the part and keep adjustable items high enough so that they do not become buried in chips.

12. TOOLING HOLES

One of the last requirements and one which is most essential in NC machining is the application of tooling holes. Tooling holes are used as an aid in manufacturing and serve no other purpose on the completed part. There are three areas where they are of benefit.

The first is in fixture manufacture where they are used for referencing other accurate dimensions. They can also be utilized when final inspection is carried out especially where coordinate measuring machines are being used.

Secondly, there are used to check fixture alignment on the NC machine. Machines of the vertical spindle variety having a floating reference zero use tooling holes to establish and set up the machine zero. By using one other tooling hole placed either on the same axis or at any exact basic position, it can be determined whether the fixture is square and parallel to the axes of the machine. (See Fig. 7).

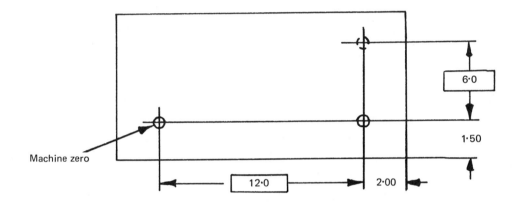

Fig. 7 Tooling holes on vertical machines used for establishing machine zero and to align the fixture.

Machines having a horizontal spindle and rotary table are more difficult to initially setup and check. If the part is to be machined on more than one side and especiajly at angles, it is critical that there be some means to align and finally check whether the fixture is square on the machine table. One method to accomplish this task is to provide four tooling holes on the largest pitch circle diameter, two being coincidental with the XY plane and the other two with the YZ plane. These allow for centering and squaring the fixture when the table is rotated through four quadrants of one revolution. There should be some way to set the Y axis. This can be

accomplished by a tooling hole square to any edge of the fixture but at a
basic dimension from the base.   (See Fig. 8).

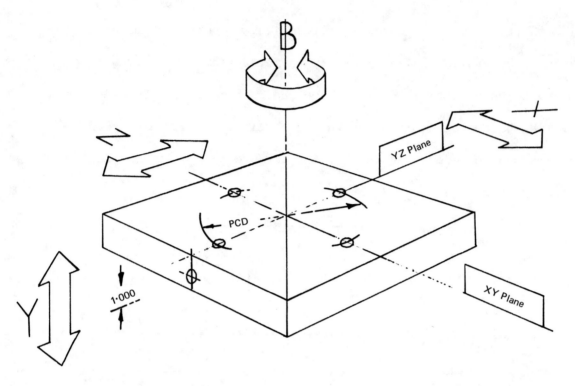

Fig. 8   Tooling holes on horizontal machines using for centering the
         fixture about the rotary axis and for squaring to the X and Y
         axes.

Finally, but more important they are an ideal way to locate the part
relative to the fixture.   Tooling holes can either be pre-machined on
the part by an auxiliary conventional operation or machined as the initial
operation on an NC machine.   When carried out by NC they become an integral
part of the operation performed by tape.

These holes may be used for a subsequent location to machine other surfaces thereby eliminating the problem of mismatch should the part have to
be removed at any time during the machine cycle.   For very accurate work
another set of tooling holes may be machined on one of these four surfaces
to locate the part so that the previously machined surfaces may be

re-machined. With this sequence, all surfaces are in the correct relation to each other. Where should they be placed? Either by accurately machining holes which are shown on the print as being drilled or by machining the holes in excess material which may be subsequently cut off.

The principles and concepts described here are not new to fixture designing. The method of applying them to NC may be a little different. By understanding what the NC machine can accomplish and having some background into what happens during the machining operation, it would be hoped that the tool designer may have a better appreciation in designing fixtures for N.C.

Creative Fixturing Plus NC Helps Draw A Better Bead on Profits

Reprinted from Modern Machine Shop, October 1977

When a contract machine shop turns out thousands of finely finished gun parts each month, profitably, chances are they can produce just about anything.

Kolar Machine, Inc., Ithaca, New York, does just that with help of a fixturing concept called multipart/multistage fixturing and an NC workcenter composed of five Moog Hydra-Point Machining Centers made by Moog, Inc., Hydra-Point Div., P.O. Box 143, Buffalo, New York 14225. Each machining center is equipped with automatic tape punch and tool changer. In addition to all the benefits typically ascribed to NC machining centers, Kolar's switch to the workcenter concept reduced rejection rates of complex, precision-machined gun, camera and business-machine parts by

Operator deburrs finished trigger housing while Moog Hydra-Point NC Machining Center continues making chips.

some 50 percent over previous conventional machining methods.

According to Mr. Larry Kolar, President of the firm, the part most typical of the company's operation is the complex trigger housing. The trigger housing is full of tapers, curves, angles, irregular and inclined surfaces, slotting and grooving—most held to ±0.001 inch tolerances. Holes range in diameter from 3/32 to 1 inch. The housing must be finely finished so that all parts, once assembled, slide and interact perfectly to eliminate malfunc-

tioning of the trigger mechanism.

To process the trigger housing is a sizeable job. The program first calls for face and side milling to square up the critical faces. All in all, there are some 419 instructions in the program calling for all 24 tools from the automatic tool changer's carousel. Some tools are recalled a second time to perform additional machining operations.

Machining any part profitably takes good planning. In the case of the trigger housing, it took sheer ingenuity. Enlisting the help of a hydraulic indexer on the table of the Moog, Kolar has developed a fixture combining multipart plus multistage setups.

Multipart setup is nothing more than loading the NC machine with several parts at once rather than machining them one at a time—this significantly reduces noncutting time which, of course, reduces labor costs, but only one plane of the work can be machined at a time.

Multipart/multistage setup combines the advantages of multipart setup with fourth axis indexing, by utilizing a multipart fixture that is designed to be mounted on and rotated by the indexer. With this approach, multiple faces of multiple parts are machined in one setup. For example, by providing access to three planes of a workpiece in one setup—setup time is cut by as much as 3:1. Otherwise you would have to work one multipart setup for each plane. Kolar developed

their own multisided fixtures for use with the hydraulic indexers purchased with their Moogs.

In production on the trigger housing, three workpieces are set up on each face of the fixture for machining in four separate phases (indexes). They utilize a mini-computer to make calculations, prepare programs and optimize tape processing.

When the machining center completes its cycle, the program halts the machine. The operator then removes completed workpieces from the fixture and reloads it with unmachined stock. What was phase one then becomes phase two and so on. Twelve parts are finished during this cycle of the operation in about half the time it would take using multipart setups alone.

Jobs at Kolar are programmed so that the Moog performs as many operations as it can on one part, at one time and with one setup. The secret is utilizing the machine's capabilities to their fullest.

The purchase of a Moog Hydra-Point 83-1000 Machining Center back in 1972 was the first venture by Kolar into NC machining centers. Since that time, four machining centers have been added: two equipped with hydraulic indexers.

Now Kolar contracts business they could not consider before, like finely finished 8-inch pump cylinders requiring circular slotting, internal hole drilling and boring.

Multipart / multistage setup combines the advantages of multipart setup with a rotary fixture adapted for a hydraulic indexer (shown in lower left). The fixture indexes, as programmed, to machine first, second, third and fourth face of the workpiece — providing access to three planes of a workpiece in one setup. In this case, three workpieces are set up on each face of the fixture for machining in four separate phases (indexes).

A robot retrieves a partially machined workpiece from the first of two Wasino lathes in a flexible machining system designed by Wasino Corp. U.S.A.

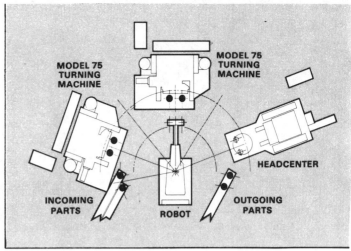

Two vertical turning machines and a HeadCenter are loaded and unloaded by the Unimate robot in this Olofsson flexible machining system.

Reprinted from Manufacturing Engineering, November 1981

Automating for Greater Gains in Productivity

Lathe automation goes beyond extending horsepower, spindle speeds, and feedrates. Here's a look at some of the things being done now and anticipated in the future

ROBERT N. STAUFFER
Senior Editor

THE CHANGES TAKING PLACE in turning machines continue to follow an evolutionary path as improvements are made in machine design, controls, workholding and workhandling equipment, and tooling. Today's NC and CNC lathes — with such advances as higher horsepower drives, faster spindle speeds, and increased feedrates — have an excellent track record in terms of increasing productivity, improving workpiece quality, and reducing machining costs.

One of the trends in this field is the proliferation in number and variety of machines now available to satisfy industry's need for efficient, productive turning. The accompanying summary of NC and CNC lathes starting on page 62 covers many of the lines which make up the current roster of machines. In most cases, the descriptive copy is limited to one model or size in that line — a preview, in effect, of the company's complete offering and particular niche in the field. Key specifications most pertinent to lathe selection are included in each writeup. The machines are listed in alphabetical order in two categories — horizontal and vertical spindle configurations. Some companies, although listed just once, offer both styles.

Automated Material Handling

Attempts to increase the productivity of turning machines are being extended beyond those approaches typically involving more power, higher speed, and more sophisticated controls. Automated material handling, including the use of programmable robots, is a prime example. In some cases, the same piece of equipment can be used for changing tools.

In a new flexible machining system recently announced by Wasino Corp., U.S.A., Hauppauge, NY, two CNC lathes are loaded and unloaded by a Fanuc Model 1 robot. The system, designed for unattended runs, is installed at Dunrite Tool & Die Inc., Farmingdale, NY.

In a typical sequence, the double-handed robot picks up a blank from a table, puts it in the spindle of the first machine, signals for the blank to be gripped, then withdraws as machining begins. When these first operations are complete, the robot retrieves the partially machined blank, substitutes a new blank from its second hand, and deposits the first piece in front of the second machine. The robot turns the part over into the correct machining position, moves it to a ready position, and waits for completion of the previous piece.

Machining cycles continue at both machines during robot moves, except for the minimal time required for workpiece placement. Finished pieces are lifted out and deposited in a hopper. The combination of robot handling and Wasino's automatic tool offset compensation to adjust for tool wear during machining, plus tool-life control and broken tool detection, make it possible to literally "push a button and walk away."

Either of two different types of measuring systems can be specified. One is an independent station employing an electronic gaging system to measure and classify finished workpieces transferred by the robot. The other consists of an in-cycle measuring sensor attached to the turret to measure OD, ID, or length of the finished parts. With both methods, the data is fed back to the CNC unit for automatic adjustment of the offsets or to stop the operation if an out-of-tolerance condition is detected.

Another flexible machining system, this one by Olofsson Corp., Lansing, MI, was installed this past summer at the Deere & Co. plant in Waterloo, IA. It consists of two Model 75 CNC vertical precision turning machines, a HeadCenter, and a Unimate Series 2000 robot.

The robot automatically loads and unloads two parts at a time from machine to machine. The system includes an orientation station which facilitates part acquisition and placement by the robot. Significant gains in productivity are possible, along with savings in labor, time, floor space, and inventory.

Lodge & Shipley Co., Cincinnati,

The end-of-arm tooling on a Prab robot reaches through the rear of a Giddings & Lewis turning center during a loading and unloading sequence.

The Acco machine-dedicated system loads and unloads parts into the chuck in a series of programmed moves. One operator can supervise three machines equipped with the loader.

OH, has designed and built a turning and chucking system using a robot for automatic demand parts handling. The system, built for Rockwell International, incorporates two L & S Numeriturn universal turning/chucking centers and a Unimate Series 4000 robot with double-handed tooling. The robot operates on a demand signal from either machine. The loading door, chuck, tailstock, and many other functions are controlled by the program. The operator remains outside the workhandling area for maximum safety, monitoring the controls and operation of the system. A third turning/chucking center can be added to the system if desired.

Dedicated Loading Systems

Dedicated loading systems used in conjunction with conveyors, postprocess gaging equipment, and other auxiliary devices are attractive for medium to high-volume production installations. Faster cycling is one of the advantages cited for the dedicated equipment.

One example of this type of system is the Accomatic® machine tool loader built by Acco Industries Inc., Industrial Lifters Div., Salem, IL. The loader's infeed chute typically holds enough parts for at least one-half hour of machine operation. Scrap rates are said to be reduced by the consistently accurate placement of parts in the machine. Parts weighing up to 160 lb (72 kg) can be handled by the equipment.

The loader is an integral part of the machine tool. If the device is temporarily down for maintenance, other machines are not affected.

Feed Force Monitoring

The removal of the operator from the machining system leads to a number of prerequisites for continuous, trouble-free operation. One of these is the need to monitor each machine for collisions and/or excessive tool wear and provide for automatic shutdown. Some machine builders now define collision barriers and incorporate such information in the software. As an added safeguard, feed force monitoring can be used to measure the force on the spindle drive.

In a system offered by Sandvik Coromant, Div. of Sandvik, Inc., Fair Lawn, NJ, the measuring principle used is based on directly registering the reaction force between the rotating spindle and the bearings attached to the housing. The system consists of two units, a sensor and an electronic unit. The sensor, which forms the actual housing for the bearings, can be installed on the main spindle. Sensor size ranges from 1 to 15″ (25.4 to 381 mm) and have force ranges from 500 to 25,000 lb (2.2 to 111.2 kN).

Better Chucks

In the past 15 to 20 years, the machine, cutting tools, and workholding devices have tended to leapfrog one another in terms of advances in technology. Higher spindle speeds, for example, came about through advances in spindle bearing designs and improved cutting tool materials.

Right now, the ball seems to be in the workholding court. The next generation chucks will have a gripping force curve approaching a flat line. That is, there will be little loss in gripping force as the chuck is rotated up to its maximum rated speed. In addition to improved safety, it will be easier to calculate chuck performance in various applications.

Modern lathe controls and turrets that hold 10 or 12 cutting tools facilitate fast machine setup and changeover from one part to another. This makes the time required to change top jaw tooling more critical. It is not uncommon for this operation to take 15 to 30 minutes. Chuck designers are tackling the problem. For example, the wedge bar design of KNCS chucks built by SMW Systems, Inc. permits top jaw changes or adjustments in one minute, or less.

Other Approaches

Up to this point in time in turning, the workpiece has been handled separately and loaded in the machine. The chuck is essentially a part of the machine. This is different from the pallet concept used with machining centers. Some development work is anticipated in fixturing of turned parts where the chuck remains with the part and may be routed to several turning centers for multiple operations.

Interfacing of equipment used for turning and other machining operations will also improve workpiece handling and flow through the plant. Round parts coming off a turning center could be fed automatically to another section of a machining line for such operations as milling, drilling, and tapping. Today, in most cases, the parts must be removed from their fixtures and loaded into a completely different system.

Automatic changeover provides significant gains in productivity. Currently, this calls for a man to change the workholding elements, tooling, and machine setup. A better approach would be to set up the chucks or fixturing devices off-line and send them randomly down a transfer system. Coupled with the use of turning tool matrices and automatic toolchanging, the entire system would lend itself to the efficient production of various types of parts or different part families on an unmanned turning center. ∎

Manufacturing Engineering *acknowledges the following individuals for their help in supplying material for this article: C. Haarlammert, production manager, Turning Machines, Cincinnati Milacron; M. Lindgren, vice president, Marketing, Giddings & Lewis Machine Tool Co.; G. Barkley, production manager, Machine Tool Systems, Acco Industries Inc.; S. Looney, national sales manager, SMW Systems, Inc.*

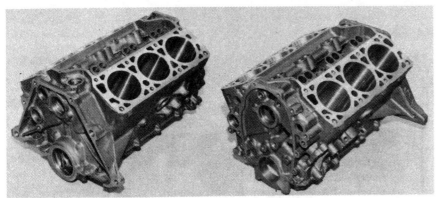

Lightweight cast-iron engine block requires special handling during machining. Blocks no longer have extra metal to save them from machining shock.

Reprinted from Tooling & Production, January 1979

Transfer line combats tool-change downtime

by **Richard P Cottrell**
VP-General Manager
Cross Fraser Div
Cross Co
Fraser, MI

and **Kurt O Tech**
Group VP
North American Operations
Cross Co
Fraser, MI

What good is a transfer line while it's down for tool change? That's the question that inspired downtime-reducing features in nearly every machine of this Cross integrated manufacturing system. The 41-machine system produces 60-degree, V-6, cast-iron engine blocks at a rate of 246/hr. The blocks are exceptionally light—with many normal manufacturing strength requirements sacrificed to maximize weight reduction. This forced a major process change, and encouraged the design of a line with emphasis on tooling considerations. Tool-change time has been cut in half.

Up to now, nearly all automobile engine blocks have been broached on the pan rail, bearing-cap seats, main-bearing half-rounds, bank faces and, frequently, the intake manifold face. The only milling performed on these surfaces has been finishing. However, the new V-6 thin-wall castings distort under the high cutting forces produced by broaching. So, on this transfer system, all these block surfaces are rough, semifinish and finish milled—with much lower cutting forces. The only broaching operation is a finishing pass on the bearing-cap seat widths, an area with very high strength.

The large amount of milling in the system led to rethinking of conventional block-milling methods. Normally, in the machines that mill the pan rail, bearing-cap seats and main-bearing half-rounds, the block is fixtured with the crankshaft axis parallel to the direction of transfer. Milling units are mounted on the sides of the machine and feed in a direction parallel to part transfer. To change these cutters, it's necessary to climb into the machine, making cutter change a time-consuming and difficult task.

Outside setup

The new transfer system solves the tool-up problem by changing both the position of the workpiece and the direction of milling, **Figure 1**. The crankshaft axis of the block is perpendicular to the direction of part transfer. The milling units mount on massive overhead bridge structures and feed at right angles to workpiece travel. The units have sufficient additional stroke so that cutters are accessible at the side of the machine during tool change. Cutter-assist devices are

provided where necessary to further reduce downtime. And this kind of thinking was continued throughout the entire system.

The setup for bank-face milling—traditionally performed by a bridge-type overhead unit parallel to the direction of transfer—is also turned 90 degrees, **Figure 2**. As on the pan-rail units, sufficient stroke was provided to present the cutters at the side of the machine. To provide optimum rigidity, tracking accuracy and way life, a design using three ways was employed. The center way is used for guiding only. This allows maximum length-to-width ratio, which reduces side loading or twisting deflection and provides stiffer tracking.

The straddle-milling operations on the main bearings are also performed in an unconventional direction to improve tool-change time. These operations have generally been performed with the block on its side, while a bank of arbor-type cutters feeds horizontally into them. Access to the cutters for tool change is im-

1. *Massive milling unit roughs the pan rail and bearing-cap seats. The milling cutters are accessible from the aisle when the milling unit moves out on overhead bridge.*

peded by the milling unit. So in the Cross system the operation is performed with the engine block mounted pan face up and the cutters feeding vertically. With the milling units above the center bed, the cutters are easily accessible from the side of the machine.

Two-post bridge

The tooling to mill the bosses on the sides of the block employs a unique fixture to speed tool change. Normally a four-post bridge-type fixture clamp is used, but tool change is difficult because the post at each corner of the fixture makes the cutters very hard to reach. So engineers developed a two-post diagonal bridge clamp. One corner on each side of this fixture is vacant, allowing fast access to the cutters during tool change.

The design also reduced the time required to change the broaching unit for the bearing-cap seat width. Normally the block is mounted on its side to perform this operation along with milling of the bulkheads. But in the Cross system, since the bulkheads are milled with the pan face up, it was possible to eliminate two rollover stations and broach the seat widths also with the pan face up.

As with the milling operations mentioned before, the broaching is performed at right angles to the direction of transfer to allow optimum tool-change access. A massive supporting base structure provides the extra rigidity required. Large openings grant ready access to the broaching unit, and a portable cutter-change assist offers further help. This hand-powered unit runs under the broaching unit, so the broach tool can be detached and placed on the portable carrier.

Milling of the front and rear faces of the block is performed in the conventional manner—with feed parallel to the direction of transfer. Here, however, the design offers a large amount of space at each side of the cutter to perform tool change, and a platform is built into the machine to make it easier to reach the tool-mounting bolts.

The machines that perform angular drilling on the bank face were also modified to reduce tool-change time. Normally, these operations are performed by single-spindle drilling units attached to floor-mounted columns on both sides of the center base. The presence of two columns hinders tool change, so in this system the columns are eliminated. The drilling units are mounted on a bridge attached to the fix-

ture. Now it's easy to reach drilling units on either side.

Controls simpler, safer

A new control console also simplifies tool change. This console contains the pushbuttons that would normally be on the main map-type control console. The pushbutton console is pendant mounted, and can be swung to either side of the machine, **Figure 3**. Thus, the person who is changing tools can use it without walking around the machine, eliminating any need for an extra person to push buttons during tool change.

The system also has a unique safety feature. Previously, in case of a jammed fixture or other obstruction, a maintenance person or operator had to attempt to free the obstruction by jogging the transfer bar with a hand button. Unfortunately, this person could easily move the bar the wrong way under power, causing severe damage. This potential hazard was eliminated by providing a handwheel that can be quickly attached

to the transfer drive mechanism. Moving the bar by hand gives the sensitivity needed to be certain of avoiding damage.

To prevent damage to the workpiece, special fixturing and holding devices must be called into play to support certain areas against the machining forces, **Figure 4**. For this, hydraulic cylinders actuate various jacks and holding devices mounted in strategic locations within the machining stations.

Another unusual aspect of this system is that it can be modified easily to manufacture aluminum engine blocks. To allow for this conversion, all machines except the cam and crank boring machines are already piped for coolant. When the conversion is made, few changes will be necessary. The anticipated tool speeds will be close to those now being used on the cast-iron block, but the feed rates will be increased on most operations.

This project, which was purchased on a single-source basis, demonstrates the advantages of this buying practice. For

2. This milling unit was specially developed to semifinish the bank faces at right angles to the direction of transfer.

...continued

3. *Special pendant-mounted console can swivel to either side of the machine.*

one thing, all equipment has a common design basis. This reduces spare-parts stocks and simplifies operator training. Also, turnkey buying allows the builder to make improvements on a system-wide basis—improvements that otherwise would probably be limited to a few machines. A prime example is the reduction in tool-change downtime, which is the highlight of this project.

4. Hydraulic jacks in this fixture prevent possible distortion of the lightweight engine block while machining. The block withstands the strain of normal operation on the road, because the stress forces come from different directions than the machining forces.

Reprinted from Manufacturing Engineering, November 1978

NC and Multipart Fixtures: Costs Down, Production Up

Two Moog NC machining centers and use of multipart jigs increase productivity on precision parts while reducing scrap rates and inventory requirements

To produce a variety of optical instruments ranging from microscopes to refractors, American Optical's Scientific Instrument Div., Buffalo, NY, manufactures more than 7000 different parts, either whole or in part. These parts require extensive precision machining to tolerances down to 0.0005" (0.013 mm), or less. NC machining and multipart fixtures are credited with reducing costs and increasing production.

Since setup time constitutes a large portion of the total part work, American Optical places a great deal of emphasis on its reduction. The company has been successful in cutting setup time by utilizing multipart setups on two Moog NC machining centers, *Figure 1.*

Jim Matuszak, supervisor of methods, is pleased with the results. "Since we went operational with the two machining centers, more than 45 different parts previously produced manually are now handled on Moogs," he says. "By tapping NC's intrinsic ability to handle multiple operations with just a single setup, we have minimized dead fixture setups in every case."

American Optical maximizes its NC machining with the special jigs. Once the machine is programmed, any number of pieces that will fit on the jig are produced in a single setup. And because of the precision of NC, rejects are down to a bare minimum.

Two Setups Eliminated. As an example of the precision machining required, Matuszak cites a 2¼" (57.2 mm) long stainless steel cylinder used as a piston guide in a tonometer. This is a noncontact instrument that propels air through the cylinder and onto the eyeball to measure intraocular pressure. Variations in air pressure are measured electronically through three slots placed equidistant on the cylinder diameter and a fourth slot placed precisely below the line of the first three. This measurement is shown on a digital readout.

"It is essential that these slots be coplanar within 0.002" (0.05 mm)," says Matuszak. "To do that manually required three different setups."

One of these setups was needed to mill a ⅛" (2.86-mm) slot solely for locating purposes, even though the stainless steel blank was supplied with a tapped hole. Then it took another setup to mill the single slot and a third one to mill the three slots.

Matuszak notes that, "Putting this piece on the Moog cuts it down to a one-setup job. It also eliminates an operation. Since we can use the tapped hole to orient the piece, we eliminate the need to mill a locating slot. And since the workpiece is not moved at all, it's a lot easier to maintain the necessary dimensions and relationships."

Production figures for this piece prove the point. The three-step manual operation required 9.2 hours to produce 100 pieces, with a scrap rate of 12%. The one-step NC operation takes 6.6 hours per 100 parts, with a 2% scrap rate. There's a good possibility that the 2% scrap is due to dull tooling not being caught in time, and that's being checked out.

Multipart Fixtures. "It makes sense to design fixturing that puts as many pieces on the table as possible," Matuszak points out. "For example, to do the drilling required on one of our

*1. TRANSFER
OF MANUAL
operations to
NC on the
Moog Hydra-Point
provides average
time savings of 35%
on direct labor
and 54% on setup.*

pieces — a turret housing — took three different jigs when done manually. On the Moog, we do four pieces per loading and complete each part in a single setup.

"A smaller part, used to hold the lens in one of our instruments, was done one at a time manually. The fixture we use now allows us to do up to 16 at a time and with just one loading."

Duplicate fixtures, *Figure* 2, allow the operator to load one fixture while parts on the other are being machined. This cuts dead machine time still further.

"One of our parts is done just one at a time on the Moog, but even that results in savings," Matuszak notes.

"There are five holes to be drilled on the part. On the manual setup each hole required dialing in a new position on the machine. With the Moog, one loading does the job. The five different hole locations are determined precisely and automatically."

Better Inventory Control. Another advantage of NC machining is inventory control. Maintaining stock on more than 7000 parts creates problems, and inventory turnover is a constant consideration.

NC machining lowers the lead time needed by American Optical to produce any given group of parts. Since parts are now produced faster and at less per-part cost, it is not necessary to

keep as many parts in inventory, or to reorder as far ahead of time as was necessary when these parts were produced manually. With a lower reorder point and smaller reorder quantities, there is less cost tied up in inventory and less shelf space required.

NC machining also reduces in-process storage. By allowing for tool changes on the machine, the number of pieces that must be stored while waiting for machine time to be available is greatly reduced. Scheduling work is easier, more predictable, and more profitable.

Matuszak says, "We're convinced that as we put more and more work on the NC machines, we'll do better and

*2. DUPLICATE FIXTURES holding multiple parts
increase machine uptime for higher production.*

better on inventory reduction.''

American Optical's justification for acquiring its second Moog NC machining center was based on some impressive figures after one year's experience with the first machine. The number of setups required to produce the amount of work handled was reduced to 19 on NC compared to 41 for manual machining. The total number of hours required to produce 100 pieces was reduced from 206 down to 133. To equal the production obtained in 2950 hours on the NC Moog would have required 4540 machine hours for manual machining. ■

Reprinted from Manufacturing Engineering, March 1978

1. CLOSE-UP OF FACE DRIVER shows drive pins used to grip and turn part. Mounting the driver in a chuck saves setup time on short runs.

2. INPUT SHAFT IS HELD and rotated with face driver shown at left. Part is turned and threaded in one seven-cut operation.

Face Drivers and NC Lathes

. . . make a productive pair for full length turning of shafts in one operation

INCREASED PRODUCTION RATES with the use of face drivers have been widely documented. Used in conjunction with NC lathes, face drivers are helping to increase the productivity of turning operations.

Lodge & Shipley, Cincinnati, OH, is a proponent of the face driver/lathe combination and has used Madison-Kosta face drivers in combination with its own NC Numeriturn turning centers. These lathes are designed for high-speed machining with great accuracy and are capable of high metal removal rates through unusual bed and spindle rigidity combined with heavy duty tailstocks. The Madison-Kosta face drivers, supplied by Madison Industries, Inc., Lincoln, R.I., provide the positive, rugged workholding system required for profitable machining.

An important feature of these face drivers is that they permit machining the entire OD of any part in one clamping. Other advantages include the ability to take heavy cuts with no slippage, high accuracy in machining operations, accurate end location, fast loading time, wide range clamping capability, and elimination of preliminary machining of end faces. Unlike chucks which have speed limitations, face drivers allow turning at much higher surface footage with no reduction in accuracy.

Operation. A face driver, *Figure* 1, consists of two main assemblies. A lo-cating shank fits on the spindle nose and has either a taper shank or flange mount. The driving head is made up of a spring-loaded center point, drive pins, and a compensating medium which permits each drive pin to adjust to irregularities in the face of the workpiece.

Face clamping with the driver is a simple two-sequence function. The part is first centered by the workpiece, establishing the precise axis of rotation.

As the tailstock engages the workpiece and axial force is applied, the center point retracts against its spring. The chisel-edged drive pins bite into the end face of the part. As the cutting tool makes its cut, torque is increased and the drive pins bite deeper into the face, assuring positive clamping.

Common methods of mounting the face driver involve the use of a tapered adapter or by flange mounting. While the latter method is preferred, an alternate method that saves setup time is to chuck the driver on the outside of the carrier body. This method is of particular benefit in the job shop or toolroom where short runs are made, and the lathe is used for both shaft turning and as a chucking machine.

Chucking the driver saves up to ½ hour in setup time, since the chuck does not have to be removed from the spindle. It takes only about 20 seconds to mount a face driver in a chuck, and there is no appreciable loss in concentricity.

Applications. Lodge & Shipley demonstrations of the face driver/lathe combination include the part shown in *Figure* 2. This is a 19 ⅞" (504.8-mm) long input shaft made of C1141 bar stock. It is turned and threaded in one operation. Total cycle time to complete the sequence of seven cuts is 7.3 minutes. Depths of cut range from 0.020 to 0.500" (0.51 to 12.70 mm). During some of the rough turning sequences, the face driver pulls the equivalent of 75-hp (55.9-kW) cuts.

A New England pump manufacturer turned his profit picture around following the application of high-speed, high horsepower turning techniques. Production requirements included machining shafts ranging in size from 1" (25.4 mm) diameter by 12" (304.8 mm) long up to 9" (228.6 mm) diameter and 10 ft (3.05 m) long. Shaft materials include 1045 steel and 304 stainless.

The previous machining procedure involved two lathe operations and a rough grind operation. The two lathe operations were required because the full length of the part could not be turned in one setup. Also, the condition of the older machines restricted their use to light depths of cut at low rpm.

The Lodge & Shipley Numeriturn system, combined with Madison-Kosta face drivers, now makes this a one-operation procedure. Nominal depths of cut range up to ⅝" (15.9 mm), using a titanium insert at 400 sfpm (121.9 m/min) and a feed of 0.035 ipr (0.89 mm/rev). The Lodge & Shipley tailstock delivers 6300 lb (28 022.4 N) of force for this specific cut. Previous machining time for this shaft was 18 minutes, not including the three separate handlings. The present machining cycle time is 4.5 minutes. ■

Presented at SME's Assemblex Conference, November 1978

Automation in Riveting Systems

By Edward M. White
The Milford Rivet & Machine Co.

Automation in riveting is mainly concerned with increasing productivity by reducing parts handling time, as well as improving safety by eliminating parts handling near the operating point of the machine. In high-level automation, multiple-head riveting and work transfer devices such as rotary tables can provide substantial increases in production rate with significant return on investment. In low-level automation, special riveting fixtures and auxiliary parts handling devices can produce dramatic results at far lower cost.

- - - - - - - - - -

INTRODUCTION

The degree of automation in riveting operations in industry today ranges widely -- from as little as with hand-loaded air presses to as much as in fully automatic assembly stations. Rivet-setting machines themselves are semi-automated in that, once the riveter is actuated, the feeding of the rivet and clinching action take place automatically. It's the handling of parts associated with riveting that makes the difference between manual and automated riveting systems.

The machine cycle time for a single clinching operation in riveting is typically about one second. But the handling of parts can take from three to ten seconds and longer, depending on the size (either large or small), number and complexity of the parts in the riveted assembly. The productivity of a riveting operation can therefore be most effectively -- and certainly, most economically -- increased by taking steps to reduce parts handling time.

In recent years, then, automation in riveting has been concerned mainly with shortening or eliminating associated manual operations. This has been accomplished mainly with riveting fixtures and auxiliary handling devices such as powered slide fixtures, rotary and linear indexing tables, and pick-and-place mechanisms. The fixtures or devices, of course, must be functionally and economically suitable to the riveting task: nature of the assembly, production, volume, return on investment.

ELEMENTS OF PARTS HANDLING TIME

The elements of parts handling time at a representative riveting station (Figure 1) are concerned with three operations:
- Removing two or more parts from containers at or near the operating point of the riveter
- Placing the parts in their proper riveting positions under the machine driver
- Removing the riveted subassembly from the machine and transferring it to a container or conveyor.

In motion-time-measurement (MTM) analysis of a specific riveting task, these handling elements can be broken down further into a number of specific motions by the operator's left and right hands. The table in Figure 2, for example, itemizes the motions involved when an operator seated at a riveting machine is handling two small parts, each jumbled in separate containers adjacent to the riveting head. The time required for the individual motions are given in time-measurement units (TMUs) -- 100,000 TMUs equal one hour. With 28 TMUs then about equal to one second, the full handling sequence in Figure 2 takes just under three seconds (no time for the machine cycle is included).

The common action of reaching for a part with either hand can be slowed down considerably when the operator must turn her body in the process of picking up the part. In the turn body chart in Figure 3, the reaching action is seen to take progressively longer than needed in Figure 2 (in which the parts are adjacent to the operating position A) as the parts container is moved back.

The type of riveting station in Figure 1 -- a single operator hand-loading parts at a single-headed riveter -- is by far the most common in industry. The production volume may be limited or the station may be on a balanced production line (which establishes the assembly cycle time), so that there is no reason for automating in order to reduce handling time.

There is also no reason for automating to improve operator safety, accomplished by eliminating the need for the operator's hands to reach in or near the operating point. The double palm switches in Figure 1, which must be simultaneously pressed to activate the machine, keep the operator's hands clear of the operating point during the cycling time.

HIGH-LEVEL AUTOMATION OF RIVETING

There are three ways of reducing the effects of parts handling time on a riveting operation:
- Combine parts handling time for two or more separate riveting operations with multiple-machine groups.
- Overlap machine cycle and parts-handling time through the use of automatic work-transfer devices.
- Reduce handling time by upgrading fixturing and machine layout.

Multiple-Machine Groups

Multiple-machine groups (Figure 4) are designed to set two or more rivets through the parts in one machine cycle -- simultaneously or, with powered fixtures, in rapid automatic sequence. The initial cost of multiple groups is usually higher than the total cost for the same number of separate riveting heads because more elaborate controls, special machine design, and fixtures are involved. Multiple-machine groups are rarely used for more than one riveting task (although the group may be disassembled and the machines regrouped or operated individually).

Extremely high production rates can be obtained with multiple-machine groups at much-reduced direct labor and overhead costs, often paying for the higher initial investment in a matter of a few months. In the two-headed machine group in Figure 4, the parts-handling time per assembly has been reduced in two ways. First, the powered slide fixture moves the roll set pins to an "out" position for maximum ease in locating the parts on the roll set pins. Secondly, the operator need only handle the parts once in the process of setting two rivets, so that the effective parts handling time per assembly is one half that of what it would be in setting two rivets in sequence on a single-headed riveter with one roll set pin. This method of reducing the unit parts handling time -- as well as overlapping the machine cycle times -- can be extended to as many as ten or more machine heads in a multiple-machine group.

Rotary Work-Transfer Devices

Most automated riveting systems are based on multiple-fixture indexing mechanisms in which fixtures containing the roll sets and parts to be riveted are automatically cycled under the rivet drivers. These automatic work-transfer devices are usually rotary (Figure 5) but may also be linear (Figure 6).

Rotary indexing tables like that in the automated riveting station in Figure 5(a) increase production rate by absorbing much of the parts handling time as part of the indexing time of the table. Several operators at a station can load the fixtures at the open fixture positions even though the table may be indexing quite rapidly. Even one operator can load multiple fixtures faster than at a single riveting station mainly because the time required to reach for and press palm buttons has been eliminated by automatic cycling of the machines.

Riveting Switches

Three different types of switches (triple, double and single from left to right in Figure 5c) are assembled in the rotary table in Figure 5(b) without any change in fixturing. In this way, the productivity of the automated station has been further increased by eliminating the need for a set-up time in replacing fixtures. The three pairs of Milford Model 56 pneumatic riveters set respective pairs of rivets, the first two pairs of which are identical and the last slightly longer. Previously, these switches were assembled by setting one or two pairs on one double-headed riveter and, if necessary, the third pair on a different double-headed machine. Clearly, there was a great deal more parts handling in the previous method -- and a substantially lower production volume.

Ejection, Controls and Non-Riveting Operations

Riveted parts may be ejected as an integral function of the fixtures or separately. The finish, size, shape, and fragility of the parts must be considered in deciding how the riveted assembly is to be ejected. It may be blown off the fixture into a discharge chute by an air jet, gently slid down a ramp onto a conveyor, or picked off by a robot pick-and-place mechanism.

The controls for operating rotary tables are usually mounted in one master control panel (right in Figure 5a), with all run-and-jog cycle features accessible to the operator. Hand-operated selectors and pushbutton switches are provided for various modes of operation. If there is more than one operator, an emergency stop switch should be installed at each operator's station. Rotary tables may be guarded by a shroud-type barrier to protect operators from the indexing fixtures.

Rotary tables may also be utilized to perform nonriveting operations on the parts to be fastened. Qualifying stations can ensure that the parts are positioned properly prior to riveting and automatically shut off the machine if they are not loaded correctly. Inspection stations can automatically determine if the parts have been fastened properly and sort assemblies into "accept" and "reject" bins.

Linear Work-Transfer Devices

Parts handling time is eliminated from the riveting cycle in much the same way on linear work-transfer devices in automated riveting stations. However, a rotary table occupies less floor space -- and usually involves a smaller investment -- for a given number of operator stations.

A wide variety of commercial drives for rotary or linear indexing stations is available in numerous sizes and types, including geneva motions, ratchet and pawl devices, and cam-operated mechanisms. The best drive depends on the types of loads and indexing speeds. It must have the proper dynamic characteristics to function reliably at a given production rate.

In addition to increasing production rate, both the rotary and linear indexing methods enhance safety by isolating the operators from where the actual riveting is done.

Riveting Brake Shoes

The brake-shoe assembly station in Figure 7(a) has been further automated over that in Figure 5 by eliminating several additional parts-handling operations. The assembly operation involves setting either seven or eight rivets to fasten the friction pads to the steel backing plates of brake shoes. The riveting station, which is adjustable for eight different riveting patterns, completes between 20 and 25 assemblies per minute. Several auxiliary operations that are usually manual have been automated: inspection at several points, unloading with a pick-and-place mechanism, and counting of the fastened brake shoes. The sequence of operations that occurs as the 12-fixture rotary table indexes clockwise in Figure 7(b) are:

- The operator places steel plates and pads over the eight (or seven) roll set pins on the fixtures passing the loading position (out of picture).
- An automatic probe checks that each pad and plate are properly located on the pins.
- Each pair of rivets is set by a pair of Model 58 pneumatic riveters as the fixture indexes through four riveting stations.
- An automatic probe checks that all rivets have been clinched in position. If not, the pick-and-place mechanism does not remove the assembly, permitting it to continue around to the operator for removal.
- The pick-and-place mechanism (right in Figure 7a) stacks five brake shoes in a container and indexes the stacks until there are four stacks of five. All 20 brake shoes are then pushed onto a conveyor for transport in a group to the next operation.

LOW-LEVEL AUTOMATION OF RIVETING

The two riveting stations described with respect to Figures 5 and 7 represent high-level riveting automation in the present state-of-the-art. With a relatively large investment (at least for a riveting station), a high production rate has been acheived in each and, with that, an extremely low unit riveting cost.

The reduction or elimination of parts handling time by upgrading fixturing and machine layout, on the other hand, can be accomplished with very little additional investment. Improved fixturing, which might be considered low-level automation, therefore receives special attention in cost reduction efforts in many firms. Assembly of a battery cable clamp (Figure 8), a drawer slide (Figure 9), and a distributor disc (Figure 10) demonstrates the value of low-level automation.

Assembling a Cable Clamp

The battery cable clamp in Figure 8(a) consists of two stamped-metal clamp handles and a very strong torsion spring. In the original assembly method, the operator preloaded the spring between the two clamp handles in a preliminary operation and then placed the three pieces together over a roll set pin. The single rivet was then set through the center of the spring and the four holes in the clamp handles. The company experienced what was considered the unacceptable production rate of 120 assemblies per hour because of the very slow preloading operation.

A powered slide fixture and nest, as shown in Figure 8(b), was then designed to eliminate entirely the preloading operation. The two clamp handles are placed in an epoxy nest and the spring is dropped into a smaller nest in the new fixture. When the machine is actuated by dual non-tie-down palm buttons, the fixture compresses the spring and brings the two clamp handles together. The roll set pin is automatically actuated through the clamp handles and spring. This in turn cycles the machine, inserting and upsetting the rivet to complete the assembly. When the pin reaches the down position, the fixture opens to release the assembly and is ready for the next riveting cycle. The result: the production rate increased to over 200 cable clamps per hour.

Assembling a Drawer Slide

In another typical case (Figure 9), a manufacturer increased productivity by eliminating handling of one part. In Figure 9(a), a shoulder rivet is clinched through the nylon roller to a stamped-metal piece to form the assembled drawer slide at top. The machine operator simply places the stamped-metal piece over the roll set pin of the Milford Model S305 riveting machine and steps on the foot pedal. The automatic vibration feeder at left in Figure 9(b) instantly feeds the nylon roller into the lower section of the jaws, and the shoulder rivet is driven down through the roller and metal piece to be clinched below. If the nylon roller were to be manually placed over the stamped-metal piece on the roll set pin, the production rate of this machine would be reduced from 1000 assemblies per hour to an estimated 600 assemblies per hour.

Note in Figure 9(b) that proper orientation of the nylon rollers requires that they turn over as they move down the slide rail from the hopper to the operating point.

Assembling a Distributor Disc

Assembly of the distributor disc at right in Figure 10(a) requires fastening a rivet and a screw machine part through each of two small identical metal shapes to opposite sides of a circular metal plate. Rather than handling the two parts separately in different operations, both are set simultaneously by the special fixed-center Milford Model 64FC riveter in Figure 10(b).

The rivet is fed from the standard hopper at right and, simultaneously, the screw machine part from the vibratory hopper at left when the operator pushes in the slide fixture (knob at bottom center) after having loaded the two metal parts and plate onto the fixture. After the first rivet and screw machine part are set in one metal shape, the operator pulls the slide out, and the fixture automatically turns 180° to bring the other metal shape into riveting orientation. When the slide is pushed back in, the machine is again actuated, and the second rivet and screw machine part are driven down through the holes in the metal shape and plate and then clinched.

Common Low-Level Methods

Other low-level methods of reducing parts handling time that are commonly used include:
- Nesting or locating fixtures -- nesting fixtures (Figure 8b) and locating fixtures (Figure 11) reduce parts handling time by helping the operator to orient the part for riveting more quickly. The simple fixture in Figure 11, in particular, is typical of many that can be made at very low cost.
- Swiveling fixtures (Figure 12) -- these help the operator to reorient the parts for multiple riveting under one head. Detents may be used to establish each riveting position, as in the small swivel fixture in Figure 12(a). The operator in Figure 12(b) turns the swivel fixture for visual location of metal-piercing rivets that are joining sections of sheet-metal ducts.
- Combined operations (Figure 13) -- the parts handling time per operation can be reduced by using the fixture on the rivet-setting

machine as an assembly point for more than one operation. In this
case, the operator is both riveting and soldering a small electrical assembly at the riveting machine.

- - - - - - - - - - - - - - - - - -

FIGURE 1

DESCRIPTION-LEFT HAND	TMU	DESCRIPTION-RIGHT HAND
Reach in direction of pieces	5.9	To one piece on pile
	6.3	Grasp one
	8.3	Shift concentration to left hand Look at piece
Reach to one piece	5.9	Move piece to pin
Grasp piece	6.3	
Move toward pin	6.0	Shift concentration to pin
	4.2	Piece to pin
	11.2	On pin
	1.7	Seat
	1.7	Let go
To pin	1.7	To box of pieces
On pin	11.2	
Seat	6.0	Shift concentration to right hand
Let go	-	Foot motion Machine cycle
	76.4	

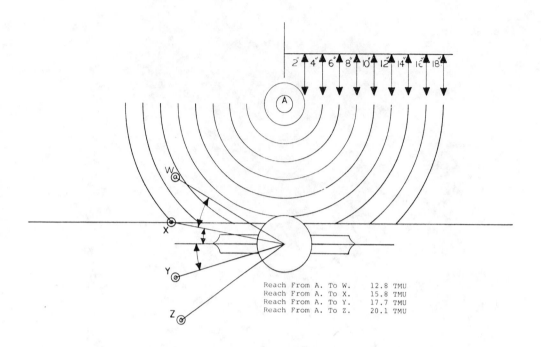

Reach From A. To W.   12.8 TMU
Reach From A. To X.   15.8 TMU
Reach From A. To Y.   17.7 TMU
Reach From A. To Z.   20.1 TMU

FIGURE 3

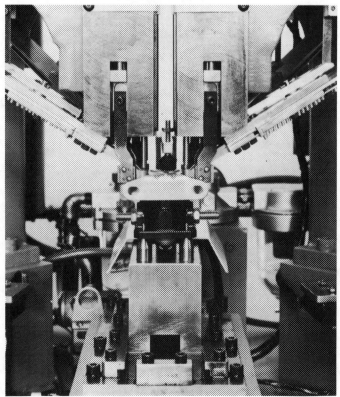

FIGURE 4

FIGURE 5a

FIGURE 5b

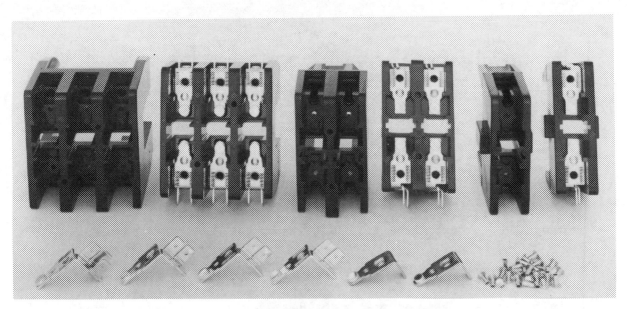

FIGURE 5c

FIGURE 6

FIGURE 7a

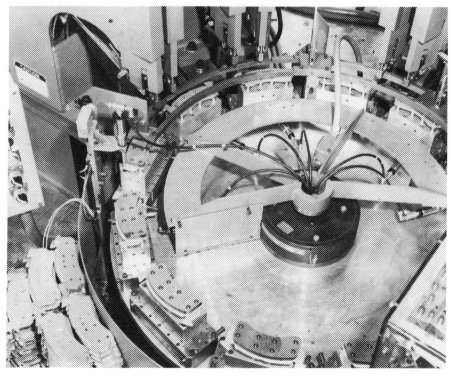

FIGURE 7b

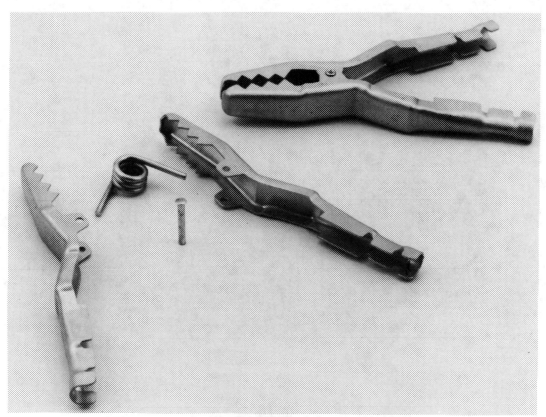

FIGURE 8a

FIGURE 8b

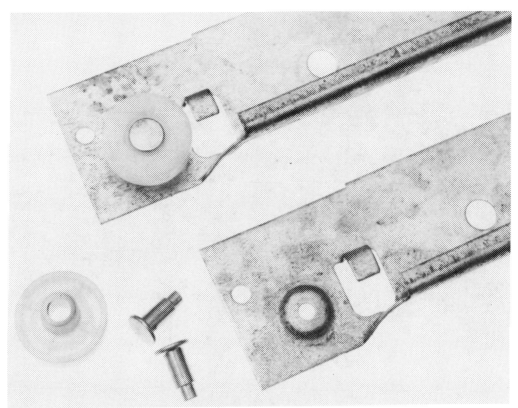

FIGURE 9a

FIGURE 9b

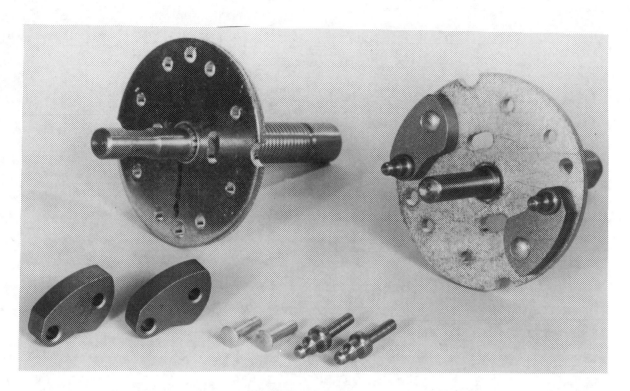

FIGURE 10a

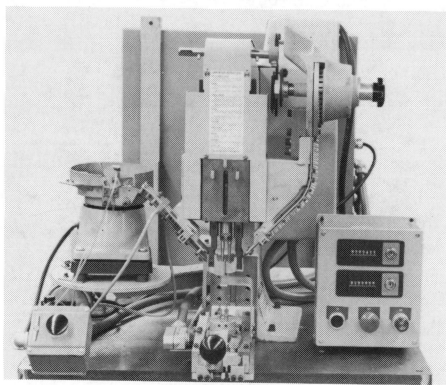

FIGURE 10b

AD78 - 760

FIGURE 11

FIGURE 12a

FIGURE 12b

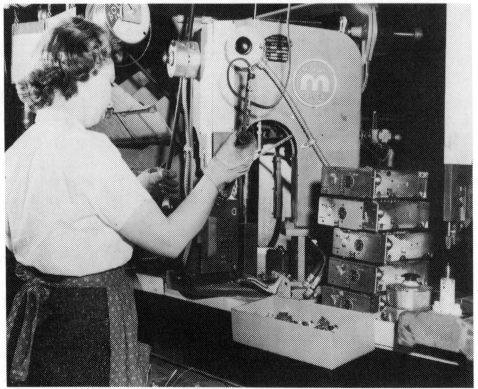

FIGURE 13

Presented at The Ford Motor Company World Wide Quality Control Conference, December 1978

Current Technology for Automatic Gaging

By Richard H. Gebelein
Moore Products Co.

Throughout the years people who have been responsible for controlling quality have looked to the inspection function for two major results:

First, the inspector must eliminate all parts which do not meet specification. Second, the inspector should furnish information about the product that will help eliminate rejects. This has not always been possible. Many checks have been performed in an arbitrary manner because it was not possible to inspect to those characteristics specified on the part print. Other checks could not be performed at all because equipment could not be made which would work in the production area. But the digital computer makes it possible to accomplish quality control inspection that simply could not be accomplished before, and the reliability and low cost of the microprocessor make it suitable for use in the factory environment. That is why digital computing technology is being used today to enhance the capability of conventional gaging equipment and to make possible new and unconventional approaches to product inspection. While this presents a set of very powerful tools to the quality control function, it creates some additional responsibilities that must be dealt with if reduced rejects and improved quality are to be achieved. Let us look at some examples which will illustrate these points.

The Automotive Industry makes tens of millions of gears each year, most of which are helical gears and are very difficult to inspect. Standard practice has developed a system where the parts will run in tight mesh with a perfect reference gear. Analog electronic circuits were used to measure the center distance between these two gears. Data about the quality of the part was extracted by interpreting variations in that center distance. The required information is shown in Figure -1-. These are the specifications on the part print of the gear.

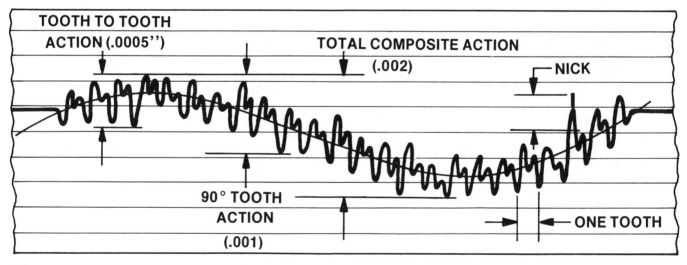

Figure 1

Analog circuits cannot make all the checks required by the part print. They also cannot detect all of the error on those checks which can be accomplished. For example, it has been possible only to determine an approximate average of errors for Total Composite Action or for Tooth-to-Tooth action. The frequency filters used to extract these characteristics also filter out a noticeable part of the error being examined. The 90° Composite Action could not be checked at all. The end result was a major compromise in accuracy which results in numerous good parts being rejected and many marginal rejects being accepted.

When digital techniques are applied to this same problem, a new level of clarity and accuracy is introduced. A very great number of precise, discreet readings are taken. These readings are not filtered in any way and so they are an accurate representation of the gear. Figure -2- shows a typical computer printout of the information gathered from such a digital inspection, including a digital plot of the same variations that were shown in Figure -1- as an analog graph.

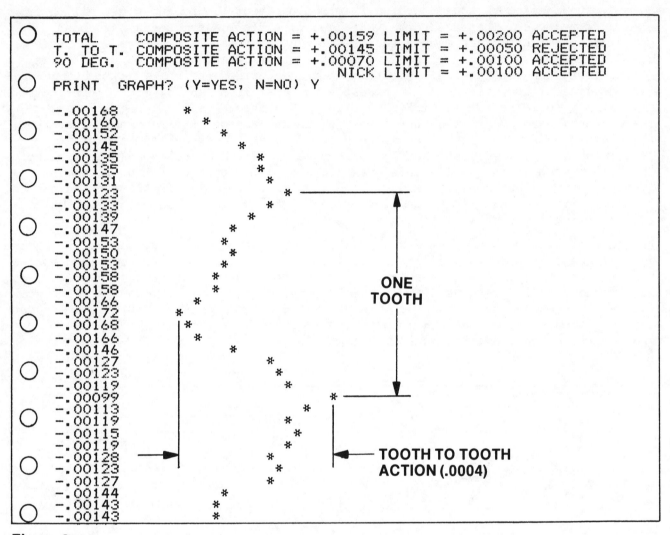

Figure 2

It is obvious that a much greater definition can be achieved. The details can be seen clearly and are represented as real and precise numbers with great amplification. These readings are manipulated in the digital program. There is virtually no limit to the examinations that can be performed. It is possible to extract the precise Total Composite Action. It is possible to extract the individual Tooth-to-Tooth Action. The 90° Tooth Action can be determined easily, acting as though a floating window, 90° in width, were moved across the graph of the part. Subtle interpretations of trends in characteristics from tooth-to-tooth can also be determined. Since none of the original readings have been attenuated, the results have high resolution and accuracy. Gears can now be inspected to the exact requirements of the part print without compromise.

This new found precision does create some new problems. It becomes essential that Product Design, Manufacturing Engineering, and Quality Control jointly consider the design and the tolerances on the part piece. How reasonable are the tolerances? Are they necessary for product performance? Do they truly cover all of the critical features of the part to make it suitable for its purpose? These decisions should be made jointly to assure optimum quality of the end product.

There are a number of inspection problems which have defied 100% inspection in the past. One example of this would be the inspection of threads. Threads have typically been inspected by applying a master thread gage which should turn onto the thread being inspected to a proper depth. This form of inspection is not practical for a 100% check in a production environment. Digital technology has solved this problem, as is illustrated in Figure -3-.

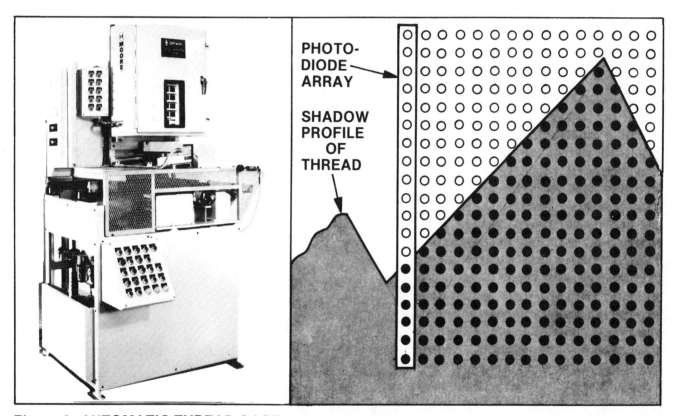

PHOTO-DIODE ARRAY

SHADOW PROFILE OF THREAD

Figure 3 AUTOMATIC THREAD GAGE

A light source is positioned to throw a shadow of the thread form onto a photo-diode array. Those diodes which are not shaded by the workpiece will generate a voltage output. Thus, a line is drawn composed of on and off signals representing one location along the thread. Motion is applied so that the diode array is optically moved along the threaded workpiece, generating a large number of lines of dots. These are recorded in the digital micro-computer and constitute a dotted picture of the threaded workpiece. The digital program analyzes that dotted picture to determine major and minor diameters, pitch variations, thread form, etc. This inspection is extremely rapid and can easily keep up with automotive production line requirements. There is no wear problem to progressively alter the readings as was the case with conventional thread gages. It is thus possible to perform a production check today which was impossible yesterday.

Once again, it is important that a team effort, including quality control personnel, decide what needs to be known about the thread. Instead of a simple functional check, you now can see the various details that are called for on the part print. Are they all necessary? Are they sufficiently complete to guarantee the successful performance of the part? The ability of the operator to apply "feel" has been removed from the inspection process. The operator can no longer forgive marginal parts, nor can he make interpretations about the relative importance of one characteristic vs. another. The computer system will make an absolute and precise judgement. It is vital that the part print accurately defines what is necessary and not simply repeat arbitrary standards that resulted from earlier limitations in the inspection process.

Passenger car wheels should not contain any significant eccentricity between the mounting hole and the tire. This will result in an uncomfortable ride as though the car were rolling along on a cam. Finding eccentricity was impossible because the manufacturing process produced a very complex shape. This same difficulty is encountered in thin walled products such as ring gears and shell bearings and in brake drums. For example, it is possible to have the exact same amount of eccentricity and ovality in two workpieces, and yet to see a radically different TIR reading when an indicator is applied. See Figure -4A- and -4B-.

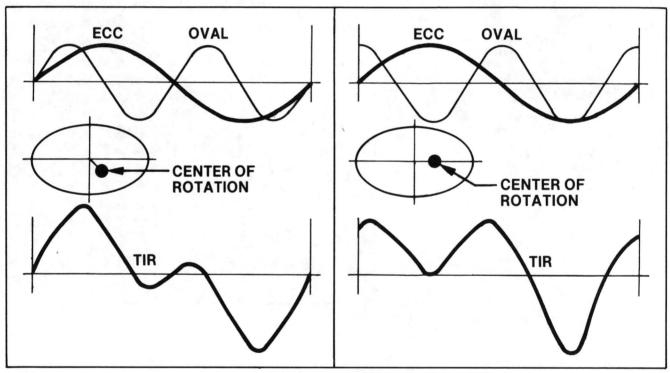

Figure 4A   **Figure 4B**

It is impossible for an operator to tell the amount of eccentricity that is present in the two TIR traces shown in that illustration. The diagram illustrates the radical differences that can occur as identical problems are mixed on a workpiece if the phase relationship varies. However, if the amount and direction of eccentricity can be determined, it can either be corrected in the workpiece or compensated for at assembly. Using digital techniques, this becomes readily possible. A Fourier Analysis is performed to find both the amplitude and the phase relation of the first harmonic. This first harmonic actually defines the amount and plane of maximum eccentricity.

The speed and reliability of microprocessor systems now makes it possible to inspect and sort such parts at line speed in hostile manufacturing environments. Photographs -4C- and -4D- show examples of automatic gages utilizing this technology.

Figure 4C WHEEL HARMONIC GAGE

Figure 4D BRAKE DRUM GAGE

All of these digital applications can incorporate data logging and statistical analysis into the inspection programs. Mass storage of data on each and every part piece is possible. Manipulation of this data is limited only by the imagination of the user and the programmer. Furthermore, programming for gages tends to be repetitive and existing programs are easily and economically adapted to the next task. Data can be printed out directly, displayed on a CRT for examination, or transmitted to a central computer for future storage and processing.

Because of consumer pressure and government regulation, it is increasingly important to inspect parts completely and accurately. Without digital technology it would not be possible to meet these challenges. And current trends in product liability make the data logging capabilities extremely valuable. These digital systems make it possible to document the quality process, and those records can be vital to warranty and to legal liability. protection.

Clearly, digital technology has proved to be one of the most valuable developments to enter the manufacturing and quality control worlds. Its future uses also appear virtually unlimited.

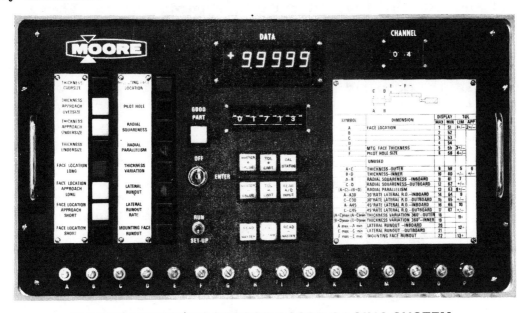

MOORE MICROPROCESSOR BASED GAGING SYSTEM

Presented at SME's Portland Tool & Manufacturing Engineering Conference, September 1981

The Development of the Coordinate Measuring Machine and Its Effect on Manufacturing Industry

By R. Gilheany
L.K. Metrology Systems Limited

In the engineering industries today there is a shortage of skilled labour and at the same time high unemployment. Looking towards the next decade, is the answer to this problem to reduce or eliminate the traditional skilled activities and substitute the machine and computer for the man?

The trend in production has been to introduce extremely sophisticated machine tools; up to a few years ago however inspection personnel had to rely on conventional equipment, some hand adjusted, such as micrometers and verniers. This equipment was invented in the middle of the last century when the developed countries were emerging from the days of the stage coach and product interchangeability was a mere dream.

Since about 1945 there has been a steady development in the consumer requirement for quality and reliability. This demand can be achieved by producing components at or close to the design optimum thus reducing the tolerancing of component parts. Conventional measuring equipment, which has considerable measurement uncertainty, is not capable of consistent control of close toleranced parts.

The introduction of the new range of machine tools - transfer lines, NC and CNC has imposed a further burden on the inspection groups. Extremely complex components can be produced on machining centres faster than the inspection floor-to-floor time.

The machine tool's prime function is to add value to a component, in the minimum time, at the requisite quality and at maximum utilisation. Therefore for the 1980's there will be requirement for:

* Countering the shortage of skilled labour.

* Improving the measurement standards.

* Rapid assessment of the products from the machine tools.

Increasingly, production engineers will have to accept the responsibilities for not only the most efficient methods of manufacture, but also the control of the quality of the product.

MEETING THE CHALLENGE OF PRODUCTION

To control production either on or off the machine there must be accurate and repeatable measurement. For too long industry has relied on measuring equipment invented in the middle of the last century, and hardly changed or modified since its first introduction, to control the products of today. It is a far cry from the days of the stage coach and the embryo locomotive to the present means of transport, and yet we are content to use the same product control equipment. The requirement of interchangeability was then a pipe dream but today is of paramount importance.

The requirements can be summarised as:

*   Improve accuracy of measurement.

*   Improve the ease and speed of measurement.

*   Reduce the time taken to record results.

*   Reduce the time taken to analyse results.

*   Improve the communication of measurement results.

Accuracy of Measurement

Measurement is never absolute; irrespective of the method employed there must always be uncertainty. The effect of environment, equipment, human influence and temperature all contribute to uncertainty.

Ease and Speed of Measurement

Advances in peripheral equipment and software development have radically changed the traditional concept of inspection. Instead of complicated setting on surface plates, sine plates, or dividing heads, components can now be placed in any attitude on the machine tables and be measured. Planned inspection programmes can be called from store; these ensure that the inspection task is performed in the most efficient manner and in a controlled pattern of movement which will not change, unless the programme is changed, no matter what time interval there is between such inspections. It follows that comparison of measured results is meaningful.

Recording Results

A time consuming element of the traditional inspection task is the clerical content. This is rarely less than 25% of the available time. The measuring machine equipment can now absorb automatically all details of measurement.

Time to Analyse Results

Many options are available to users, ranging from teletype to visual display. All have the ability to process measurement information and to display this in many forms. A full analysis of the results can be computer assessed or visually edited.

Communicating the Results

Improving the communication of results will be the next major development. Economic influences will, in the next decade, dictate the development. The cost of raw materials will inevitably rise, and management must reduce the stock holding and lead times for production.

One of the bottle necks of production is the decision making delay on the acceptance of parts produced incorrectly to specification. Invariably design has to be consulted and in multi-national companies where the design offices are divorced from production, there is considerable delay in transmitting information about the fault and in the reply. During this communicating process, misinterpretation can occur and lead either to an incorrect decision or even longer delay. The requirement for improved communication will become one of the most important in the manufacturing industries.

Measuring Machines

The introduction of measuring machines into the manufacturing industries of the developed countries has been a growth industry during the last 15 years. In the mid 1960's a major expenditure in an inspection organisation was for conventional mechanical measuring equipment with sorties into multi-gauging fixtures with pneumatic or electronic gauging heads, today measuring machines costing up to £100,000 can be justified.

Primarily the justification is to counteract the shortage of skilled labour. However, the use of the machine escalates, and as the associated software developments keep pace with industrial demands, the savings show a sharp increase resulting from:

a) Improved through-put; savings of up to 400% have been recorded.

b) No requirement for jigs and fixtures.

c) No drafting of jigs and fixtures.

d) No calibration requirements for c).

e) No storage of jigs or fixtures.

f) Automatic analysis and printout of measurement results.

In addition to the justification arising from the savings there is a further advantage which is difficult to quantify, but nevertheless important. Machine tools will have considerably reduced downtime resulting from waiting for inspection acceptance. The planned sequence of inspection can be timed in precisely the same manner as the cutting operations on the manufacturing tool.

Measurement from a co-ordinate machine is rarely disputed; it is accepted by production and costly re-inspection of products is eliminated. As the use of the machines has proliferated, there has been a constant dialogue between supplier and customer. One feature has, of recent years, been dominant - how to improve the accuracy and repeatability. The problems of accuracy and its assessment have not been fully understood and in many applications accuracy is mistaken for precision. In the simplest interpretation:

"Accuracy is the agreement of a measurement with a true value".
"Precision is the repeatability of a set of measurements".

Development of Co-ordinate Measuring Machines

The co-ordinate measuring machines (CMM) have now become an accepted method of inspection control. The developments of the machine configurations have followed designs (see Fig. 1).

a) Cantilever

b) Moving Bridge

c) Overhead Gantry

d) Horizontal Arm

It will be recognised that the CMM's configuration has related itself to that of the machine tools. The majority of the machining centres now have horizontal spindles, which combined with a rotary table gives both accessability and versatility, so it is with the CMM. The horizontal arm CMM similar to the LK Metre Four (see Fig.2) is the most adaptable for integrating into the machining lines.

Improving Accuracy

Some years ago we concentrated our design on the use of granite as the most accurate datum from which machines could be produced. Our reasons for this decision were:

Granite can be lapped to extreme accuracy.
Air bearings average out errors and are not sensitive to texture.
Granite has been stress relieved for millions of years and is very stable.
No wear due to air bearings.
The stiffness/weight ratio of granite is the same as steel and better than cast iron.
Air bearings have no friction forces to cause elastic deflections or hysteresis errors.

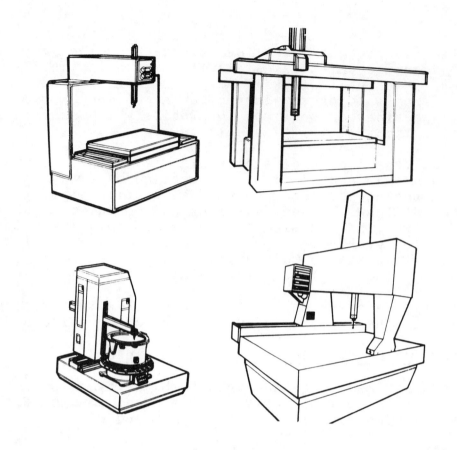

Figure 1

Figure 2

The coefficient of linear thermal expansion of granite is 5×10^6 per $^{\circ}$C. (12×10^6 per $^{\circ}$C for steel and 22×10^6 for aluminium), which is important when considering bending due to thermal gradients. Granite has a large conduction path for heat which results in reduced thermal gradient.

The self-damping capacity of granite is 15 times better than cast iron.

Air bearings are self-cleaning

No corrosion problems with granite.

No burrs raised on granite faces due to accidental damage.

No seizure of air bearings on the granite if air is left off for long periods.

It was of interest to read an article in Precision Engineering January 1979 on the design and construction of an ultra precision diamond turing machine and to note that for the machine base and independent metrology unit granite has been chosen.

Development of Co-ordinate Measuring Machine Software

Although over the last five years Co-ordinate Measuring Machine designs may have developed to the styles as shown in Fig. 1, each design has its own attributes and is suited to varying types of components. It is safe to say however that the general developments of the mechanics of the machines, is small compared to the revolution taking place in the Co-ordinate Measuring Machine peripheral equipment area. It is apparent that as Industry accepts the use of Co-ordinate Measuring Machines as standard practice, its familiarity increases its demands on the capabilities of the Co-ordinate Measuring Machine.

As in all walks of life the advent of the Computer has had a tremendous effect, and it has provided us with means of meeting Industries demands upon the inspection system.

A Chronological Summary of the Co-ordinate Measuring Machine Peripheral Development

A 'peripheral' is a piece of equipment which when it is interfaced, or connected to a Co-ordinate Measuring Machine improves it's overall capability.

A Co-ordinate Measuring Machine can be loosely described as a mechanised surface plate, with the first machines well over a decade ago, showing the position of the sensing head on three digital displays the head or probe being manually moved around the component. The results were then transferred onto paper by the inspector using the basic Co-ordinate Measuring Machine, improved inspection throughput because using one piece of equipment, widths, depths, hole positions, hole sizes can be inspected, including information in two related axes from one inspection sequence, which with conventional inspection techniques, i.e. surface plates, height gauges, micrometers etc., is not possible.

Over conventional methods time savings of up to 50% were envisaged. It was apparent however that errors were made and time was lost by the inspector writing down his results, and in order to alleviate this problem, a teletype was interfaced to the machine so that data could be automatically printed when required. This straight away improved throughput time by a further 10%. Again it was realised that the teletype had a means of creating paper tapes and reading them, so that with simple modifications, instructions as to how a job should be inspected with nominals and tolerances were printed, together with the result. (See Fig. 3a and 3b). This ensured 100% inspection with documented results, although the decision as to whether or not the feature was in error was a decision left to the operator, as a teletype is not an 'intelligent' device. The fact that a tape was used, further improved the flexibility of the machine, for a tape written with sufficient information, allowed an inspector with no previous knowledge of the component, to inspect it, merely by following the instructions.

As mentioned previously the teletype is not intelligent, and the next demand from industry was to make the Co-ordinate Measuring Machine make the decision of in or out of tolerance and report the result accordingly. This required the introduction of some form of intelligence, namely a computer. More than eight years ago a Digital Equipment PDP8 computer was interfaced to a teletype and Co-ordinate Measuring Machine. (See Fig. 4). Not only did it make decisions about being in or out of tolerance, but it would allow the calculation of diameter size and central co-ordinates from 3 points, release results in either Cartesian or Polar form, as well as take into account misalignment in 2 axes of the component in parallel to the surface of the Co-ordinate Measuring Machine surface plate. These few additions made possible by the computer not only improved the capability of the Co-ordinate Measuring Machine by improved throughput by a further 10% allowing savings of up to 70% over conventional inspection methods.

With the introduction of the computer people now became critical of the speed of the teletype as a reporting device, its read and write speeds of 10 characters per second (cps) although fast compared to humans, was very slow compared to other peripherals appearing on the commercial market.

To meet these ends, Visual Display Units, (V.D.U.'s) were utilised, allowing input of inspection schedule programs on cassette, at a speed of up to 1000 cps, and also using a printer capable of printing results at up to 180 cps. See Fig. 5a, 5b and 5c. To handle these new peripherals a decision was taken to use a new model of computer, which also had enhancements, manufactured by Hewlett Packard. All of the improvements which gave improved speeds, greater reliability and flexibility, were obviously reflected in costs. The first teletype cost approximately £3,000.00, now the system costs were up six or seven times, but the advantages made such increases justifiable.

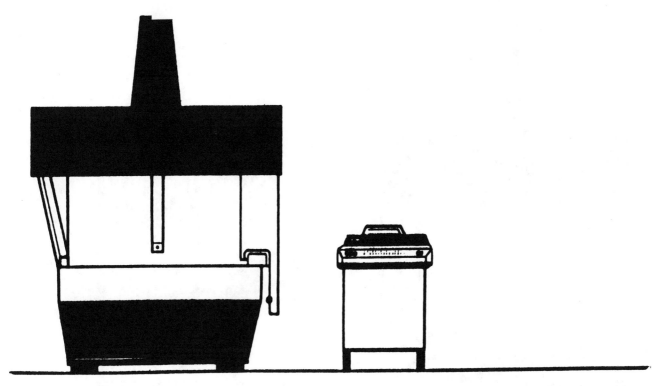

Figure 3A   **Data Recording**

Figure 3B

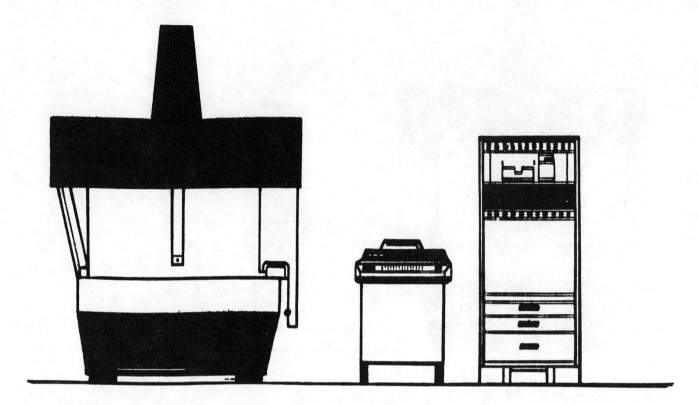

Computer Assisted Inspection

Figure 4

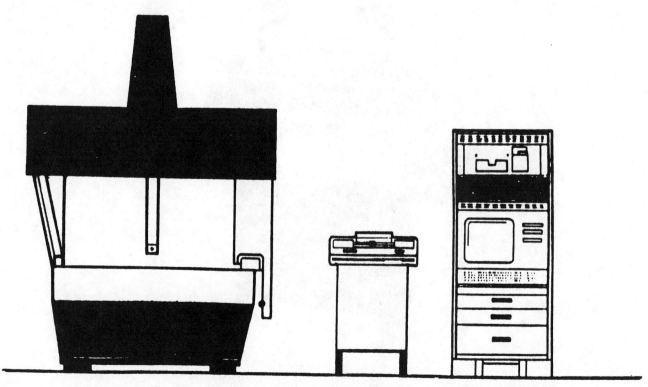

CNC Inspection with Display

Figure 5A

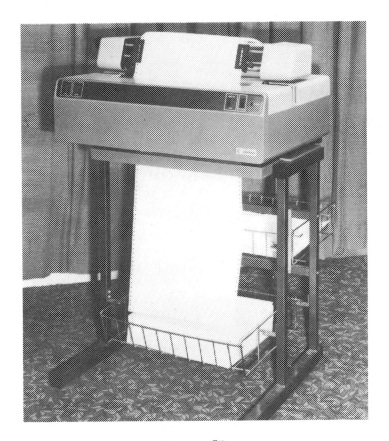

Figure 5B

Figure 5C

In 1977 a fresh look was needed with regard to the inspection capabilities with computers. From the original PDP8 system, extra facilities had been added, and added and in fact there was a serious problem of the system becoming unmanageable. It was therefore decided to redesign and completely develop a new software system, with enhancements but more simplicity of schedule commands for the inspector. Furthermore the decision was also taken to give the control of the Co-ordinate Measuring Machine to the computer, allowing completely automated inspection, thus leaving the inspector free to inspect components more suited to conventional techniques, or free to schedule more inspection programs for other types of components.

The sort of capability then achieved, besides automated inspection were as follows:

    i.   Automatic setup to the component without the need for jigs and fixtures, an obvious cost saver.

    ii.   Checks for hole positions and size, with more than 3 points, allowing checks on roundness.

    iii.   Inspection of flatness.

    iv.   Depths, widths, lengths, radii.

    v.   Tolerancing with unilateral, bilateral or true position.

    vi.   Inspection of angles, including inspection of planes, or lines.

    vii.   Inspection of compound bores and the intersection of their axes in space.

All of these extra capabilities greatly enhanced the versatility of the machine, enlarging it's scope of inspection.

The fact that we had a machine capable of automatic motion, or manual motion under drive control, also improved the repeatability of the Co-ordinate Measuring Machine as it guaranteed that different people, when using the machine under manual drive control would move the machine at the same speed, as this was a preset condition. Thus reducing one of the worst problems in inspection, i.e. the repeatability of different personnel with the same gauge on the same job.

To the present day the system has been enhanced allowing inspection on 3 or 4 axis Co-ordinate Measuring Machines. On 4 axis machines with rotary tables drum type components can be easily inspected.

We now offer several different types of peripherals on our machines, the selection of which system being dependant on the requirements of the customer. The requirements need not necessarily be the type or complexity of component, e.g. how would a fully automatic machine be of any benefit in a 'one-off' or low batch tool room type operation, but also other parameters such as environment, e.g. different types of systems are more robust, and the type of personnel is also very important when selecting a system.

The range starts from a low-cost system based on the Hewlett Packard 85A Calculator, for manual machines only, and goes to Hewlett Packard or Digital Equipment Computer Systems with Floppy Discs or hard discs for fully automated systems with growth potential.

The software too is now capable of offering inspection to the standards of BS 308 including MMC condition checks for holes, slots and faces. At present work is being finished on MMC fits for groups and patterns of holes with documentation relating to which features may need machining to achieve an acceptable condition, thus reducing the need for costly inflexible gauging equipment.

Graphical representation on plotters is available, indicating on preprinted sheets whether a component is in or out of tolerance on the feature concerned. Roundness can also be plotted. (See Fig. 6a and 6b).

FUTURE TRENDS AND DEVELOPMENTS

Digitising of a Surface

Possibly the greatest development effort within the company is being devoted to "Scanning". This is where using standard inspection "touch" triggers or electronic probes, we can follow the contour of an unknown surface, and digitise it. If the shape or profile is a known shape we can check against nominals and tolerances, and plot the results accordingly. See Fig.7a,7b and 7c. The state of the system now on release to customers, allows not only digitisation of the surface but can also 'best fit' this shape against a known form indicating how much motion and twist is required to do this. Perhaps this is best described by inspecting a Turbine Blade. A slice through the blade can be scanned, the results being relative to a 'stacking point' or datum. We can then find by how much we must move the datum and twist the blade in order to get the best fit between the blade profile and its nominal shape and tolerance. A second option is that it may be preferred to have a blend between the acceptable error on stacking point, twist and profile. To this end the system is also capable of calculating an error averaged over these three parameters, i.e. not allowing perfect profile and stacking point, with maximum twist, when for the performance of the blade a general average would be more preferable.

A blade was used in this case but profiling can be applied to any shape.

NC Tape Generation

With our advancements in digitising we have become aware of the simple step from inspection to production.

We are at present working with several computer aided Design and Manufacturing Systems (CAD/CAM), to bridge this gap.

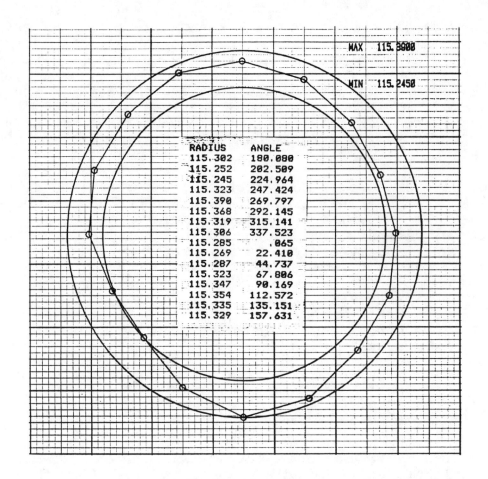

RADIUS	ANGLE
115.302	180.080
115.252	202.509
115.245	224.964
115.323	247.424
115.390	269.797
115.368	292.145
115.319	315.141
115.306	337.523
115.285	.065
115.269	22.410
115.287	44.737
115.323	67.806
115.347	90.169
115.354	112.572
115.335	135.151
115.329	157.631

MAX 115.3900

MIN 115.2450

Figure 6A

Figure 6B

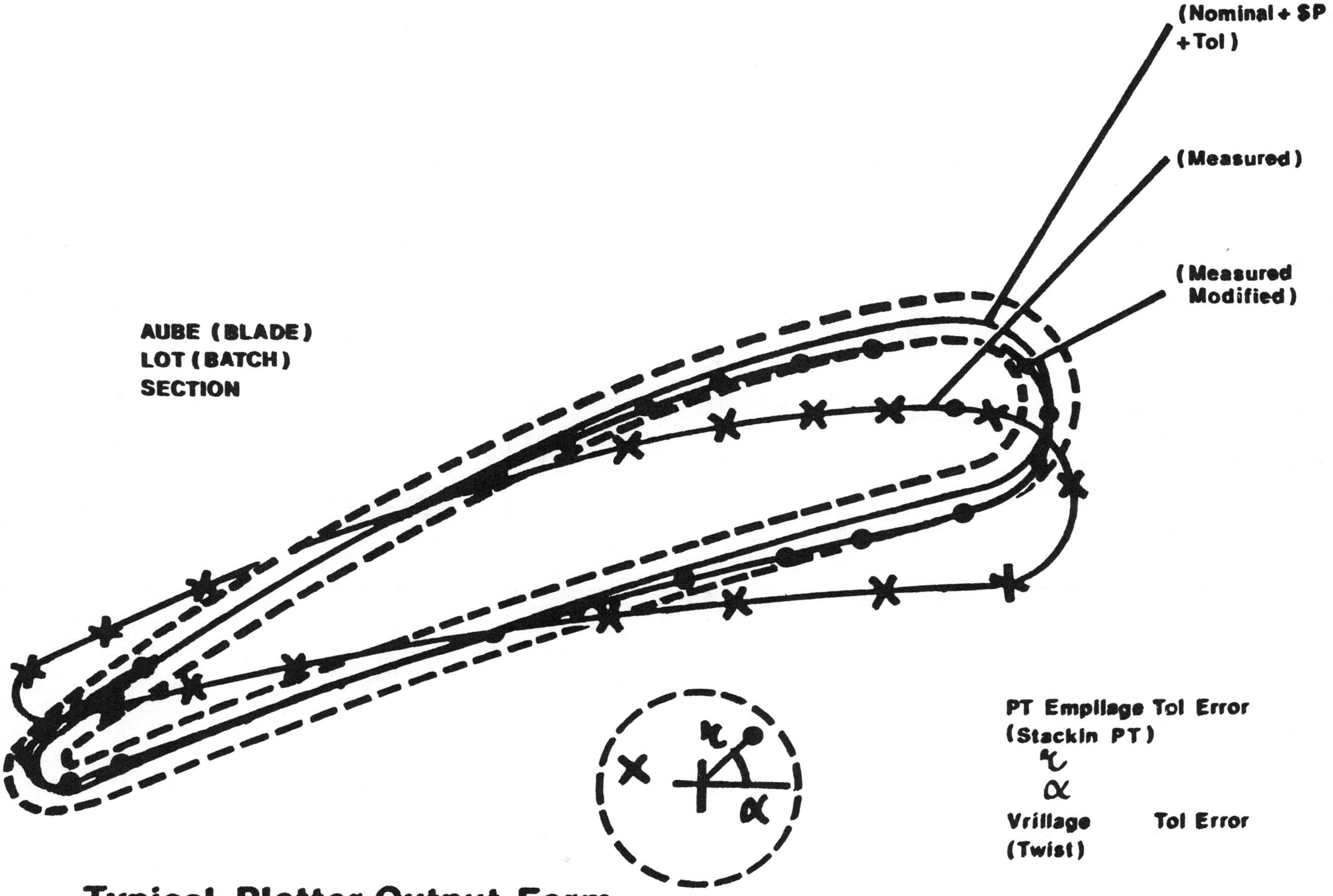

(Nominal + SP +Tol)

(Measured)

(Measured Modified)

AUBE (BLADE)
LOT (BATCH)
SECTION

PT Emplilage Tol Error
(Stackin PT)

Vrillage   Tol Error
(Twist)

Typical Plotter Output Form
These can be displayed singly or concurrently

Figure 7A

BLADE PROFILE OPTIMISATION CONCEPT.

Theoretical Input

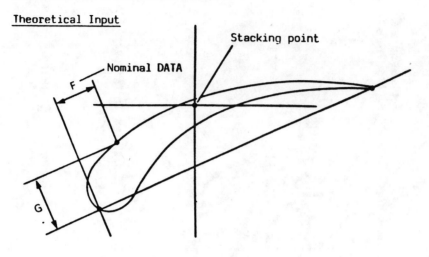

Actual Scanning data.

Superimposed data.

Best fit.

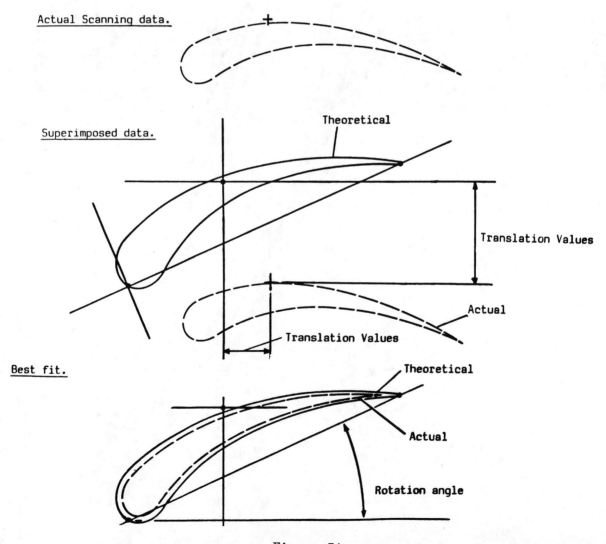

Figure 7A

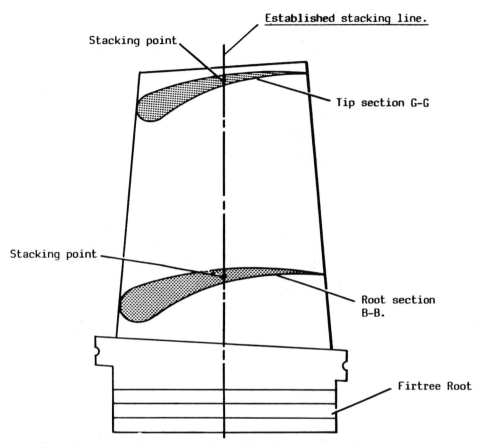

Establishment of stacking line utilizing tip & root sections.

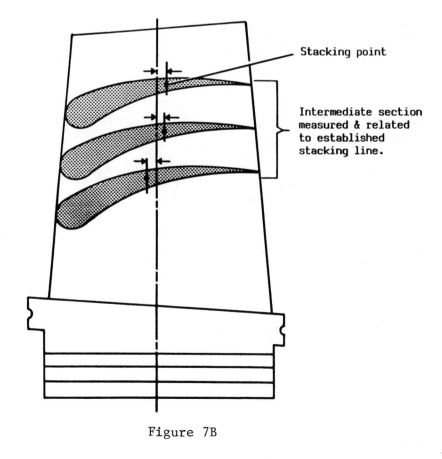

Figure 7B

For example, if you have a model of a shape for a tool or die, for example, you know the model is correct but you do not know its shape, we can digitise the surface, either as points on the surface, or as centreline data, by using a sensor equivalent in size to an acceptable cutting tool. The data can then be passed to the CAM system and utilised to produce an NC tape for a machine tool, thus giving the capability to inspect an unknown model or shape and then reproduce it on an NC machine. Work is also being conducted into allowing us to run both the CMM and the CAM system on the same computer.

Flexible Machining Lines

In the UK and other countries there is considerable planning activity towards FMS and inspection. The first positive move which involved my Company (LK Group of Companies) came from the Czechoslovakian Government Controlled Institute of Machine Tools and Machining (VUOSO) at Prague. In 1980 a flexible system started to control the entire process from raw material castings to fully inspected finished parts. The main activity of a very limited number of operators will be the loading of pallets onto the conveyor lines, and in observation but not intervention.

The two CMM's selected for this application are from the LK Metre Four range. (See Fig. 2).

The machines will automatically receive, locate, and inspect the products at any stage of the machining cycle as determined by part programmes and experience.

The amount of inspection called for by the programme will be developed from the on-going inspection of parts, from the stored data the process capabilities will be established. Control, therefore will be from practical experience.

The CMM's will signal to the central computer the status of the parts inspected, pass or fail, so that corrective action, if required, can be introduced.

Either, or both, Metre Four machines can be in action at any one time. Should one machine be idle it can be programmed for other work before reverting to the main programme.

The Metre Four has two rotary tables on the extended X axis of the machine. One rotary table is on-line to the pallet system feeding the machines. The second table can be manually operated off-line, for example to investigate production problems of control, without disturbance of the central system.

The CMM's have a capability for measuring 400 mm^3 components.

The CMM's are integrated into a flexible production system incorporating two banks of four parallel controlled horizontal machines with in-built rotary tables. The machines have 10 racked containers for 144 tools in each. At the start of the operation up to 1440 tools are placed in the racked containers. It follows that with the availability of this selection of machine tools the standard 3 axis probe used in the measuring machine would not have the capability of measuring all the attributes produced by the range of cutting tools. The requirement also applies to the multi axis probe.

Therefore, to satisfy the requirement an automatic probe changer has been developed (Fig 8), this is located on the Metre Four table.

The CMM can now be programmed for control of all attributes produced by the range of machine tools and cutters.

The mechanical developments of the CMM has been paralleled by the developments in both hardward and software.

Robotics

Work is also being perused into control of robots, which are achieving higher standards of repeatability and accuracy for not only automatic loading onto CMM's but also for inspection of components using non-contact type sensors (see Fig. 9).

It will be seen that components can now be conveyor moved to the inspection centre, be measured, and if correct resume movement in the system and if in error transmit signals for corrective action. All data can be stored and statistically appraised for future action.

One obstacle still remains for a system to be totally automatic. This is the subjective inspection, the measurement of geometric features such as roundness, squareness of features to each other, and the assessment of surface finish by measurement, or the "eye ball" look at a component for "fitness for purpose".

In the high technological manufacture and control, for example aerospace, there is also a requirement for component identification of control and traceability. Traditionally this has always been interpreted that the identification must be the inspection stamp of an inspector. Changes will have to be introduced to permit machine identification and this be referenced to approved capability studies.

The inspection organisations now have measuring technology which is equal to the manufacturing technologies. The developments in measuring equipment with the use of micro processors or computers have, during the last decade, been phenomenal, we look forward to the 80's with confidence.

Fig. 8

Production Line
Quality Control Analysis

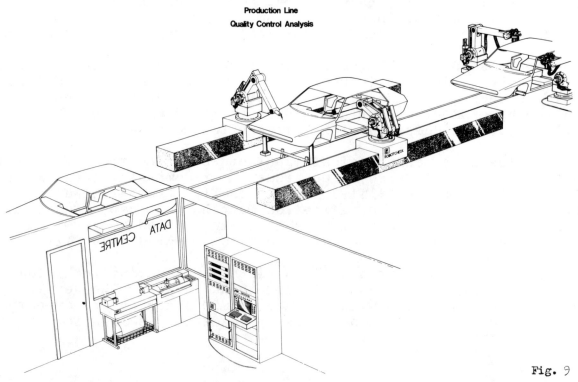
Fig. 9

Presented at SME's Portland Tool & Manufacturing Engineering Conference, September 1981

Calibration—The Key to Precision Machining

By Glenn Herreman
Hewlett-Packard Company

There are some basic principles of metrology that must be understood by mechanical inspectors, calibrators, quality and manufacturing engineers at all levels before reliable measurements can be made.

From the use of micrometers, calipers and indicators to measuring instruments and measuring machines as well as control of a manufacturing process, Geometry plays a very important role in the ultimate end result--the product.

The discussion will cover such items as straightness, squareness, Abbe's Principle and their effect on the accuracy of a manufacturing or measuring process.

The "Six Degrees of Freedom" of machine and instrument slides and their effect on process control will also be discussed. Statistical data plots will be shown as examples of accuracy and repeatability problems. When your measurement process is in control, corrective action can be taken with confidence to improve the manufacturing process. The result is a quality product.

The inspector and engineer, through study of basic principles of metrology and related experiences must acquire a high level of expertise to solve day to day problems. This expertise will establish the credibility necessary to work successfully with others toward mutual goals.

GEOMETRY IN METROLOGY--PARTNERS IN PROCESS CONTROL

Process control is our means to continuous on time production and higher profits. It should therefore be foremost on the minds of Quality and Manufacturing Engineers. Process control is a broad term, covering many areas and disciplines. This paper will deal with the discipline of producing and inspecting parts in a metal cutting environment. Through this paper we hope to give some insight into a few problems that we feel, from our experience and contacts, are least understood and most often overlooked when a process is not in control.

These problem areas have to do with:

1. Machines used to produce the parts- be it cutting, punching, grinding, turning, etc.

2. The measuring equipment used to inspect these parts such as measuring machines, optical comparators, toolmakers microscopes, surface plates, etc.

3. The geometry of some turned and ground parts that cannot be accurately or adequately measured with standard inspection equipment.

We will begin this discussion with the machine tools as they are the foundation of all that follows. If your manufacturing equipment does not meet the requirements necessary to produce the parts to specification, you are in trouble from the start.

KEY WORDS - Six degrees of freedom
- Orthoganality
- Abbe's Principle
- Flatness
- Repeatability
- Circular Geometry

I. MACHINE GEOMETRY

Each axis of a machine slide has six degrees of freedom (Fig. 1) acting simultaneously on the movement of the slide as it moves from position A to position B. The freedom of movements are working independently on each axis. These degrees of freedom are defined as:

1. Position: The freedom of the slide to move to a position within a band which is less than or greater than the command position. This will be accuracy and repeatability.
2. Straightness, Horizontal: The side to side movement of the slide.
3. Straightness, Vertical: The up and down movement of the slide.
4. Pitch: An angle movement, front to back in line with the slide travel.
5. Yaw: An angle movement side to side.
6. Roll: A side to side angular rotation movement around the screw.

THE SIX DEGREES OF FREEDOM

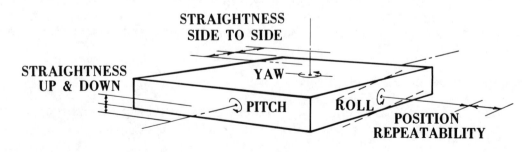

Fig. #1

There is one other degree of freedom that should be included as a seventh degree of freedom - Squareness. The 90 degree movement of one axis relative to another axis:

X to Y; X to Z_1; X to Z_2; Y to Z_1; and Y to Z_2. The Laser Interferometer will measure 5 of the 6 degrees of freedom. The Laser system will not measure roll. The horizontal measurement of roll can be accomplished with a precision or electronic level.

Vertical roll is another problem and particularly important as it applies to a Coordinate Measuring Machine using a 90 degree probe holder. You can imagine the problem of having a 90 degree probe pointing at some arbitrary reference point at the upper end of the travel and then because of vertical roll, rotate to some other point as the Z axis slide moves down to other positions. If the 90 degree probe is oriented in the X axis direction, rotation due to Z axis Roll will cause error displacements in Y axis. All of this will go unnoticed, and if questioned, will be difficult to check.

It is not necessary to measure all of the six degrees of freedom to qualify a machine tool or measuring machine. You can determine the collective effects of pitch and Yaw at the cutting tool by measuring positioning accuracy at different places on the machine. For example, measuring position accuracy of the X axis slide at two levels above the table will give an indication if there is a pitch problem. Similarly, measuring the X axis at two positions along the Y axis will give an indication if there is a Yaw problem. If there appears to be a problem, Pitch and Yaw measurements can then be made independently to give you the information necessary to correct the problem.

To illustrate the "Freedom of Position", I would like to use the following example of the calibration of an NC Mill equipped with software to accommodate backlash correction and linear error correction (Fig. 2). The example shows position data of six data points taken bidirectionally. The X axis on the graph indicates the position of the data @1.125" increments along the axis. The Y axis on the graph indicates the error, plus or minus, from the command position. The center line is the average of the data points and the outer limit lines indicate the boundary limits of the +3 Sigma probability curve. In this example we see two sets of good data separated by some offset that indicates a backlash problem. When we can measure this backlash as shown, we can put the necessary backlash correction into the machine software and effectively eliminate the backlash as shown in Fig. 3. We have made a big improvement in the machine precision, but we still have to correct for positioning accuracy.

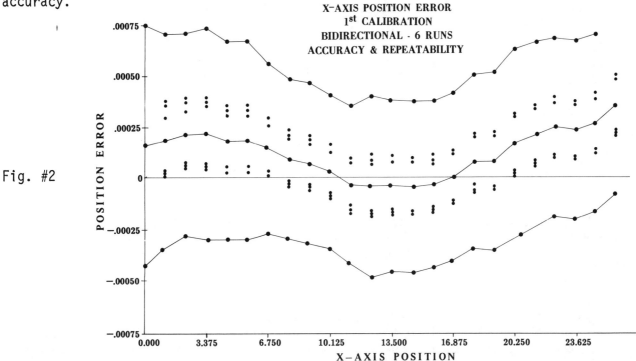

Fig. #2

X-AXIS POSITION ERROR
1st CALIBRATION
BIDIRECTIONAL - 6 RUNS
ACCURACY & REPEATABILITY

With the data shown, we know where to compensate for the X axis positioning errors (Fig. 4). It is not perfect because there are a fixed number of corrections that must be distributed over all three axis. In this example we end up with a straight curve which a programmer can adjust to zero simply by putting in a linear correction per inch.

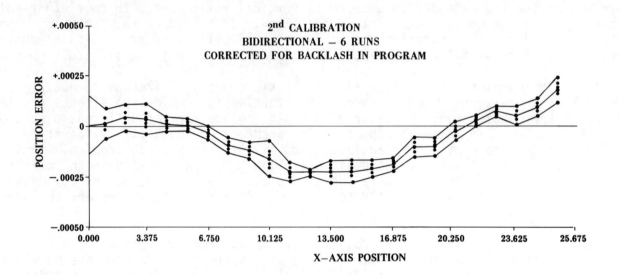

Fig. #3

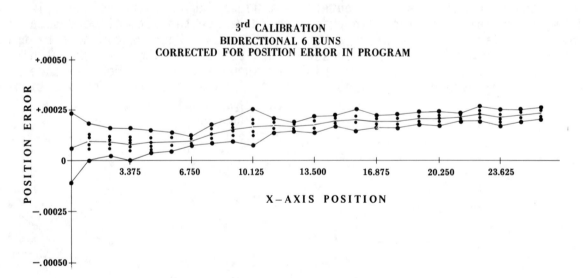

Fig. #4

In the next example, I want to illustrate what happens when the X - Y axes are not square to each other and the X axis does not move in a straight line. We will also look at the problems encountered when the Z axis is not square to X or Y axes. For the purpose of this oversimplified example we will define positioning accuracy as perfect.

The example (Fig. 5), shows a rectangular part with one hole in the lower left corner, three holes in line along the upper surface with one hole along the same X axis, but at a different level. The tool axis is identified by the special target circle.

The first operation will be to drill or inspect at the target position, (Fig. 6), this is our reference. We next move the Y axis stage to position the part as shown in Fig. 7. What we do not know is that the Y axis is not square to the X axis. Therefore, without moving the X axis stage, we have an X axis position error because of X - Y Squareness.

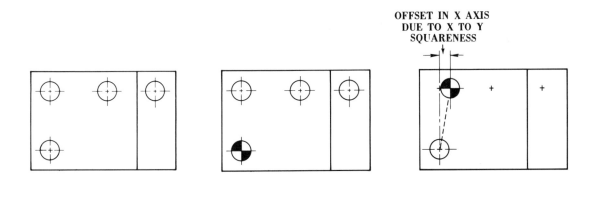

Fig. #5   Fig. #6   Fig. #7

Our next move will be as shown in Fig. 8, and by definition the X axis travel is not straight. This out-of-straight condition in the X axis causes a position error movement in the Y axis, yet the Y axis stage has not moved.

As we continue and move the stage in X axis to the position shown in Fig. 9, the X axis straightness moves back to its original orientation.

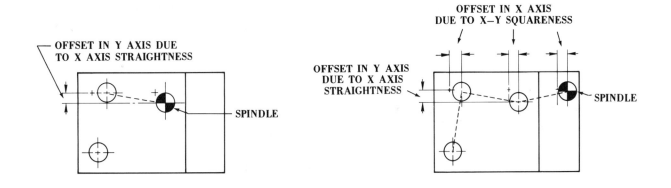

Fig. #8   Fig. #9

Our next area of interest is the Z axis geometry, Fig. 10. We see the apparent axis and actual axis intersecting the part on the lower level at the same position. When we raise the spindle to clear the part and then move the part over by a specified distance under the apparent axis, (Fig. 11), the apparent axis and the actual axis are no longer in coincidence at the part. The Z axis being out of square will cause the spindle to contact the part at a point other than the specified position. In this example, the X axis error caused by the Z axis out-of-square would be added to the X axis error caused by the Y axis out-of-square. The Z axis out-of-squareness may be oriented in any 360 degrees and could effect both X and Y axes.

If the operation is the inspection of a good part, the stage will position the part correctly under the apparent axis which is different than the actual out-of-square optical axis. In this case you could easily measure a good part to be reject. None of the errors in this example could be attributed to positioning inaccuracies of the X-Y slides, they were all caused by bad geometry. These problems are not hypothetical. The examples illustrate an on-going problem with measuring and manufacturing stages and/or slides.

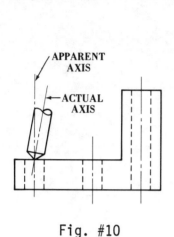

Fig. #10

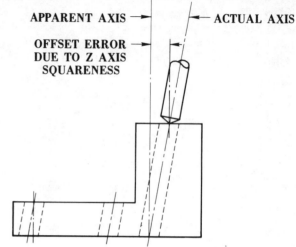

Fig. #11

II. ABBE'S PRINCIPLE

Another geometry problem is known as Abbe's Principle or Abbe Offset. To paraphrase Dr. Abbe, "--the most precise measurement can only be made when the measuring device is in line with the part being measured--". A good example of Dr. Abbe's principle is the standard micrometer (Fig. 12). The anvil, spindle and measuring thimble are in line with each other as well as being in line with the part being measured. An example of a measuring device that violates Abbe's Principle is the vernier caliper, (Fig. 13). It is a handy shop and inspection tool but again the offset jaws can deflect and give inaccurate readings.

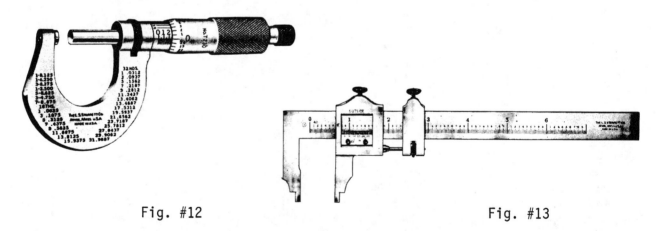

Fig. #12   Fig. #13

To extend these simple examples into the machine shop and inspection areas, we do not have to look very far. The knee type milling machine (Fig. 14) is a good example. Please do not misunderstand the intent of these examples; the knee type mill is the backbone of many shops, they are extremely versatile and we would be lost without them. My intent is honorable, I simply want to make you aware of some of the machine problems so you can better trouble shoot production problems. The problem is primarily with the X axis, because of the long table overhang from the short bearing support surface of the saddle. In addition the X axis gibs are not locked during table traverse. The geometry of the table movement is a predictable arc as the table traverses in the X axis. Abbe's Principle comes into effect when you consider that the scale and reader are placed approximately 4 to 6 inches below the table surface and the cutting tool is working approximately 6 to 8 inches above the table surface. The fact that the table is traversing in an arc should give us a clue. We know that the cordal length between a given angle will increase as we move further away from the center of rotation. With this understanding we can recognize that the distance traversed at the scale and reader will be less than the distance traversed by the work passing under the cutting tool. The work is not in line with the master. We have seen these errors as high as .020". More typically the errors are in the .005" to

KNEE TYPE MILL

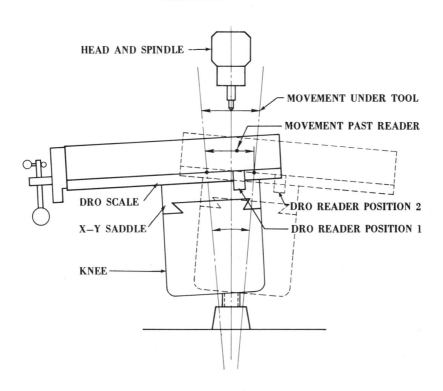

Fig. #14

range. Another word of caution -- be sure the Z axis knee is securely locked in place. If the knee is not locked, there will be a point at about mid-travel where the scale and reader will move as one unit because the knee rocks from one side to the other as the table weight moves past center.

Another Abbe Offset problem is illustrated by the example of an NC punch press, Fig. 15. This illustration shows the Laser Interferometer set up to check the X axis at the screw and also at a position close to the turret, approximately 24" away from the lead screw. The first calibration is at the lead screw to determine its accuracy. The second calibration is at the turret, with the Y axis at maximum distance from the turret, in this case 24".

NC PUNCH PRESS

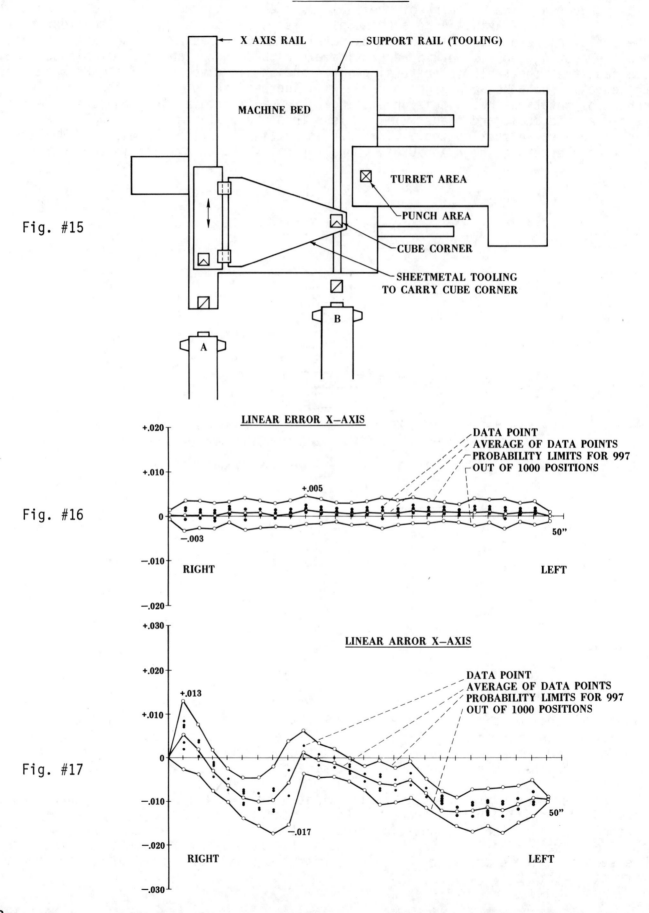

Fig. #15

Fig. #16

Fig. #17

The +3 Sigma data plot (Fig. 16) taken bidirection at the lead screw shows a total error band of .008". This data plot indicates very good accuracy and repeatability. However, due to out-of-straightness, or Yaw motion of the X axis slide, we see a different error curve when we look at the results of the same positioning data taken 24" from the X axis slide (Fig. 17). Now instead of a total error band of .008" we find we have a total error band of .030". With this kind of documentation the fabrication supervisor can determine which parts to run on this machine and which parts to farm out.

III. SURFACE PLATES

Surface plates are the master reference surface for most shops and inspection areas. They are presumed to be flat because they are a reference surface. I sometimes wonder if we have a mental block when it comes to considering that measuring equipment might wear out just like a cutting tool. Surface plates seem to fall into this category.

There are two important parameters concerning the effective precision of a surface plate. One is flatness and the other is repeatability. You can be in trouble if either of these parameters are out of specification.

Flatness: "-All of the surface shall be contained within two parallel planes equal to the tolerance for the given grade of plate-".

Repeatability The ability to repeat the same reading within size and grade tolerance on a known flat surface when approached from any direction on the plate.

The isometric plot of a surface plate (Fig. 18), shows a typical irregular surface, illustrating high and low areas and also areas of radical change. With this type of documentation at the surface plate, the user has a better idea where to work when inspecting precision parts. This type of data plot is available when the calibration data is input to a Desktop Computer and plotted on an X-Y plotter. The numeric data plot (Fig. 19) is also available, showing actual deviations of the isometric plot.

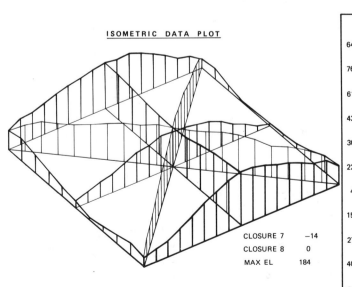

ISOMETRIC DATA PLOT

CLOSURE 7   −14
CLOSURE 8   0
MAX EL   184

Fig. #18

NUMERIC DATA PLOT BASE PLANE REFERENCE

```
64  89  92  105 120 131 140 151  158 151 154 139  118  95  68  57  30
    92                                                           65
76      110                      156                      76     38
            108              CLOSURE 0          88
61          110                  150              95            18
               118                            108
43             128           142           125                  6
                  136                 128
30                134 138 132                                   0
                   142   143       CLOSURE −14
22  37  52  61  84 101 108 125 136 119 112 105  88  61  42  17  6
                     127 134
4                    120   140 136                             8
                       114            140
15                113           156           136             16
                     100                 126
27          87                  170               108         34
               74                                 112
40      74                   178                   108   56
    63                                                96
30  44  55  79  104 130 151 173 184 184 173 159 148 130 109 101 64
                    MAX EL 184
```

Fig. #19

Figure 20 will illustrate what can happen when the surface plate is not flat. In this example the plate is convex. We would adjust the part for a zero indicator reading at each end. This would take place at the low area of the plate. When you move the indicator to the middle of the part to check for straightness, the indicator transfer base has now moved to a higher area of the surface plate, raising the indicator above the part. Assuming the part is straight the indicator would show a deviation equal to the plate curvature. Since we are working with geometry that cannot be seen with the naked eye, our only reference is what we see on the indicator and we could easily attribute this error to the part in question.

Repeatability is another application of Abbe's Principle and is a very important part of surface plate measuring accuracy. As the transfer stand (Fig.21) moves over the surface plate, the hills and valleys will cause the stand to pitch and move through an arc. There will be some error in the Z axis at the stand, but the error will be amplified at the indicator tip depending on its distance from the stand (overhang), and which side of the slope the stand happens to be resting on.

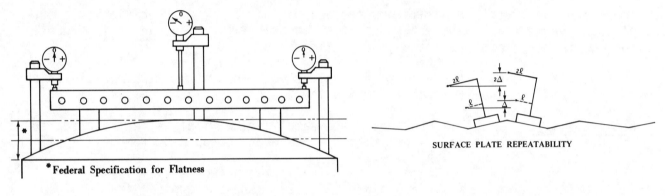

*Federal Specification for Flatness

SURFACE PLATE REPEATABILITY

Fig. #20   Fig. #21

IV. CIRCULAR GEOMETRY

Key Words: Roundness   Parallelism
               Lobing   Runout
               Effective Diameter   Out-of-flat
               Concentricity   Cylindricity
               Squareness

Roundness Geometry is another insidious type of problem. It is possible to have mating parts that will be accepted when measured with a two point measuring devise, yet the parts will not assemble. You cannot "see" the problem and it is difficult to measure without expensive and sensitive measuring equipment such as a Roundness Measuring machine.

I will use Fig. 22 to illustrate. The part, produced in the production shop, has a hole .3750" +.0005", - .0000. The hole, inspected with a production "air plug" accepted the part at .3755". Because the parts were for "pilot run", they had to go to Pilot Run inspection for documentation. Pilot Run used a plain plug set, .3750" Go- .3755" No Go, rather than an air plug which they didn't have. The .3750" Go plug would not enter--the parts were rejected. Why? Because of Roundness Geometry. The plot, Fig. 23, shows the hole to have 7 lobes and a 2 point measuring device cannot detect odd lobes in a hole or on a shaft. The effective diameter of a hole is smaller than the measured size and the effective diameter of a shaft is larger than the measured size.

.375 +.0005
     −.0000

Fig. #22

EQUIPMENT CONNECTIONS

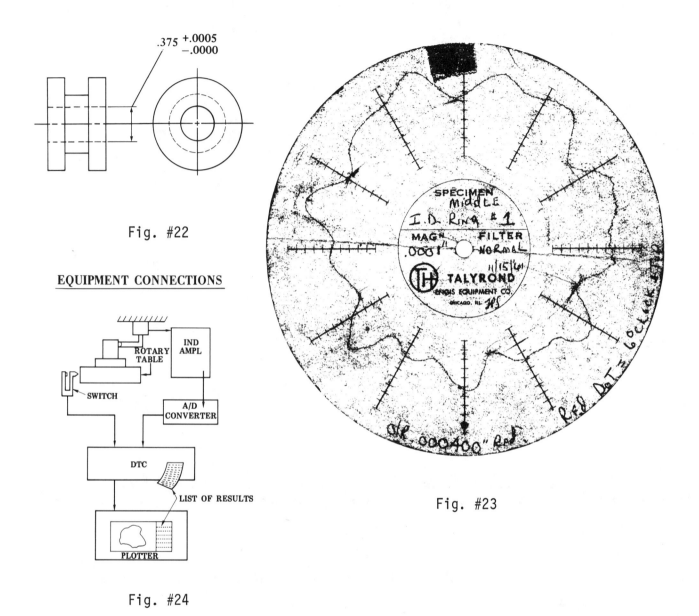

Fig. #23

Fig. #24

With the use of Geometric Tolerancing, Roundness Geometry has suddenly become Circular Geometry with many more ramifications. The polar recorder is no longer adequate. To facilitate the additional information required for Geometric Tolerancing, we have digitized our analog system Fig. 24 to work through a Desktop Computer (DTC) and plotter to plot traces in their proper orientation with centers identified and pertinent data calculated and printed alongside the plots, Fig. 25. We can also see the orientation of each trace center which indicates that the shaft at "3" is eccentric and the positions of the centers for traces 4 & 5 indicate the shaft is slightly bowed. Trace #5 is also 3 lobed.

The DTC also speeds up the operation. Rather than align the part to within ± one meter division for center and tilt, we only have to be within ± five meter divisions for center and tilt, the DTC aligns the part reference to spindle zero and plots both reference traces on chart center.

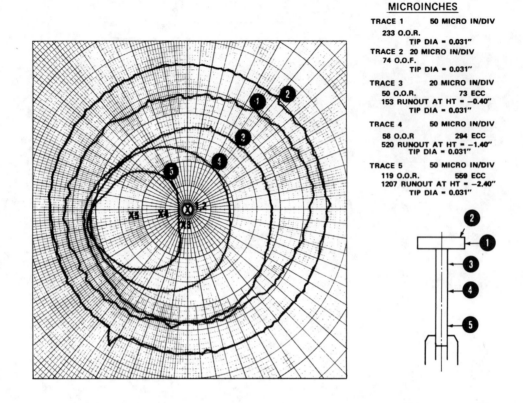

MICROINCHES

TRACE 1	50 MICRO IN/DIV
233 O.O.R.	
	TIP DIA = 0.031"
TRACE 2	20 MICRO IN/DIV
74 O.O.F.	
	TIP DIA = 0.031"
TRACE 3	20 MICRO IN/DIV
50 O.O.R.	73 ECC
153 RUNOUT AT HT = −0.40"	
	TIP DIA = 0.031"
TRACE 4	50 MICRO IN/DIV
58 O.O.R	294 ECC
520 RUNOUT AT HT = −1.40"	
	TIP DIA = 0.031"
TRACE 5	50 MICRO IN/DIV
119 O.O.R.	559 ECC
1207 RUNOUT AT HT = −2.40"	
	TIP DIA = 0.031"

Fig. #25

SUMMARY

To summarize, there are many geometry problems relating to machine, instrument, inspection and production parts that you should be aware of. The problem areas are:

1) Six degrees of freedom of all slides

2) The effect that the geometry freedom of one axis might have on the positioning accuracy of another axis including the squareness of one axis to another.

3) The Abbe' Principle or Abbe" Offset.

4) Surface Plate Flatness and Repeatability.

5) Circular Geometry

CONCLUSION

Our experiences from many calibrations and discussions with many people indicate to us a need for more awareness of the problems discussed in this paper. It is for this reason that this paper has been oriented to some of the basics. Accurate and properly functioning equipment are necessary to achieve process control.

REFERENCES

1. "NMTBA Definition and Evaluation of Accuracy and Repeatability for Numerically Controlled Machine Tools." National Machine Tool Builders Association, 7901 Westpark Drive, McLean, Virginia, 22101.

2. Ernest Abbe, Professor, Physicist, Germany 1840-1905.

3. Repeatability: "Check Surface Plate Flatness", R.J. Rahn, Research Director, Rahn Granite Surface Plate Co., Federal Specification GGG-P-463C, Sept. 1973.

4. Sir George Airey, Astronomer, "Engineering Metrology", K.J. Hume (pg. 62).

5. "Improving Circular Geometry Measurements", F.R. Berry, Hewlett-Packard Co., 35th ASQC Midwest Conference, 1980, Tulsa, Oklahoma.

BIBLIOGRAPHY

1. "Fundamentals of Dimensional Metrology". Ted Busch - Delmar Publishers, Inc.

2. "How to Calibrate Surface Plates in the Plant". B.C. Moody - The Tool Engineer Oct. 1955.

3. "Calibration of Surface Plates". Laser Measurement Systems Application Note 156-2. Hewlett-Packard Co.

4. "Calibration of Machine Tools. Laser Measurement Systems Application Note 156-4. Hewlett-Packard Co.

ACKNOWLEDGEMENT

This paper is based on inputs from many people, printed material, experiments and experience over many years. My present associates, Frank Berry and Cecil Dowdy, and past associates have all been generous contributors.

I also want to acknowledge the many learned Sales people and Manufacturers who have also contributed to my learning curve.

Methodology Classification (LSC Code)

Methodology Class.	Function	Industrial/Business
320	40	434
720	70	435
760		439

1. *TRANSAXLE TEST STAND performs tests automatically under computer control. Testing is aborted if any serious deviations from the test pattern are noted by the computer.*

Reprinted from Manufacturing Engineering, March 1978

Selecting Clamps via Matrix Pinboard

. . . is part of the automation in computer controlled transaxle test stands

Pneumatically operated clamps and locating pins actuated by a preset matrix pinboard facilitate the automatic testing of automotive transaxles on a series of newly developed computer controlled test stands. Five of the stands, built by Scans Associates, Inc. of Livonia, MI, were recently installed at the Chrysler Corp. transmission plant in Kokomo, IN. The stands, one of which is shown in *Figure* 1, are used in testing transaxles for Plymouth Horizon and Dodge Omni automobiles.

Test Program. The tests consist of driving the input shaft at a controlled acceleration rate and measuring the corresponding output speed to check that the shift points are at the right

**2. TRANSAXLE
IS CLAMPED**
*to head plate by two
or three rotary clamps,
depending on
model being tested.*

speed and that there is no excessive clutch slippage. In addition, output pressures are measured frequently at seven test points and checked against acceptable limits. The shift shaft on the transaxle is also actuated to assure that the park, reverse, neutral, and drive positions function properly and that the park/neutral switch produces the correct signals.

The various transaxle models require different differential ratios. A magnetic pickup is used to compare the number of ring gear teeth to a 60-tooth gear on the output shaft to verify that the correct differential is installed in the model being tested.

Setup. Transaxles are delivered to the test stands on J-bars and lowered into place ready for positioning and automatic connection to electrical, water, air, and oil lines. A handwheel can be used to line up the teeth in a driver adapter with those on the transaxle drive shaft. The test unit is then advanced to the head plate where it is

positioned with two locating pins.

Next, driving pins are advanced automatically into two driving holes in the back gear. This gives a 1:1 driving ratio through the transaxle. Typical test speeds range up to 4000 rpm. The power source is a 30-hp (22.4-kW) electric motor driving through an Eaton Dynamatic transmission which provides a variable input speed.

After the transaxle is properly located, *Figure* 2, the head plate clamps are activated, a sealing block is advanced to cover the transaxle cavity, and probes are attached to pick up pressure signals from various internal sections of the unit being tested. Pressures monitored are those that are controlled by the transaxle valve body.

Clamping Features. Five rotary type clamps are provided on the mounting plate, but only two or three are used for any particular transaxle model. Rotary actuation up to 90° provides the necessary clearance for mounting and removing the transaxle. The

mounting plate has a number of cutouts to accommodate different model configurations.

An AMP matrix pinboard located in the operator's control console, *Figure* 3, is used to select the appropriate clamps and locating pins for each model. The pinboard incorporates bus-type continuous contact springs which are arranged on X and Y coordinates separated by a perforated insulating board and are attached to a plastic laminate front panel. Inserting a shorting pin completes the electrical connection between the two sets of contact strips. Any two circuits can be joined by inserting pins where the contact strips cross.

The clamps are pneumatically operated on 50 psi (344.7 kPa) air. The main requirement is to maintain concentricity between the test stand driving gear and the transaxle input spline.

Computer Control. Each test stand has its own Data General 1210 computer. A Data General 1220 host com-

*3. MATRIX PINBOARD
is used to program
automatic actuation of clamps
and locating pins.*

puter is interfaced with the test computers to accumulate test results and print out management reports indicating trends in transaxle quality, production conditions, etc.

After clamping and sealing are complete, the test computer controls the testing cycle according to its program. Digital displays enable the operator to observe test results. At the end of a test, an *Accept* or *Reject* light is lit, and the operator depresses an *Accept* or *Reject* pushbutton to signal the computer that the test is complete. The

operator has other options, such as recycling the test and overriding the computer's decision under some circumstances. He also has pushbuttons and switches to supply necessary data to the computer.

After a test is complete, test results are transferred to the host computer — usually while the next transaxle is being tested. The information is used to update the test results file and the reject statistics. Management reports are printed out by the host computer periodically.

Scans provided the software for the host computer along with the program for transferring data from the test computers to the host computer. Actual test programs are provided by the customer.

Counting plus and minus limits imposed on the various performance features being checked, there are more than 300 reasons why a transaxle can be rejected. Units that pass the testing cycle satisfactorily carry a high degree of reliability and assurance of top performance. ∎

CHAPTER SEVEN

TOOL HOLDING DEVICES

Reprinted from The Carbide Journal, September-October 1977

N/C Tooling Update

Presented at the
1977 Advanced Carbide Tooling Seminar
by

Ralph Johnson
Chief Engineer
Ex-Cell-O Corp., Tool & Abrasive Products Div.

Within the last several years considerable progress has been made by the metal working industry in developing and standardizing tooling for use on N/C machines. This has been a joint effort involving personnel from machine tool builders, machine tool users and cutting tool manufacturers working through such organizations as—Society of Carbide and Tool Engineers, National Machine Tool Builders Association, Cemented Carbide Producers Association, Cutting Tool Manufacturers Association and American National Standards Institute. Among the more noted standards adopted are . . .

1) "V" flange tool shanks for use on N/C machining centers.
2) Qualified tool holders for N/C turning machines.
3) Metric cartridges for use on both machining centers and turning machines.

In addition, various quick change systems have been devoted or existing systems expanded or modified for application on N/C machines.

Probably the one item which has had the most significant impact on the market is the ANSI "V" flange tool shank for use in machining centers. Need for a standard shank can readily be appreciated by reviewing Figure 1 which shows only a few of the many tool shanks used today. The number of different shanks in use may easily exceed thirty varieties and sizes.

Caterpillar Tractor was one of the first users to take the initiative in solving this problem. Because of the varieties required to tool numerous makes of machine tools, tool holder inventories were reaching astronomical figures with little or no interchangeability between machines. A Caterpillar study resulted in a holder known in the industry as the "Cat Shank".

Ex-Cell-O Workcenter Division was the first U.S. builder to furnish Caterpillar and other U.S. users a machine utilizing this shank. These machines were delivered three years ago. Ex-Cell-O was the first builder to offer the "V flange shank" as standard equipment. This shank is now being adopted by ANSI.

CV-50 NMTB TAPER SHANK
BASIC TOOL HOLDER
DIMENSIONS

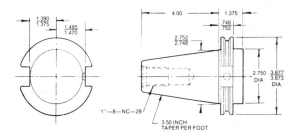

Figure 2. CV-50 NMTB taper shank basic tool holder dimensions.

This design features a "V" groove on the flange which is gripped by the machine tool automatic tool change arm. The standard encompasses shanks with #30, 40, 45, 50 and 60 NMTB tapers. The rear of the shank has a standard NMTB thread that accepts a retention knob or a pull back stud used to retain the holder in the machine spindle. With the retention knob removed, the holder may be used on older conventional boring and milling machines with threaded draw bars. The drive slots in the flange are foolproofed by being machined to different depths from the centerline. The tool may be orientated in both the tool carrier and spindle. This orientation allows the use of rectangular or irregular shaped boring bars and tools of a larger diameter than the distance between the sockets in the tool carrier. This facilitates the use of more positions in the tool carrier. It also ensures consistent tool orientation as it is presented to the work piece. Offset boring bars for back spotfacing or counterboring larger diameters than the entrance hole may be used. The machine table positions the part hole offcenter from the spindle and the bar is fed through the hole. The machine table when repositions the part on centerline and the backboring is accomplished. When boring is completed the spindle stops and orien-

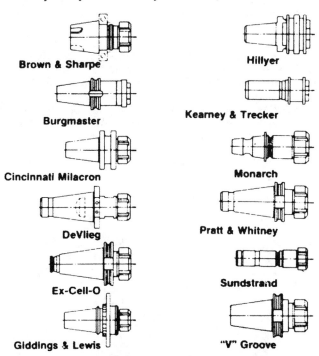

Brown & Sharpe

Burgmaster

Cincinnati Milacron

DeVlieg

Ex-Cell-O

Giddings & Lewis

Hillyer

Kearney & Trecker

Monarch

Pratt & Whitney

Sundstrand

"V" Groove

Figure 1. Basic collet chucks (special for each machine).

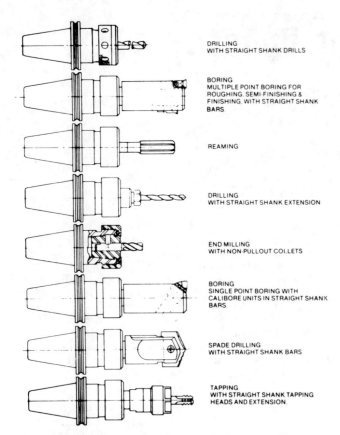

DRILLING
WITH STRAIGHT SHANK DRILLS

BORING
MULTIPLE POINT BORING FOR
ROUGHING, SEMI-FINISHING &
FINISHING, WITH STRAIGHT SHANK
BARS.

REAMING

DRILLING
WITH STRAIGHT SHANK EXTENSION

END MILLING
WITH NON-PULLOUT COLLETS

BORING
SINGLE POINT BORING WITH
CALIBORE UNITS IN STRAIGHT SHANK
BARS.

SPADE DRILLING
WITH STRAIGHT SHANK BARS

TAPPING
WITH STRAIGHT SHANK TAPPING
HEADS AND EXTENSION.

Figure 3. Preset collet holder typical applications.

tates the tool. The reverse procedure is used to withdraw the tool.

We can all appreciate how this standard shank will ultimately reduce tool inventories. The next step in reducing tool inventories is a sensible approach to tooling the machine using standard tooling wherever possible. Several tool holder manufacturers are now offering a

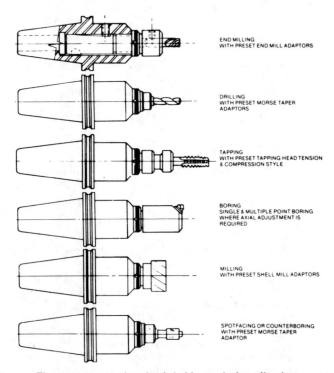

END MILLING
WITH PRESET END MILL ADAPTORS

DRILLING
WITH PRESET MORSE TAPER
ADAPTORS

TAPPING
WITH PRESET TAPPING HEAD TENSION
& COMPRESSION STYLE

BORING
SINGLE & MULTIPLE POINT BORING
WHERE AXIAL ADJUSTMENT IS
REQUIRED

MILLING
WITH PRESET SHELL MILL ADAPTORS

SPOTFACING OR COUNTERBORING
WITH PRESET MORSE TAPER
ADAPTOR

Figure 4. Automotive shank holder typical applications.

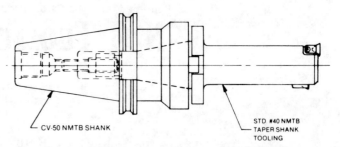

CV-50 NMTB SHANK

STD #40 NMTB
TAPER SHANK
TOOLING

Figure 5. NMTB taper holders typical application.

basic assortment of "V" flange holders to accomplish this.

Figure 3 illustrates a collet chuck which is probably the most versatile holder used in tooling machining centers.

The basic use of the collet chuck is for drilling. It may also be used for end milling merely by changing the collet. In addition standard "off the shelf" straight shank tooling for reaming, boring, tapping, spade drilling & drilling with extensions may be used in the collet chucks.

Another popular holder used on machining centers is the automotive straight shank holder, (Figure 4).

This is used where axial adjustment of the tool is required. Standard "off the shelf" automotive shank tooling may be used for drilling, tapping, end milling, boring, and face milling.

The NMTB taper holder (Figure 5) has a "V" flange shank and accepts other tooling with standard NMTB shanks.

This holder facilitates using tools already in inventory. If the user had a special & costly boring bar or tool that was used before obtaining a machining center, he may now use it. When using this approach, consideration should be given to the length and weight of the assembly. It should be within the limits of good tool design and application practice.

Most tool holder manufacturers may also provide other standard "V" flange holders such as . . .

Morse Taper Adaptors .... Figure 6
End Mill Holders .... Figure 7
Tapping Heads .... Figure 8

Another useful approach in tooling a machining center is the use of bar blanks. These may often be used on special applications and result in reduced cost and shorter lead time. Bar blanks are available in both straight shank & "V" flange types (Figure 10).

These may be used to make special boring bars, spade drill holders, block type tools and extended length end mill, drilling and counterboring holders as shown in Figure 11.

Various boring applications using standard brazed and solid carbide tools, bar with indexable inserts and

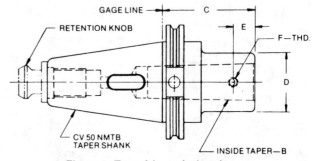

GAGE LINE — C
RETENTION KNOB
E
F—THD.
D
CV 50 NMTB
TAPER SHANK
INSIDE TAPER—B

Figure 6. Tang drive reducing sleeves.

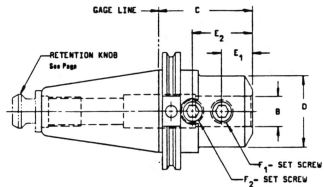

Figure 7. End mill holders.

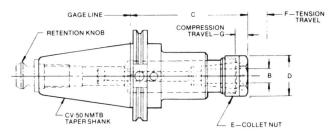

Figure 8. Tapping heads.

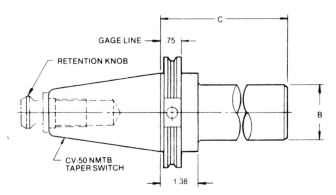

Figure 9. Blank bars are tough hardened and finish ground all over, and ready to be machined to accept either Calibore precision boring units, indexable tool cartridges, round or square tool bits, block type tools, spade drills, or other precision units.

bar blanks with single point tools are shown in Figure 12.

Qualified tool holders which have been standardized are covered by ANSI B94-45 - 1973 standard. Briefly this standard covers dimensional specifications, styles and designations for precision indexable insert holders. With the increased usage of N/C horizontal and vertical turning and chucking machines the need for tool holders manufactured to closer controlled tolerances arose. The standard encompasses some 257 different sizes and styles and assures the tool user can obtain an interchangeable tool from a large number of sources.

When close tolerances in the order of .0005 must be held, tool adjustment is a major problem. In order to accomplish this Ex-Cell-O has designed the "Bore-set" (Figure 12). This is a micro-adjustable adapter with an integral "V" flange shank calibrated for fine adjustment of the tool radius. Tool adjustments as fine as .0002 are not uncommon.

Qualified or controlled dimensions are held to a ±.003 tolerance on length and width as shown in (Figure 14).

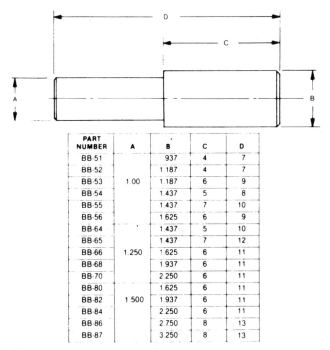

PART NUMBER	A	B	C	D
BB-51		937	4	7
BB-52		1.187	4	7
BB-53	1.00	1.187	6	9
BB-54		1.437	5	8
BB-55		1.437	7	10
BB-56		1.625	6	9
BB-64		1.437	5	10
BB-65		1.437	7	12
BB-66	1.250	1.625	6	11
BB-68		1.937	6	11
BB-70		2.250	6	11
BB-80		1.625	6	11
BB-82	1.500	1.937	6	11
BB-84		2.250	6	11
BB-86		2.750	8	13
BB-87		3.250	8	13

Figure 10. Blank bars.

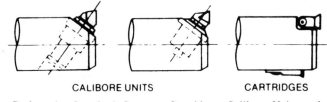

CALIBORE UNITS   CARTRIDGES

Pockets for Standard Compact Cartridges, Calibore Units and Calibore II Top Adjustable Units to perform multiple operations can easily be machined in these blanks.

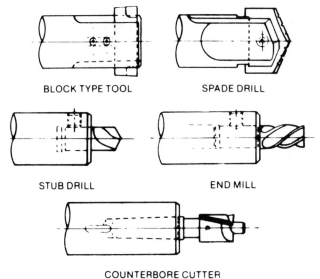

BLOCK TYPE TOOL   SPADE DRILL

STUB DRILL   END MILL

COUNTERBORE CUTTER

Applications shown represent but a few of the many uses of these bars.

Figure 11.

These qualified holders eliminate the use of the consuming presetting of tools and reduce set-up time. Tool size adjustments may be accomplished by use of the "tool offset" feature normally available on most N/C machines.

421

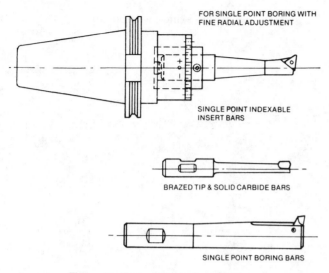

FOR SINGLE POINT BORING WITH
FINE RADIAL ADJUSTMENT

SINGLE POINT INDEXABLE
INSERT BARS

BRAZED TIP & SOLID CARBIDE BARS

SINGLE POINT BORING BARS

Figure 12. Bore set typical applications.

Figure 13. ANSI standard qualified tool holders.

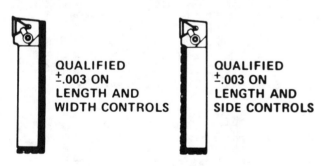

QUALIFIED
±.003 ON
LENGTH AND
WIDTH CONTROLS

QUALIFIED
±.003 ON
LENGTH AND
SIDE CONTROLS

Figure 14. Right hand tooling.

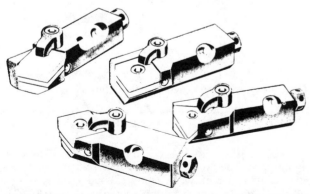

Figure 15. New indexable insert metric cartridges ANSI - ISO STD.

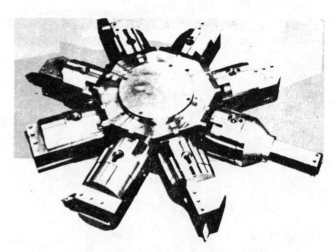

Figure 16. Typical application for a 8 turret position N/C lathe.

Another standard to develop in recent years was the precision indexable insert cartridge B94-48 - 1977 assigned (not published). This is a pure metric standard and covers styles, methods of mounting and dimensional tolerances for cartridges to ensure interchangeability between suppliers. The standard encompasses some 19 different lead angles in both positive and negative types in five different sizes. Typical metric cartridges are shown in Figure 15.

These cartridges are used primarily in special tooling applications. They may be designed into boring bars to accomplish the use of indexable inserts in simple single point boring applications or they are used in multiple step boring and chamfering operations. Metric cartridges may be adjusted axially and radially in order to provide minor tool size change. Metric cartridges may also be used on complicated boring head and tool block set-ups. With good design application they may eliminate many of the special cartridges and with their multiple usage, perishable tool inventories may be reduced substantially.

Figure 17. Turret principle applied to a two spindle vertical turning machine.

Several "quick change" tooling systems for N/C turning and chucking machines have been developed. These systems enable users to change tooling set-ups on N/C machines with a minimum of down time. There are several such systems on the market at present, one of these is the Ex-Cell-O "OGESTO" system. One typical application for a 8 turret position N/C lathe is shown in Figure 16.

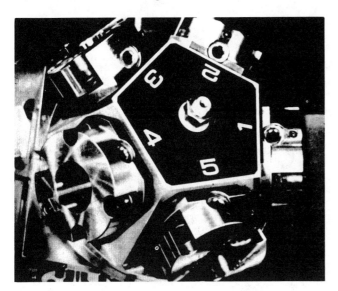

Figure 18. Five position turret adaptor.

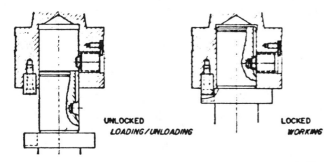

Figure 19. The Ogesto quick change adaptor is held in the holder socket with the locating locking plunger. The plunger is actuated by turning it with an Allen wrench, 1/2 to 3/4 turn, "right" to lock, "left" to unlock.

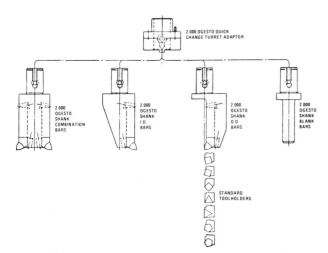

Figure 20. Typical bars used to turn, face, bore, chamfer and contour parts.

Figure 17 shows the same principle applied to a two spindle vertical turning machine with a five position turret.

The system is quite simple and consists of a turret adaptor Figure 18 and a mating straight shank bar. A few turns of the locking screw releases or clamps the bar as desired.

Tools on a five position turret can be changed in five minutes or less. Figure 19 shows how the system works.

Figure 20 shows typical bars used to turn, face, bore, chamfer, and contour parts as required. Here again, note the use of qualified tool holders and metric cartridges.

All that has been discussed exists and is in use today. Now a look into the future.

Machine tool and cutting tool technology are both moving forward.

One recent development which promises to revolutionize the through boring of close tolerance holes is the application of "bore-lap" tools to N/C machines (Figure 21). This system, under development by Ex-Cell-O assures repeated hole to hole size within .0002. "Bore-Laps" utilize diamond or Borazon materials and

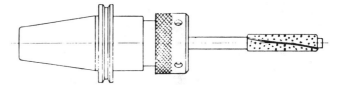

Figure 21. Bore-Lap tool.

Figure 22. Circular interpolation of the pallet enables the center bore of the workpiece to be milled by the 18,000 rpm spindle. The slots are milled through linear interpolation of the workpiece.

Figure 23. This 12 x 20 x 3-inch aluminum test workpiece (left) was milled in four hours on the high speed Ex-Cell-O 208 Workcenter machine. Similar parts took as long as ten hours using conventional methods.

Figure 24. The principal parts of the tool holder are, (L to R): retention knob, plunger, ball bearings, tool shank, spring and collet. To the right of the parts is an assembled tool holder and a 1-1/4 inch solid carbide cutting tool used in high speed milling.

have been known to last upwards of 100,000 pieces before replacement was required. These tools may be used in collet holders or end mill holders and finish the hole removing .001-.0015 stock in seconds.

High speed machining is the latest technical capability to be applied to workcenter machines. Ex-Cell-O recently furnished a machine capable of spindle speeds to . . . 18,000 RPM. This machine was shipped to a West Coast airframe manufacturer.

Figure 22 shows the machine and aluminum billet.

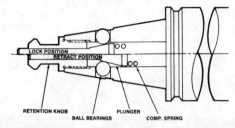

LOCK POSITION
RETRACT POSITION

RETENTION KNOB   BALL BEARINGS   PLUNGER   COMP. SPRING

Figure 25. Tool holder shows the location of the plunger, bearings and springs in both lock and retract positions.

Figure 23 shows the airframe part partially and finish machined.

The metal removal rate of 30 cubic inches per minute was achieved using a newly developed carbide two flute end mill. Figure 24 shows the holder both assembled and disassembled.

This holder embodies a unique self locking safety feature to insure the holder does not come out of the spindle should the retention mechanism fail. Figure 25 shows how the spring loaded cone plunger seats locking balls into a recess in the machine spindle. When the retention mechanism moves forward to release the holder, it depresses the plunger releasing the locking balls allowing the holder to be removed from the spindle by the tool change. These are but a few of the things under development. With the rapid progress being made on carbide coatings, ceramics and new cutting tool materials one can expect more high speed machining applications resulting in increased productivity.

The All-new Kenturn System

Kennametal has come up with a new toolholder system – one that gives promise of greatly reduced downtime in numerically controlled turning applications

Reprinted from Manufacturing Engineering, July 1976

NUMERICALLY CONTROLLED LATHES are great for increasing your plant productivity — as long as they're kept running. The moment they stop, they're losers. Trouble is, virtually all NC lathes have to be down for a considerable amount of time for setup and tool changing. Both problems are solved to a greater or lesser degree with the new Kenturn tooling system, a product of Kennametal.

Designed for Speed. Tool changing in a conventional system requires that the operator stop his lathe, and change his insert. This operation can be complicated by heat in the insert and toolholder or by inaccessibility — or by both. Whatever the problem, the new system is designed to beat it through removal of the toolholder itself. Further, changing of the toolholder can be accomplished while the lathe is in operation, depending on the turret position and the operation being performed. Since the associated Kennametal turret is a 12-station type instead of the customary 6-station unit, it's possible to provide a given job with a complete set of backup tooling, should it be deemed necessary. Similarly — and probably more likely — the lathe can be tooled for two different jobs. With this capability, any change in jobs requires only a change in programming. The lathe itself is ready to go.

How It Locks. Each toolholder is equipped with a cylindrical shank designed to fit a sleeve threaded in the turret. Radial orientation is provided by a dowel pin — press fitted to the turret — that fits a groove in the toolholder.

Once it's inserted in the sleeve, and given proper radial orientation, the toolholder is locked in position with an actuating screw. Function of this screw is to advance a 5/8-inch (15.9-mm) diameter ball. This ball, in turn, contacts three others spaced at 120-degree intervals. When forced outward in this manner, the three smaller balls (they measure 1/2 inch or 13.7 mm in diameter) engage the bearing surface of the sleeve. The resulting locking force is in the order of 3000 lbs (13.3 kN). The ratio between the force exerted by the operator in tightening the assembly and the final locking force is 14:1.

Total System Configuration. While it's designed with NC turning in mind, the Kenturn system can also be used on other machining installations, transfer lines being an outstanding example. When used in a lathe, however, it does require use of the Kennametal turret, a unit that can be specified when new equipment is purchased. At the same time, it's possible to retrofit turrets to existing NC equipment. Approximate cost for a typical retrofit is $4100. Cost of the individual toolholders is approximately $100.

All of the holders have common F and C dimensions with qualification to within ±0.003 inch (0.076 mm). Thus the system provides for complete interchangeability. Equally important, it allows an NC lathe to become something of a machining center, in that one or two sets of tools can be mounted in the turret, another set can be in the rack at the machine, and still another being set up in the crib. With CNC control, it's possible to preprogram the necessary tool changes as a function of the machine. ∎

DEVELOPED BY KENNAMETAL, system provides for the rapid fire changing of complete toolholders – not just the inserts.

TOOLHOLDER LOCKING SYSTEM makes use of a replaceable sleeve and three hardened balls that engage the sleeve. ▶

TWELVE-STATION TURRET enables the operator to make two complete setups at a time.

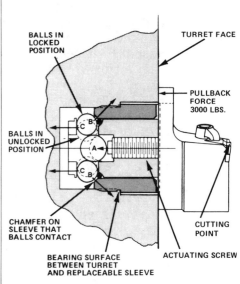

425

The Quick Change Chuck System

by Michael R. Cannon
Boeing Commercial Airplane Co.

Introduction

With the increases in production rates at Boeing in the 1975 time period, the Company was keenly interested in making labor saving improvements in manufacturing. The Manufacturing Research and Development organization recognized that a substantial amount of lost time could be saved by accelerating the process of changing drills in hand-held drill motors. With this objective in mind, an effort was begun to investigate the marketplace for a faster means of changing drills.

The closest available chucking systems consistent with this objective were keyless chucks produced by several manufacturers. These chucks were evaluated and found to have deficiences in three areas:

(1) Under severe drilling conditions the drills could become unchucked, thereby posing a threat to both personnel and part quality.

(2) Drill runout (TIR) tended to be higher with these chucks making it more difficult to accurately start the hole without drill running.

(3) The larger size of these chucks worsened access problems when drilling around airframe structures.

The results from the keyless chuck evaluation initiated an internal effort to develop a faster chucking system that did not have these deficiences.

The earliest prototype of a Quick Change Chuck System at Boeing was a crude adaptation of a quick release air hose coupling. This first prototype was useful for demonstration purposes but, of course, was impractical. This early activity laid the groundwork for a follow-on design centering around the use of a hexagonal shaped adaptor brazed onto the shank of the drill. The chuck consisted of an internally broached body to provide torque drive, and a ball bearing which seated in a groove in the adaptor by means of a spring-loaded sleeve to retain the drill in the chuck. The hexagonal drive quick change chuck concept was met with enthusiasm in a 1976 factory pilot program conducted in the 7/7/7 Division in Renton, Wash. The users of the chuck could change drills faster than with key chucks, drill closer to assembly structure, and enjoyed the reduced weight of the new chuck which improved the balance of pistol grip motors.

With the pilot program considered a success, the development effort shifted to optimizing the producibility of the system. In the course of this effort problems were encountered in maintaining dimensional accuracy of the hexagonal stock used for the adaptor, and also in accurately broaching the chuck. It was determined that the resolution of these problems would increase the manuacturing costs of the new system to levels seriously eroding the derived savings.

Later that year, Messrs. Earl Hill, Mark Soderberg and Richard Benson of Boeing redesigned and patented the quick change chuck around a cylindrical configuration rather than hexagonal shapes. This new design permitted the use of high speed screw machines and production grinding equipment to drive down the manufacturing costs of the chucking system.

Protoype chucks were developed for pistol grip and angle motors and successfully tested in production. A total cost savings of over $500,000/yr was estimated with the use of this system at Boeing Commercial.

Pistol Grip Quick Change Chuck

The pistol grip quick change chuck is a direct replacement for conventional 1/4 in. three jaw chucks (Figure 1). The pistol grip chuck is operated by extending a spring-loaded sleeve and inserting a drill with an integral quick change adaptor (Figure 2). The chuck contains three retaining balls which engage in the detent holes of the drill adaptor when the sleeve is released. These three balls provide the torque driving mechanism and serve to retain the drill in the chuck (Figure 3). The chuck is approximately 1-3/4 in. long, 3/4 in. diameter, and weighs 2-1/4 oz. The quick change adaptor is joined to the drill shank by either the silver brazing or interference fit process, and must meet the minimum torque requirements of NAS 965 (Figure 4). The drill adaptor is heat treated to R_c 40-50 to prevent detent distortion when high drilling torque loads are experienced, such as on hole exit breakthrough. The concentricity of the adaptor is closely controlled by grinding the adaptor to final size $(\frac{.3020 \text{ in.}}{.3010 \text{ in.}})$ after assembly.

Time studies conducted at Boeing have determined that drills can be changed seven seconds faster with the quick change chuck than with conventional key type chucks. At Boeing, where there are approximately 5,500 assembly mechanics performing a total of over 360,000 drill changes per week this, factor alone translates into 700 hours of saved labor per week.

In addition to labor savings, there are several other significant benefits resulting from the use of the quick change chuck:

(1) Decreased Drill Motor Requirements

It is common practice for assembly mechanics to have several drill motors set up at their work station with the variety of tools necessary to perform their assigned tasks. This is frequently done because it is faster to change drill motors than to change tools with a conventional chuck. This is a costly practice in terms of increasing the value of drill motor inventories and the associated maintenance costs for these tools. The Quick Change Chuck System eliminates the need for this practice due to the rapidity in which tools can be changed.

(2) Extended Drill Life

Drill shanks routinely become damaged when slippage occurs in conventional chucks. This necessitates removal of shank scoring, typically by sanding,

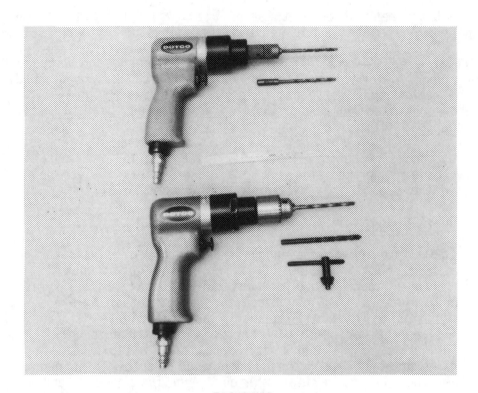

FIGURE 1
PISTOL GRIP QUICK CHANGE
VS. CONVENTIONAL CHUCK

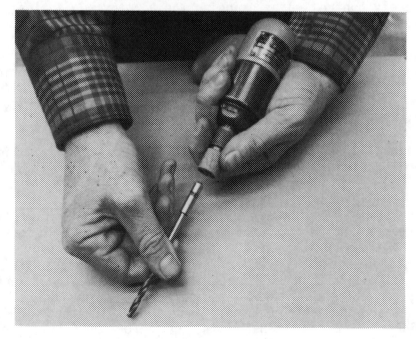

FIGURE 2
PISTOL GRIP CHUCK OPERATION

QUICK CHANGE DRILL SYSTEM
PISTOL GRIP MOTORS

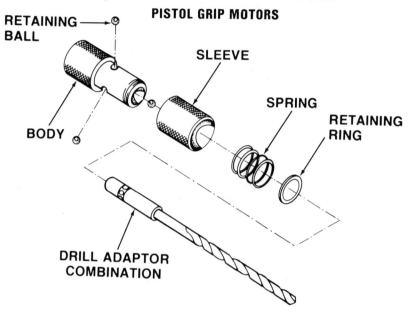

FIGURE 3
EXPLODED DIAGRAM

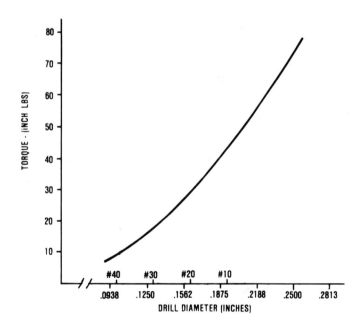

FIGURE 4
MINIMUM TORQUE REQUIREMENTS

or scrapping the drill when the shank is excessively damaged. The positive locking drive of the quick change drill by the retaining balls in the chuck has eliminated this problem, thereby lowering drill attrition.

(3) Improved Drilling Accessibility

The smaller size of the chuck allows assembly mechanics to drill closer to surrounding structure thereby permitting the use of shorter, less expensive drills (Figure 5).

(4) Reduced Part Damage

A common problem with conventional drill chucks is skin damage on drill breakthrough caused by the chuck jaws scoring the skin surface. The quick change system eliminates this problem due to the conical surface of the adaptor at the drill/adaptor interface.

(5) Decreased Chuck Cost

The quick change chuck is approximately 50% the cost of conventional drill chucks in comparable order quantities, and is less susceptible to damage than conventional chucks

(6) Reduced Drill Motor Weight

The quick change chuck is approximately four ounces lighter than a conventional 1/4 in. drill chuck. This weight reduction improves the balance of the drill motor and lessens operator fatigue.

Angle Motor Quick Change Chuck

Angle drill motors are extensively used in the aerospace industry to perform drilling operations where space constraints do not permit the use of pistol grip motors. The angle motor Quick Change Chuck System replaces the threaded shank drills formerly used in these motors (Figure 6).

The operating principle of the angle motor chuck is similar to the pistol grip chuck except that the chuck collar is rotated, rather than extended, to open the chuck (Figure 7). The ball detent drive concept is identical to the pistol grip version (Figure 8).

Changing threaded shank drills requires the use of a hexagonal socket wrench to prevent the motor from rotating while another wrench is used to grip the threaded adaptor and remove the drill. In many cases assembly mechanics will improvise on this changing process and use pliers to remove the drill. This common practice frequently damages the wrench flats on the threaded adaptor and can cause distortion contributing to increased drill runout.

With the severe drilling torque loads experienced with low speed, high torque angle motors, the threaded adaptor often seats with sufficient force to cause shearing at the thread relief when an attempt is made to remove the drill. This occurrence causes the unit to be withdrawn from service and sent to

FIGURE 5
IMPROVED DRILLING ACCESS

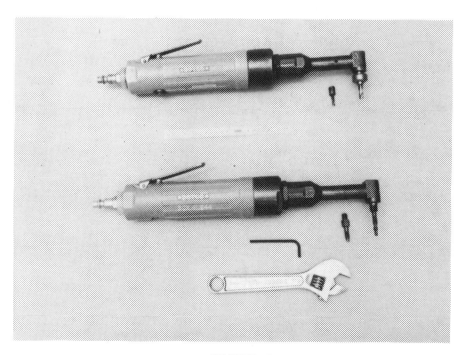

FIGURE 6
ANGLE MOTOR QUICK CHANGE
VS. THREADED SHANK DRILLS

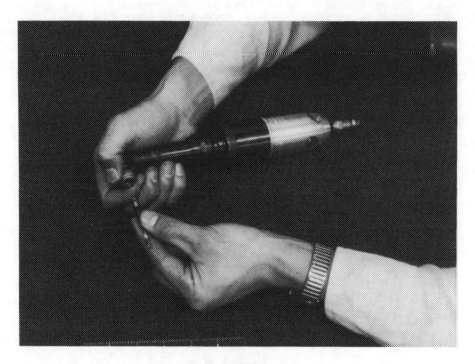

FIGURE 7
ANGLE MOTOR CHUCK OPERATION

QUICK CHANGE DRILL SYSTEM
90° POWER VANE MOTOR

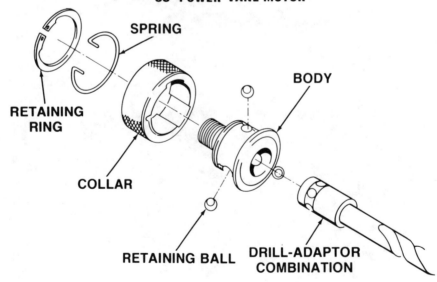

FIGURE 8
EXPLODED DIAGRAM

maintenance to remove the broken shank from the motor. The angle motor quick change chuck precludes the occurrence of these problems.

Time studies have determined that the use of the angle motor quick change chuck saves over eight seconds per drill change. The cylindrical design of the quick change adaptor is easier to manufacture than the threaded shank, thereby making the new system less expensive to procure. The same advantages of reduced drill motor requirements and extended drill life as explained for the pistol grip quick change also apply.

Miscellaneous Tools

Assembly mechanics are required to use a variety of tools in hand-held drill motors to perform tasks such as reaming, countersinking, deburring, sanding and spot facing. To derive the full benefit of the quick change chuck concept at Boeing, these tools have also been converted to the quick change system (Figure 9).

To accommodate tools used in very low quantities where adaptation to the quick change configuration by the manufacturer is not practical, a colleting quick change adaptor was developed (Figure 10). Tools are inserted into this adaptor and secured by means of closing an internal collet around the shank of the tool by tightening a collet nut.

Costs and Impacts

There are currently three sources for quick change drills:

(1) Precision Twist Drill and Machine Co.
    Crystal Lake, IL

(2) Semco Twist Drill & Tool Co., Inc.
    Santa Maria, CA

(3) U. S. Tool Grinding Co.
    St. Louis, MO

Quick change drills are procured by Boeing at an average increase of 40 cents above standard aircraft general purpose drill costs. Currently the combined manufacturing output of the suppliers is approximately 160,000 drills/wk.

The considerable size of the drill distribution system at Boeing, both in terms of sheer volume (120,000 resharpens/wk) and the wide variety of drill sizes and types used, necessitated implementing the quick change drill system in planned phases. From estimated usage figures supplied by Manufacturing, long range procurement requirements were established. Implementation quantities for each phase of the program were determined as were the sustaining quantities of new drills required to keep the "pipeline" full. Inventory levels were established based on procurementlead times, vendor performance, and drill return rates from the factory. Temporary inventory build up was a natural side effect of the implementation since until each phase was implemented a dual inventory of standard drills was maintained.

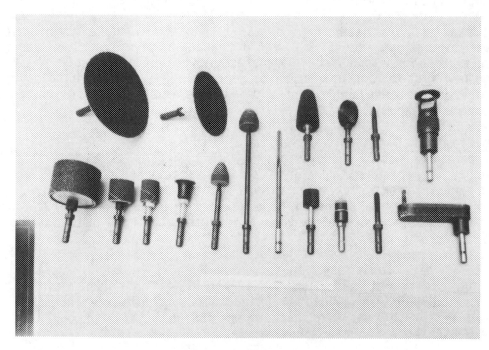

FIGURE 9
QUICK CHANGE MISCELLANEOUS TOOLS

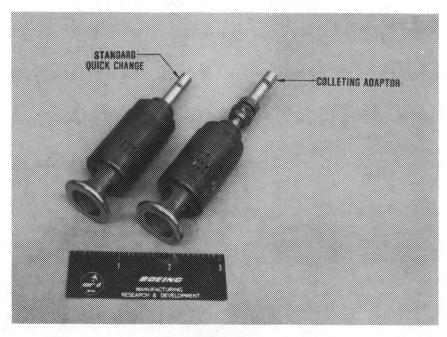

FIGURE 10
QUICK CHANGE ADAPTED TOOL
VS. COLLETING ADAPTOR

The true accuracy of usage estimates was determined after all standard drills had been removed from an area that was implemented with the quick change system. For this reason, tool rooms and tool boxes were purged of standard drills as soon as practical. Once this had been accomplished, actual usage figures were discernable, and could be compared with the original usage estimates on which procurement plans were based.

Another advantage of phasing the implementation is that the used standard drills returned from an implemented area can be used to sustain the requirements of areas not yet implemented with the quick change system. When the implementation is completed, the last remaining standard drills can either be converted to the quick change system, or sold on the open market.

SUMMARY

When Boeing decided to kick-off the quick change drill program in 1977 there was one supplier for these tools. That same year Manufacturing Research and Development initiated an effort to develop an Automatic Quick Change Drill Assembly Machine to augment the drill supply and provide an in-house capability of supplying emergent drill requirements (Figure 11). Currently Boeing Commercial is in the final phases of implementing the Quick Change Drill System and sources are well established.

The use of this new time saving drill chucking system at Boeing has been well received by factory personnel performing drilling operations, and helps meet the goal of reducing manhour expenditures in manufacturing operations through increased productivity.

FIGURE 11
AUTOMATIC QUICK CHANGE DRILL
ASSEMBLY MACHINE

INDEX